McGraw-Hill Networking & Telecommunications

Build Your Own

TRULOVE · *Build Your Own Wireless LAN with projects*

Crash Course

LOUIS · *Broadband Crash Course*
VACCA · *i-Mode Crash Course*
LOUIS · *M-Commerce Crash Course*
SHEPARD · *Telecom Convergence,* Second Edition
SHEPARD · *Telecom Crash Course*
LOUIS · *Telecom Management Crash Course*
BEDELL · *Wireless Crash Course*
KIKTA/FISHER/COURTNEY · *Wireless Internet Crash Course*

Demystified

HARTE/LEVINE/KIKTA · *3G Wireless Demystified*
LAROCCA · *802.11 Demystified*
MULLER · *Bluetooth Demystified*
EVANS · *CEBus Demystified*
BAYER · *Computer Telephony Demystified*
HERSHEY · *Cryptography Demystified*
TAYLOR · *DVD Demystified*
HOFFMAN · *GPRS Demystified*
SYMES · *MPEG-4 Demystified*
CAMARILLO · *SIP Demystified*
SHEPARD · *SONET/SDH Demystified*
TOPIC · *Streaming Media Demystified*
SYMES · *Video Compression Demystified*
SHEPARD · *Videoconferencing Demystified*
BHOLA · *Wireless LANs Demystified*

Developer Guides

GUTHERY · *Mobile Application Development with SMS*
RICHARD · *Service and Device Discovery: Protocols and Programming*
VACCA · *i-Mode Crash Course*

Professional Telecom

SMITH/COLLINS · *3G Wireless Networks*
BATES · *Broadband Telecom Handbook,* Second Edition
COLLINS · *Carrier Class Voice over IP*
HARTE · *Delivering xDSL*
HELD · *Deploying Optical Networking Components*
MINOLI · *Ethernet-Based Metro Area Networks*
BENNER · *Fibre Channel for SANs*
BATES · *GPRS*
MINOLI · *Hotspot Networks: Wi-Fi for Public Access*
LEE · *Lee's Essentials of Wireless*
BATES · *Optical Switching and Networking Handbook*

WETTEROTH · *OSI Reference Model for Telecommunications*
SULKIN · *PBX Systems for IP Telephony*
RUSSELL · *Signaling System #7,* Fourth Edition
SAPERIA · *SNMP on the Edge: Building Service Management*
MINOLI · *SONET-Based Metro Area Networks*
NAGAR · *Telecom Service Rollouts*
LOUIS · *Telecommunications Internetworking*
RUSSELL · *Telecommunications Protocols,* Second Edition
MINOLI · *Voice over MPLS*
KARIM/SARRAF · *W-CDMA and cdma2000 for 3G Mobile Networks*
BATES · *Wireless Broadband Handbook*
FAIGEN · *Wireless Data for the Enterprise*

Reference

MULLER · *Desktop Encyclopedia of Telecommunications,* Third Edition
BOTTO · *Encyclopedia of Wireless Telecommunications*
CLAYTON · *McGraw-Hill Illustrated Telecom Dictionary,* Third Edition
PECAR · *Telecommunications Factbook,* Second Edition
RUSSELL · *Telecommunications Pocket Reference*
RADCOM · *Telecom Protocol Finder*
KOBB · *Wireless Spectrum Finder*
SMITH · *Wireless Telecom FAQs*

Security

HERSHEY · *Cryptography Demystified*
NICHOLS · *Wireless Security*

Telecom Engineering

SMITH · *Cellular System Design and Optimization*
ROHDE/WHITAKER · *Communications Receivers,* Third Edition
SAYRE · *Complete Wireless Design*
OSA · *Fiber Optics Handbook*
LEE · *Mobile Cellular Telecommunications,* Second Edition
BATES · *Optimizing Voice in ATM/IP Mobile Networks*
RODDY · *Satellite Communications,* Third Edition
SIMON · *Spread Spectrum Communications Handbook*
SNYDER · *Wireless Telecommunications Networking with ANSI-41,* Second Edition

BICSI

Network Design Basics for Cabling Professionals
Networking Technologies for Cabling Professionals
Residential Network Cabling
Telecommunications Cabling Installation

Optical Communications Essentials

Optical Communications Essentials

Gerd Keiser

McGraw-Hill

New York Chicago San Francisco Lisbon London Madrid
Mexico City Milan New Delhi San Juan Seoul
Singapore Sydney Toronto

The McGraw-Hill Companies

Library of Congress Cataloging-in-Publication Data

Keiser, Gerd,
Optical communications essentials / Gerd Keiser.
p. cm.
Includes bibliographical references and index.
ISBN 0-07-141204-2 (alk. paper)
1. Optical communications. I. Title.

TK5103.59.K42 2003
621.382'7—dc21 2003051171

Copyright © 2003 by The McGraw-Hill Companies, Inc. All rights reserved. Printed in the United States of America. Except as permitted under the United States Copyright Act of 1976, no part of this publication may be reproduced or distributed in any form or by any means, or stored in a data base or retrieval system, without the prior written permission of the publisher.

1 2 3 4 5 6 7 8 9 0 DOC/DOC 0 9 8 7 6 5 4 3

ISBN 0-07-141204-2

The sponsoring editor for this book was Stephen S. Chapman, the editing supervisor was David E. Fogarty, and the production supervisor was Sherri Souffrance. It was set in Century Schoolbook by Macmillan India Ltd. The art director for the cover was Handel Low.

Printed and bound by RR Donnelley.

This book was printed on recycled, acid-free paper containing a minimum of 50% recycled, de-inked fiber.

McGraw-Hill books are available at special quantity discounts to use as premiums and sales promotions, or for use in corporate training programs. For more information, please write to the Director of Special Sales, Professional Publishing, McGraw-Hill, Two Penn Plaza, New York, NY 10121-2298. Or contact your local bookstore.

Information contained in this work has been obtained by The McGraw-Hill Companies, Inc. ("McGraw-Hill") from sources believed to be reliable. However, neither McGraw-Hill nor its authors guarantee the accuracy or completeness of any information published herein, and neither McGraw-Hill nor its authors shall be responsible for any errors, omissions, or damages arising out of use of this information. This work is published with the understanding that McGraw-Hill and its authors are supplying information but are not attempting to render engineering or other professional services. If such services are required, the assistance of an appropriate professional should be sought.

To Ching-yun and Nishla for their support and understanding during the writing of this book

Contents

Preface

The tremendous growth of the optical communications field has resulted in a large group of people who need to understand the fundamental concepts of the technology without going into extensive technical details. They range from individuals with a minimum technical knowledge to practicing engineers in specialized disciplines who need to get a quick broad view of related technical specialties. This book addresses the needs of these people. In addition, the book can be used as a reference for optical communications engineers, since it contains a great amount of practical information. The introductory nature of the topics will enable the reader to apply the basic concepts of the technology, comprehend the contents of articles appearing in popular trade journals and magazines, answer product sales questions, understand the job functions of various people in the field, and assess the value of new developments.

The audience for the book includes technical sales and marketing personnel, business managers in telecommunications, investment people, assembly technicians, equipment installers, component researchers and designers, cable installation engineers, cable plant design engineers, test and measurement technicians, network operations personnel, troubleshooting and repair personnel, optical communications engineers, and engineers in specialized disciplines (e.g., circuit designers, materials engineers, test equipment designers) who are new to optical communications and need a technical overview of the field.

The book consists of 20 chapters that allow for an easy reading of any particular topic. There are ample illustrations and photographs throughout each chapter to thoroughly explain the concepts without going into a great deal of complex technical detail. A minimum amount of mathematics is used in order to enable a wide range of people to understand the material. Advanced topics and concepts are presented in indented sections in each chapter for those who want to delve deeper into specific topics. Those readers who not interested in the particulars can skip these topics without losing continuity. Thus the book can be read at an introductory level in which only the concepts are examined or at a higher level by means of the tutorial topics if the reader wants greater technical depth.

A special feature of the book is access via the Internet to an abbreviated version of a simulation tool from VPIphotonics for noncommercial educational use. This version is called VPIplayer and contains predefined component and link configurations that allow interactive concept demonstrations. Results are shown in a format similar to the displays presented by laboratory instruments. VPIplayer can be downloaded free from www.VPIphotonics.com. In addition, at www.PhotonicsComm.com there are numerous interactive examples of optical communication components and links related to topics in this book that the reader can download and simulate.

Gerd Keiser

Acknowledgments

Special thanks go to Rudolph Moosburger and Arthur Lowery of VPIphotonics for enabling the use of VPIplayer and the associated modeling and simulation demonstrations. The concept of this book is due to Evie Bennett of William Frick and Company, who pointed out the need for a book that nonengineering people working in the optical communications field can understand. Numerous people have helped directly or indirectly with various aspects of this work. Among them are Sharmila Vidyadhara of Agilent Technologies; Hon Tsang of Bookham Technologies; Fred Quan of Corning; Jean-François Cauchon and Thierry Champagne of DICOS Technologies; Carl Larsen of Domaille Engineering; Alain Poirier of EXFO; Jeff Sitlinger of Fitel Interconnectivity Corp.; Luciano “Lucky” Covati of General Dynamics; Frank Fu, Sunny Sun, and Lily Wu of Koncent; Bill Emkey and Curtis Johnson of kSARIA; Ping Shum of the Nanyang Technological University, Singapore; Lars Rønn, Jacob Philipsen, and Daniel Joseph Barry of NKT Integration; Amir Chitayat and Ronit Levin Tzairi of Sagitta; Robert E. Orr of Sherman & Reilly; Jessica Held of Vermeer; and Hooman Jazaie of VPIphotonics. This book especially benefited from the expert editorial guidance of Stephen S. Chapman and David E. Fogarty of McGraw-Hill, who provided quick responses to all my questions and requests. As a final personal note, I am grateful to my wife Ching-yun and my daughter Nishla for their patience and encouragement during the time I devoted to writing this book.

Optical Communications Essentials

Chapter

1

Basic Concepts of Communication Systems

Ever since ancient times, people continuously have devised new techniques and technologies for communicating their ideas, needs, and desires to others. Thus, many forms of increasingly complex communication systems have appeared over the years. The basic motivations behind each new one were to improve the transmission fidelity so that fewer errors occur in the received message, to increase the transmission capacity of a communication link so that more information could be sent, or to increase the transmission distance between relay stations so that messages can be sent farther without the need to restore the signal fidelity periodically along its path.

Prior to the nineteenth century, all communication systems operated at a very low information rate and involved only optical or acoustical means, such as signal lamps or horns. One of the earliest known optical transmission links, for example, was the use of a fire signal by the Greeks in the eighth century B.C. for sending alarms, calls for help, or announcements of certain events. Improvements of these systems were not pursued very actively because of technology limitations at the time. For example, the speed of sending information over the communication link was limited since the transmission rate depended on how fast the senders could move their hands, the receiver was the human eye, line-of-sight transmission paths were required, and atmospheric effects such as fog and rain made the transmission path unreliable. Thus it turned out to be faster, more efficient, and more dependable to send messages by a courier over the road network.

The invention of the telegraph by Samuel F. B. Morse in 1838 ushered in a new epoch in communications—the era of electrical communications. In the ensuing years increasingly sophisticated and more reliable electrical communication systems with progressively larger information capacities were developed and deployed. This activity led to the birth of free-space radio, television, microwave, and satellite links, and high-capacity terrestrial and undersea wire lines for sending voice and data (and advertisements!) to virtually anywhere in the world.

However, since the physical characteristics of both free-space and electric wire-based communication systems impose an upper bound on the transmission capacities, alternative transmission media were investigated. A natural extension was the use of optical links. After extensive research and development on the needed electrooptical components and the glass equivalent of a copper wire, optical fiber communication systems started to appear in the 1970s. It is this technology that this book addresses.

To exchange information between any two devices in a communication system, some type of electric or optical signal which carries this information has to be transmitted from one device to the other via a communication channel. This channel could consist of a wire, radio, microwave, satellite, infrared, or optical fiber link. Each of the media used for such communication channels has unique performance characteristics associated with it. Regardless of its type, the medium degrades the fidelity of the transmitted signal because of an imperfect response to the signal and because of the presence of electrical and/or optical noise and interference. This can lead to misinterpretations of the signal by the electronics at the receiving end. To understand the various factors that affect the physical transfer of information-bearing signals, this chapter gives a basic overview of fundamental data communication concepts. With that as a basis, the following chapters will describe how information is transferred using lightwave technology.

1.1. Definitions

We start by giving some concepts and definitions used in data communications and the possible formats of a signal. The signal format is an important factor in efficiently and reliably sending information across a network.

A basic item that appears throughout any communications book is the prefix used in metric units for designating parameters such as length, speed, power level, and information transfer rate. Although many of these are well known, a few may be new to some readers. As a handy reference, Table 1.1 lists standard prefixes, their symbols, and their magnitudes, which range in size from 10^{24} to 10^{-24}. As an example, a distance of 2×10^{-9} m (meters) $= 2$ nm (nanometers). The three highest and lowest designations are not especially common in communication systems (yet!), but are included in Table 1.1 for completeness.

Next let us define some terms and concepts that are used in communications.

- *Information* has to do with the content or interpretation of something such as spoken words, a still or moving image, the measurement of a physical characteristic, or values of bank accounts or stocks.
- A *message* may be considered as the physical manifestation of the information produced by the source. That is, it can range from a single number or symbol to a long string of sentences.
- The word *data* refers to facts, concepts, or instructions presented as some type of encoded entities that are used to convey the information. These can include

TABLE 1.1. Metric Prefixes, Their Symbols, and Their Magnitudes

Prefix	Symbol	Decimal	Magnitude	Multiple
yotta	Y			10^{24}
zetta	Z			10^{21}
exa	E			10^{18}
peta	P		Quadrillion	10^{15}
tera	T		Trillion	10^{12}
giga	G	1,000,000,000	Billion	10^{9}
mega	M	1,000,000	Million	10^{6}
kilo	k	1,000	Thousand	10^{3}
centi	c	0.01	Hundredth	10^{-2}
milli	m	0.001	Thousandth	10^{-3}
micro	μ	0.000001	Millionth	10^{-6}
nano	n	0.000000001	Billionth	10^{-9}
pico	p		Trillionth	10^{-12}
femto	f		Quadrillionth	10^{-15}
atto	a			10^{-18}
zepto	z			10^{-21}
yocto	y			10^{-24}

arrays of integers, lines of text, video frames, digital images, and so on. Although the words *data* and *message* each have a specific definition, these terms often are used interchangeably in the literature since they represent physical embodiments of information.

- *Signals* are electromagnetic waves (in encoded electrical or optical formats) used to transport the data over a physical medium.

A block diagram of an elementary communication link is shown in Fig. 1.1. The purpose of such a link is to transfer a message from an originating user, called a *source*, to another user, called the *destination*. In this case, let us assume the end users are two communicating computers attached to different local area networks (LANs). The output of the information source serves as the message input to a transmitter. The function of the *transmitter* is to couple the message onto a *transmission channel* in the form of a time-varying signal that matches the transfer properties of the channel. This process is known as *encoding*.

As the signal travels through the channel, various imperfect properties of the channel induce impairments to the signal. These include electrical or optical noise effects, signal distortions, and signal attenuation. The function of the *receiver* is to extract the weakened and distorted signal from the channel, amplify it, and restore it as closely as possible to its original encoded form before *decoding* it and passing it on to the message destination.

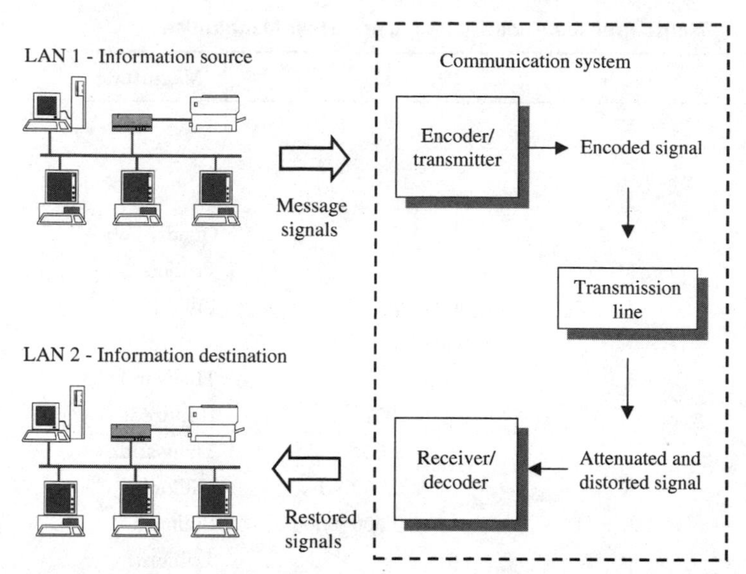

Figure 1.1. Block diagram of a typical communication link connecting separate LANs.

1.2. Analog Signal Formats

The signals emitted by information sources and the signals sent over a transmission channel can be classified into two distinct categories according to their physical characteristics. These two categories encompass analog and digital signals.

An *analog signal* conveys information through a *continuous and smooth variation in time* of a physical quantity such as optical, electrical, or acoustical intensities and frequencies. Well-known analog signals include audio (sound) and video messages. As examples,

- An optical signal can vary in color (which is given in terms of its wavelength or its frequency, as described in Chap. 3), and its intensity may change from dim to bright.
- An electric signal can vary in frequency (such as the kHz, MHz, GHz designations in radio communications), and its intensity can range from low to high voltages.
- The intensity of an acoustical signal can range from soft to loud, and its tone can vary from a low rumble to a very high pitch.

The most fundamental *analog signal* is the periodic *sine wave*, shown in Fig. 1.2. Its three main characteristics are its amplitude, period or frequency, and phase. The *amplitude* is the size or magnitude of the waveform. This is generally designated by the symbol A and is measured in *volts*, *amperes*, or *watts*, depending on the signal type. The *frequency* (designated by f) is the

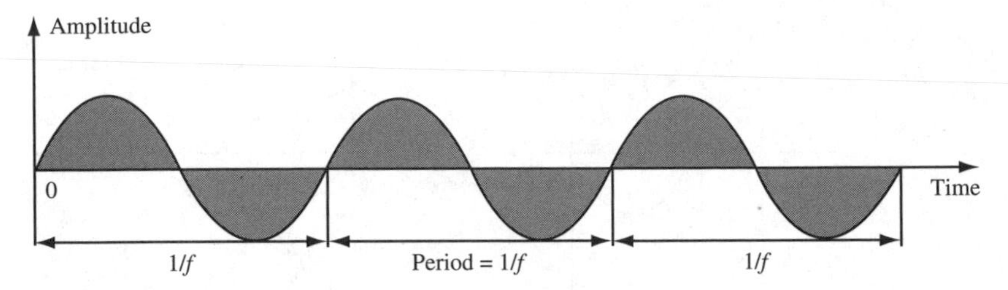

Figure 1.2. Characteristics of a basic sine wave.

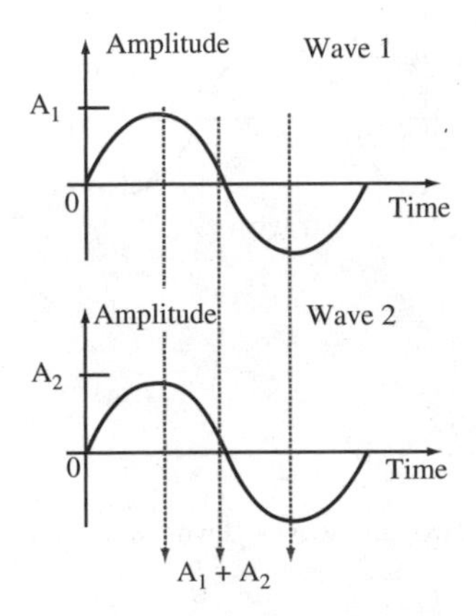

Figure 1.3. Two in-phase waves will add constructively.

number of cycles per second that the wave undergoes (i.e., the number of times it oscillates per second), which is expressed in units of *hertz* (Hz). A hertz refers to a complete cycle of the wave. The *period* (generally represented by the symbol T) is the inverse of the frequency, that is, period $= T = 1/f$. The term *phase* (designated by the symbol ϕ) describes the position of the waveform relative to time 0, as illustrated in Fig. 1.3. This is measured in *degrees* or *radians* (rad): $180° = \pi$ rad.

If the crests and troughs of two identical waves occur at the same time, they are said to be *in phase*. Similarly, if two points on a wave are separated by whole measurements of time or of wavelength, they also are said to be in phase. For example, wave 1 and wave 2 in Fig. 1.3 are in phase. Let wave 1 have an amplitude A_1 and let wave 2 have an amplitude A_2. If these two waves are added, the amplitude A of the resulting wave will be the sum: $A = A_1 + A_2$. This effect is known as *constructive interference*.

Figure 1.4 illustrates some *phase shifts* of a wave relative to time 0. When two waves differ slightly in their relative positions, they are said to be *out of phase*. As an illustration, the wave shown in Fig. 1.4*c* is 180° (π rad) out of phase with the wave shown in Fig. 1.4*a*. If these two waves are identical and have the same amplitudes, then when they are superimposed, they cancel each other out, which is known as *destructive interference*. These concepts are of

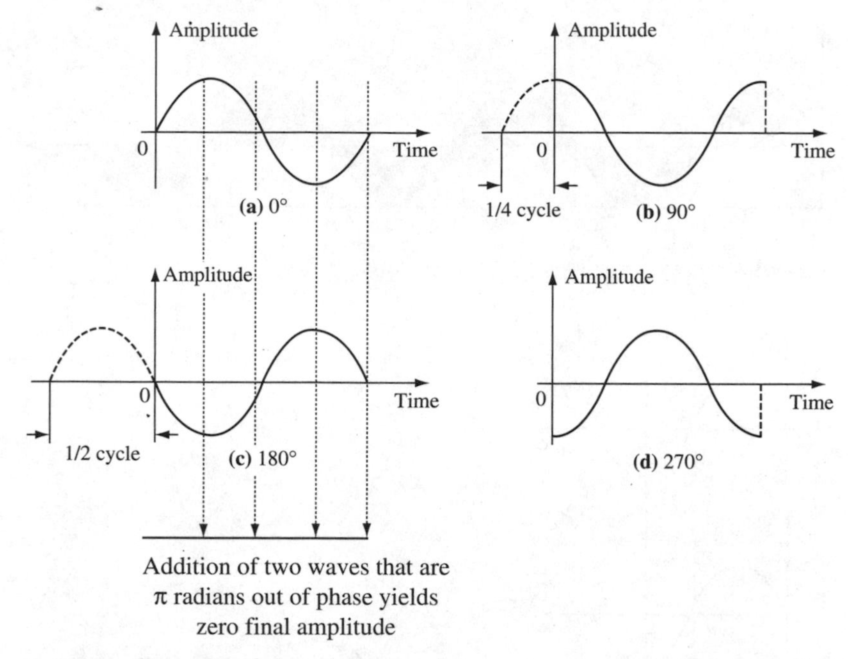

Figure 1.4. Examples of phase differences between two sine waves. Two waves that are 180° out of phase will add destructively.

importance when one is considering the operation of optical couplers, as described in Chap. 8.

Example

- A sine wave has a frequency $f = 5\,\text{kHz}$. Its period is $T = 1/5000\,\text{s} = 0.20\,\text{ms}$.
- A sine wave has a period $T = 1\,\text{ns}$. Its frequency is $f = 1/(10^{-9}\,\text{s}) = 1\,\text{GHz}$.
- A sine wave is offset by ¼ cycle with respect to time 0. Since 1 cycle is 360°, the phase shift is $\phi = 0.25 \times 360° = 90° = \pi/2$ rad.

Two further common characteristics in communications are the frequency spectrum (or simply spectrum) and the bandwidth of a signal. The *spectrum* of a signal is the range of frequencies that it contains. That is, the spectrum of a signal is the combination of all the individual sine waves of different frequencies which make up that signal. The *bandwidth* (designated by B) refers to the width of this spectrum.

Example If the spectrum of a signal ranges from its lowest frequency $f_{\text{low}} = 2\,\text{kHz}$ to its highest frequency $f_{\text{high}} = 22\,\text{kHz}$, then the bandwidth $B = f_{\text{high}} - f_{\text{low}} = 20\,\text{kHz}$.

1.3. Digital Signal Formats

A *digital signal* is an ordered sequence of *discrete symbols* selected from a finite set of elements. Examples of digital signals include the letters of an alphabet,

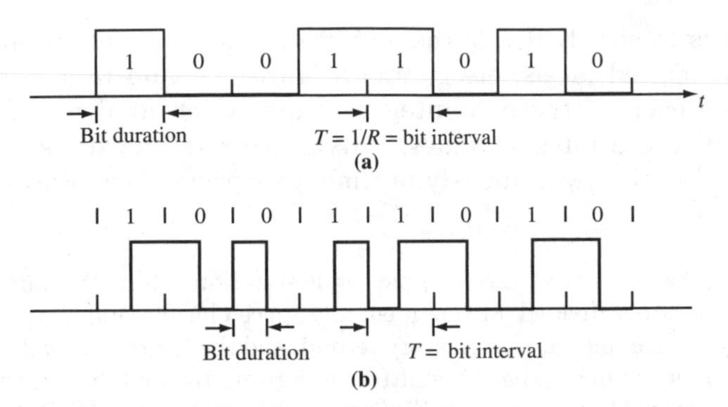

Figure 1.5. Examples of two binary waveforms showing their amplitude, period, and bit duration. (*a*) The bit fills the entire period for 1 bit only; (*b*) a 1 bit fills the first half and a 0 bit fills the second half of a period.

numbers, and other symbols such as @, #, or %. These discrete symbols are normally represented by unique patterns of pulses of electric voltages or optical intensity that can take on two or more levels.

A common digital signal configuration is the *binary* waveform shown in Fig. 1.5. A binary waveform is represented by a sequence of two types of pulses of known shape. The information contained in a digital signal is given by the particular sequence of the presence (a *binary one*, or simply either *one* or 1) and absence (a *binary zero*, or simply either *zero* or 0) of these pulses. These are known commonly as *bits* (this word was derived from binary digits). Since digital logic is used in the generation and processing of 1 and 0 bits, these bits often are referred to as a *logic one* (or *logic* 1) and a *logic zero* (or *logic* 0), respectively.

The time slot T in which a bit occurs is called the *bit interval*, *bit period*, or *bit time*. (Note that this T is different from the T used for designating the period of a waveform.) The bit intervals are regularly spaced and occur every $1/R$ seconds (s), or at a rate of R *bits per second* (abbreviated as bps in this book), where R is called the *bit rate* or the *data rate*. As an example, a data rate of 2×10^9 bits per second (bps) $=$ 2 Gbps (gigabits per second). A bit can fill the entire bit interval or part of it, as shown in Fig. 1.5*a* and *b*, respectively.

A block of 8 bits often is used to represent an encoded symbol or word and is referred to as an *octet* or a *byte*.

1.4. Digitization of Analog Signals

An analog signal can be transformed to a digital signal through a process of periodic sampling and the assignment of quantized values to represent the intensity of the signal at regular intervals of time.

To convert an analog signal to a digital form, one starts by taking instantaneous measures of the height of the signal wave at regular intervals, which is called *sampling* the signal. One way to convert these analog samples to a digital format is to simply divide the amplitude excursion of the analog signal into N equally spaced levels, designated by integers, and to assign a discrete binary word to each of these N integer values. Each analog sample is then assigned one of these integer values. This process is known as *quantization*. Since the signal varies continuously in time, this process generates a sequence of real numbers.

Example Figure 1.6 shows an example of digitization. Here the allowed voltage-amplitude excursion is divided into eight equally spaced levels ranging from 0 to V volts (V). In this figure, samples are taken every second, and the nearest discrete quantization level is chosen as the one to be transmitted, according to the 3-bit binary code listed next to the quantized levels shown in Fig. 1.6. At the receiver this digital signal is then demodulated. That is, the quantized levels are reassembled into a continuously varying analog waveform.

Nyquist Theorem Note that the equally spaced levels in Fig. 1.6 are the simplest quantization implementation, which is produced by a *uniform quantizer*. Frequently it is more advantageous to use a *nonuniform quantizer* where the quantization levels are roughly proportional to the signal level. The companders used in telephone systems are an example of this.

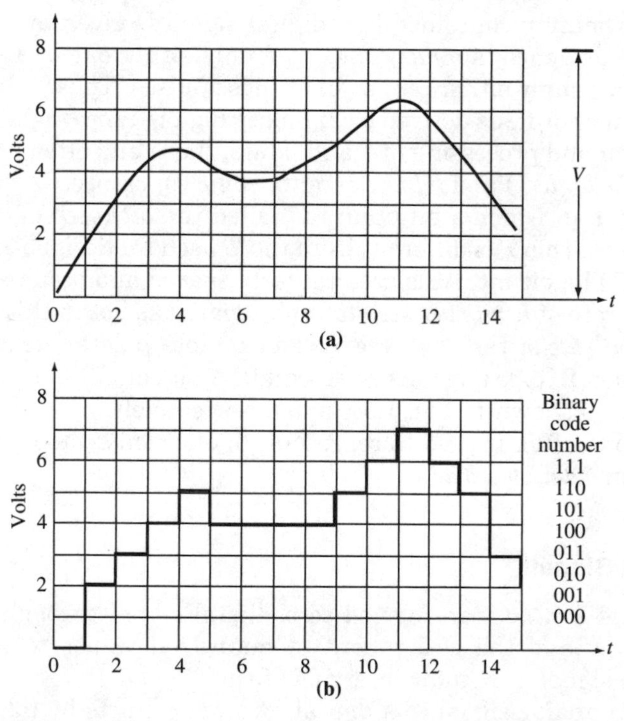

Figure 1.6. Digitization of analog waveforms. (*a*) Original signal varying between 0 and V volts; (*b*) quantized and sampled digital version.

Intuitively, one can see that if the digitization samples are taken frequently enough relative to the rate at which the signal varies, then to a good approximation the signal can be recovered from the samples by drawing a straight line between the sample points. The resemblance of the reproduced signal to the original signal depends on the fineness of the quantizing process and on the effect of noise and distortion added into the transmission system. According to the *Nyquist theorem*, if the sampling rate is at least 2 times the highest frequency, then the receiving device can faithfully reconstruct the analog signal. Thus, if a signal is limited to a bandwidth of B Hz, then the signal can be reproduced without distortion if it is sampled at a rate of $2B$ times per second. These data samples are represented by a binary code. As noted in Fig. 1.6, eight quantized levels having upper bounds $V_1, V_2, \ldots, V$ can be described by 3 binary digits ($2^3 = 8$). More digits can be used to give finer sampling levels. That is, if n binary digits represent each sample, then one can have 2^n quantization levels.

1.5. Electromagnetic Spectrum

To understand the distinction between electrical and optical communication systems and what the advantages are of lightwave technology, let us examine the spectrum of electromagnetic (EM) radiation shown in Fig. 1.7.

1.5.1. Telecommunication spectral band

All telecommunication systems use some form of electromagnetic energy to transmit signals from one device to another. Electromagnetic energy is a combination of electrical and magnetic fields and includes power, radio waves, microwaves, infrared light, visible light, ultraviolet light, x rays, and gamma rays. Each of these makes up a portion (or band) of the electromagnetic spectrum. The fundamental nature of all radiation within this spectrum is the same in that it can be viewed as electromagnetic waves that travel at the speed of light, which is

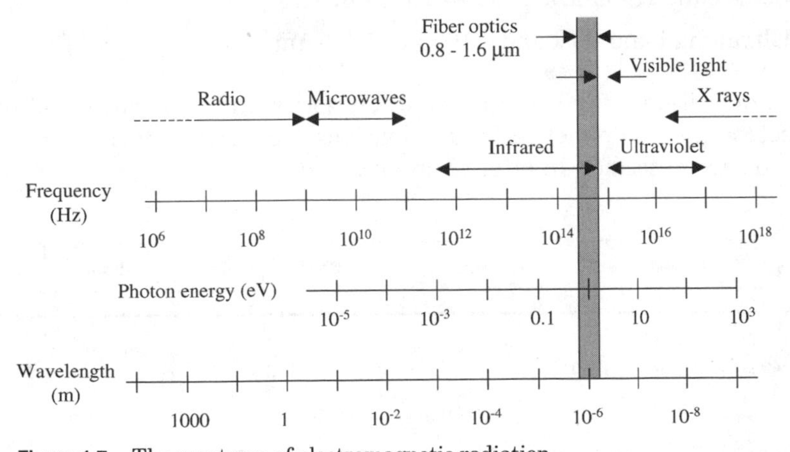

Figure 1.7. The spectrum of electromagnetic radiation.

about c = 300,000 kilometers per second (3×10^8 m/s) or 180,000 miles per second (1.8×10^5 mi/s) in a vacuum. Note that the speed of light in a material is less than c, as described in Chap. 3.

The physical property of the radiation in different parts of the spectrum can be measured in several interrelated ways. These are the length of one period of the wave, the energy contained in the wave, or the oscillating frequency of the wave. Whereas electric signal transmission tends to use frequency to designate the signal operating bands, optical communications generally use *wavelength* to designate the spectral operating region and *photon energy* or *optical power* when discussing topics such as signal strength or electrooptical component performance. We will look at the measurement units in greater detail in Chap. 3.

1.5.2. Optical communications band

The optical spectrum ranges from about 5 nm (ultraviolet) to 1 mm (far infrared), the visible region being the 400- to 700-nm band. Optical fiber communications use the spectral band ranging from 800 to 1675 nm.

The International Telecommunications Union (ITU) has designated six spectral bands for use in intermediate-range and long-distance optical fiber communications within the 1260- to 1675-nm region. As Chap. 4 describes, these band designations arose from the physical characteristics of optical fibers and the performance behavior of optical amplifiers. As shown in Fig. 1.8, the regions are known by the letters O, E, S, C, L, and U, which are defined as follows:

- Original band (O-band): 1260 to 1360 nm
- Extended band (E-band): 1360 to 1460 nm
- Short band (S-band): 1460 to 1530 nm
- Conventional band (C-band): 1530 to 1565 nm
- Long band (L-band): 1565 to 1625 nm
- Ultralong band (U-band): 1625 to 1675 nm

The operational performance characteristics and applications of optical fibers, electrooptic components, and other passive optical devices for use in these bands are described in later chapters.

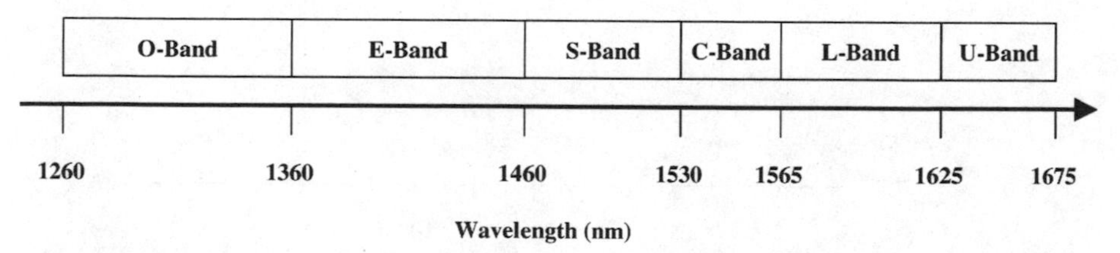

Figure 1.8. Definitions of spectral bands for use in optical fiber communications.

1.6. Transmission Channels

Depending on what portion of the electromagnetic spectrum is used, electromagnetic signals can travel through a vacuum, air, or other transmission media. For example, electricity travels well through copper wires but not through glass. Light, on the other hand, travels well through air, glass, and certain plastic materials but not through copper.

1.6.1. Carrier waves

As electrical communication systems became more sophisticated, an increasingly greater portion of the electromagnetic spectrum was utilized for conveying larger amounts of information faster from one place to another. The reason for this development trend is that in electrical systems the physical properties of various transmission media are such that each medium type has a different frequency band in which signals can be transported efficiently. To utilize this property, information usually is transferred over the communication channel by superimposing the data onto a sinusoidally varying electromagnetic wave, which has a frequency response that matches the transfer properties of the medium. This wave is known as the *carrier*. At the destination the information is removed from the carrier wave and processed as desired. Since the amount of information that can be transmitted is directly related to the frequency range over which the carrier operates, increasing the carrier frequency theoretically increases the available transmission bandwidth and, consequently, provides a larger information capacity.

To send digital information on a carrier wave, one or more of the characteristics of the wave such as its amplitude, frequency, or phase are varied. This kind of modification is called *modulation* or *shift keying*, and the digital information signal is called the *modulating signal*. Figure 1.9 shows an example of *amplitude shift keying* (ASK) or *on/off keying* (OOK) in which the strength (amplitude) of the carrier wave is varied to represent 1 or 0 pulses. Here a high amplitude represents a 1 and a low amplitude is a 0.

Thus the trend in electrical communication system developments was to employ progressively higher frequencies, which offer corresponding increases in bandwidth or information capacity. However, beyond a certain carrier frequency, electrical transmission systems become extremely difficult to design, build, and operate. These limitations, plus the inherent advantages of smaller sizes and lower weight of dielectric transmission materials

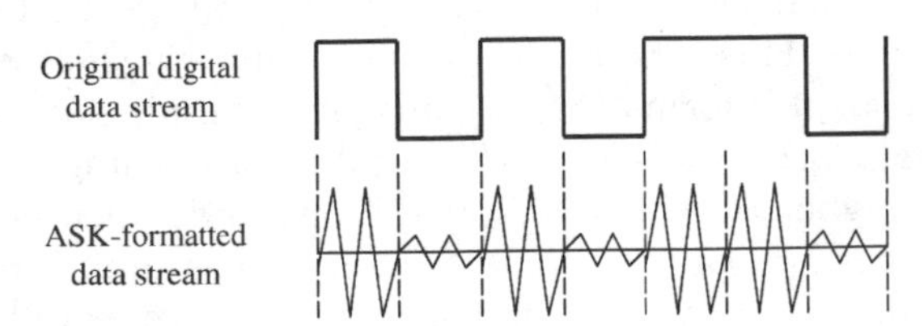

Figure 1.9. Concept of carrier waves.

such as glass, prompted researchers to consider optical fiber transmission technology.

1.6.2. Baseband signals

Baseband refers to the technology in which a signal is transmitted directly onto a channel without modulating a carrier. For example, this method is used on standard twisted-pair wire links running from an analog telephone to the nearest switching interface equipment.

The same baseband method is used in optical communications; that is, the optical output from a light source is turned on and off in response to the variations in voltage levels of an information-bearing electric signal. As is described in Chap. 6, for data rates less than about 2.5 Gbps, the light source itself can be turned on and off directly by the electric signal. For data rates higher than 2.5 Gbps, the optical output from a source such as a laser cannot respond fast enough. In this case an external device is used to modulate a steady optical output from a laser source.

1.7. Signal Multiplexing

Starting in the 1990s, a burgeoning demand on communication network assets emerged for services such as database queries and updates, home shopping, video-on-demand, remote education, audio and video streaming, and video conferencing. This demand was fueled by the rapid proliferation of personal computers coupled with a phenomenal increase in their storage capacity and processing capabilities, the widespread availability of the Internet, and an extensive choice of remotely accessible programs and information databases. To handle the ever-increasing demand for high-bandwidth services from users ranging from home-based computers to large businesses and research organizations, telecommunication companies worldwide are implementing increasingly sophisticated digital multiplexing techniques that allow a larger number of independent information streams to share the same physical transmission channel. This section describes some common techniques. Chapter 2 describes more advanced methodologies used and proposed for optical fiber transport systems.

Table 1.2 gives examples of information rates for some typical voice, video, and data services. To send these services from one user to another, network providers combine the signals from many different users and send the aggregate signal over a single transmission line. This scheme is known as *time-division multiplexing* (TDM). Here N independent information streams, each running at a data rate of R bps (bits per second), are interleaved electrically into a single information stream operating at a higher rate of $N \times R$ bps. To get a detailed perspective of this, let us look at the multiplexing schemes used in telecommunications.

Early applications of fiber optic transmission links were mainly for large-capacity telephone lines. These digital links consisted of time-division multiplexed

TABLE 1.2. Examples of Information Rates for Some Typical Voice, Video, and Data Services

Type of service	Data rate
Video on demand/interactive TV	1.5 to 6 Mbps
Video games	1 to 2 Mbps
Remote education	1.5 to 3 Mbps
Electronic shopping	1.5 to 6 Mbps
Data transfer or telecommuting	1 to 3 Mbps
Video conferencing	0.384 to 2 Mbps
Voice (single channel)	64 kbps

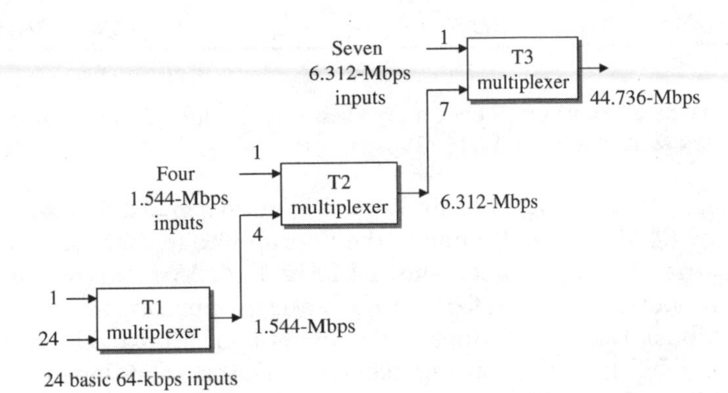

Figure 1.10. Digital transmission hierarchy used in the North American telephone network.

64-kbps voice channels. The multiplexing was developed in the 1960s and is based on what is known as the *plesiochronous digital hierarchy* (PDH). Figure 1.10 shows the digital transmission hierarchy used in the North American telephone network. The fundamental building block is a 1.544-Mbps transmission rate known as a DS1 rate, where DS stands for *digital system*. It is formed by time-division multiplexing 24 voice channels, each digitized at a 64-kbps rate (which is referred to as DS0). *Framing bits*, which indicate where an information unit starts and ends, are added along with these voice channels to yield the 1.544-Mbps bit stream. At any level a signal at the designated input rate is multiplexed with other input signals at the same rate.

DSn versus Tn In describing telephone network data rates, one also sees the terms T1, T3, and so on. Often people use the terms Tn and DSn (for example, T1 and DS1 or T3 and DS3) interchangeably. However there is a subtle difference in their meaning. DS1, DS2, DS3, and so on refer to a *service type*; for example, a user who wants to send information at a 1.544-Mbps rate would subscribe to a DS1 service. T1, T2, T3, and so on refer to the technology used to deliver that service over a physical link. For example,

TABLE 1.3. Digital Multiplexing Levels Used in North America, Europe, and Japan

Digital multiplexing level	Number of 64-kbps channels	Bit rate, Mbps		
		North America	Europe	Japan
0	1	0.064	0.064	0.064
1	24	1.544		1.544
	30		2.048	
2	96	6.312		6.312
	120		8.448	
3	480		34.368	32.064
	672	44.736		
4	1920		139.264	
	4032	274.176		
	5760			397.200

the DS1 service is transported over a physical wire or optical fiber using electrical or optical pulses sent at a T1 = 1.544-Mbps rate.

The system is not restricted to multiplexing voice signals. For example, at the DS1 level, any 64-kbps digital signal of the appropriate format could be transmitted as one of the 24 input channels shown in Fig. 1.10. As noted there and in Table 1.3, the main multiplexed rates for North American applications are designated as DS1 (1.544 Mbps), DS2 (6.312 Mbps), and DS3 (44.736 Mbps). Similar hierarchies using different bit rate levels are employed in Europe and Japan, as Table 1.3 shows. In Europe the multiplexing hierarchy is labeled E1, E2, E3, and so on.

1.8. SONET/SDH Multiplexing Hierarchy

With the advent of high-capacity fiber optic transmission lines in the 1980s, service providers established a standard signal format called *synchronous optical network* (SONET) in North America and *synchronous digital hierarchy* (SDH) in other parts of the world. These standards define a synchronous frame structure for sending multiplexed digital traffic over optical fiber trunk lines. The basic building block and the first level of the SONET signal hierarchy are called the *Synchronous Transport Signal—Level 1* (STS-1), which has a bit rate of 51.84 Mbps. Higher-rate SONET signals are obtained by byte-interleaving N of these STS-1 frames, which are then scrambled and converted to an *optical carrier—level N* (OC-N) signal. Thus the OC-N signal will have a line rate exactly N times that of an OC-1 signal. For SDH systems the fundamental building block is the 155.52-Mbps *Synchronous Transport Module—Level 1* (STM-1). Again, higher-rate information streams are generated by synchronously multiplexing N different STM-1 signals to form the STM-N signal. Table 1.4 shows commonly used SDH and SONET signal levels.

TABLE 1.4. Commonly Used SONET and SDH Transmission Rates

SONET level	Electrical level	SDH level	Line rate, Mbps	Common rate name
OC-1	STS-1	—	51.84	—
OC-3	STS-3	STM-1	155.52	155 Mbps
OC-12	STS-12	STM-4	622.08	622 Mbps
OC-48	STS-48	STM-16	2,488.32	2.5 Gbps
OC-192	STS-192	STM-64	9,953.28	10 Gbps
OC-768	STS-768	STM-256	39,813.12	40 Gbps

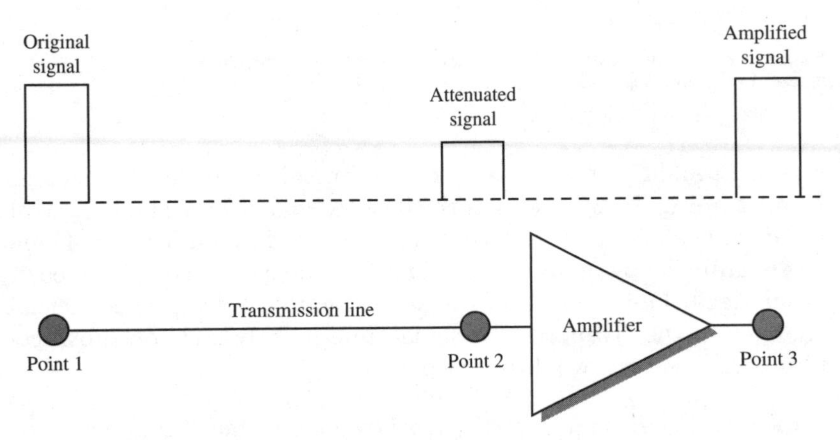

Figure 1.11. Amplifiers periodically compensate for energy losses along a channel.

1.9. Decibels

Attenuation (reduction) of the signal strength arises from various loss mechanisms in a transmission medium. For example, electric power is lost through heat generation as an electric signal flows along a wire, and optical power is attenuated through scattering and absorption processes in a glass fiber or in an atmospheric channel. To compensate for these energy losses, amplifiers are used periodically along a channel to boost the signal level, as shown in Fig. 1.11.

A convenient method for establishing a measure of attenuation is to reference the signal level to some absolute value or to a noise level. For guided media, the signal strength normally decays exponentially, so for convenience one can designate it in terms of a logarithmic power ratio measured in *decibels* (dB). In unguided (wireless) media, the attenuation is a more complex function of distance and the composition of the atmosphere. The dB unit is defined by

$$\text{Power ratio in dB} = 10 \log \frac{P_2}{P_1} \tag{1.1}$$

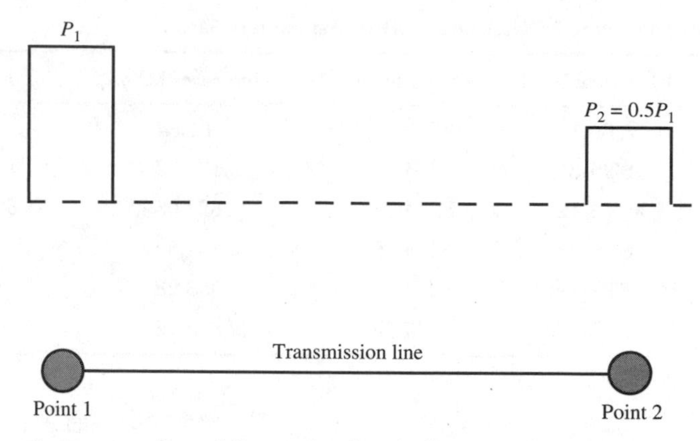

Figure 1.12. P_1 and P_2 are the electrical or optical power levels of a signal at points 1 and 2.

where P_1 and P_2 are the electrical or optical power levels of a signal at points 1 and 2 in Fig. 1.12, and log is the base-10 logarithm. The logarithmic nature of the decibel allows a large ratio to be expressed in a fairly simple manner. Power levels differing by many orders of magnitude can be compared easily when they are in decibel form. Another attractive feature of the decibel is that to measure changes in the strength of a signal, one merely adds or subtracts the decibel numbers between two different points.

Example Assume that after a signal travels a certain distance in some transmission medium, the power of the signal is reduced to one-half, that is, $P_2 = 0.5\,P_1$ in Fig. 1.12. At this point, by using Eq. (1.1) the attenuation or loss of power is

$$10 \log \frac{P_2}{P_1} = 10 \log \frac{0.5P_1}{P_1} = 10 \log 0.5 = 10(-0.3) = -3\,\text{dB}$$

Thus, -3 dB (or a 3-dB attenuation or loss) means that the signal has lost one-half of its power. If an amplifier is inserted into the link at this point to boost the signal back to its original level, then that amplifier has a 3-dB gain. If the amplifier has a 6-dB gain, then it boosts the signal power level to twice the original value.

Table 1.5 shows some sample values of power loss given in decibels and the percentage of power remaining after this loss. These types of numbers are important when one is considering factors such as the effects of tapping off a small part of an optical signal for monitoring purposes, for examining the power loss through some optical element, or when calculating the signal attenuation in a specific length of optical fiber.

Example Consider the transmission path from point 1 to point 4 shown in Fig. 1.13. Here the signal is attenuated by 9 dB between points 1 and 2. After getting a 14-dB boost from an amplifier at point 3, it is again attenuated by 3 dB between points 3 and 4.

TABLE 1.5. Representative Values of Decibel Power Loss and the Remaining Percentages

Power loss, dB	Percentage of power left
0.1	98
0.5	89
1	79
2	63
3	50
6	25
10	10
20	1

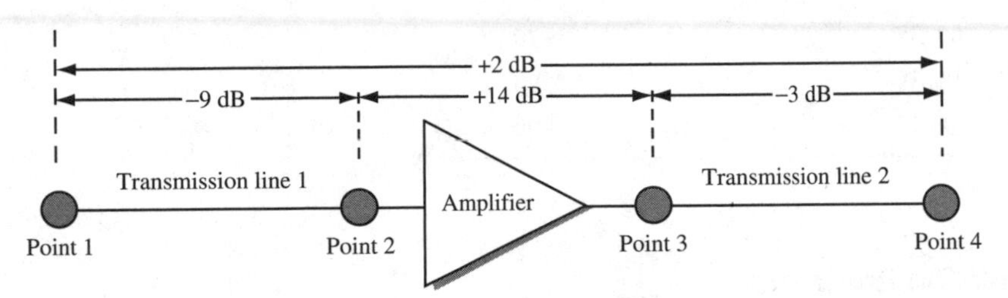

Figure 1.13. Example of attenuation and amplification in a transmission path.

Relative to point 1, the signal level in decibels at point 4 is

$$\begin{aligned}\text{dB level at point 4} &= (\text{loss in line 1}) + (\text{amplifier gain}) + (\text{loss in line 2})\\ &= (-9\text{ dB}) + (14\text{ dB}) + (-3\text{ dB}) = +2\text{ dB}\end{aligned}$$

Thus the signal has a 2-dB gain (a factor of $10^{0.2} = 1.58$) in power in going from point 1 to point 4.

Since the decibel is used to refer to ratios or relative units, it gives no indication of the absolute power level. However, a derived unit can be used for this. Such a unit that is particularly common in optical fiber communications is the *dBm*. This expresses the power level P as a logarithmic ratio of P referred to 1 mW. In this case, the power in dBm is an absolute value defined by

$$\text{Power level, dBm} = 10 \log \frac{P\text{ mW}}{1\text{ mW}} \tag{1.2}$$

An important rule-of-thumb relationship to remember for optical fiber communications is 0 dBm = 1 mW. Therefore, positive values of dBm are greater than 1 mW, and negative values are less than this. Table 1.6 lists some examples of optical power levels and their dBm equivalents.

TABLE 1.6. Some Examples of Optical Power Levels and Their dBm Equivalents

Power	dBm equivalent
200 mW	23
100 mW	20
10 mW	10
1 mW	0
100 μW	−10
10 μW	−20
1 μW	−30
100 nW	−40
10 nW	−50
1 nW	−60
100 pW	−70
10 pW	−80
1 pW	−90

1.10. Simulation Programs

Computer-aided modeling and simulation software programs are essential tools to predict how an optical communication component, link, or network will function and perform. These programs are able to integrate component, link, and network functions, thereby making the design process more efficient, less expensive, and faster. The tools typically are based on graphical interfaces that include a library of icons containing the operational characteristics of devices such as optical fibers, couplers, light sources, optical amplifiers, and optical filters, plus the measurement characteristics of instruments such as optical spectrum analyzers, power meters, and bit error rate testers. To check the capacity of the network or the behavior of passive and active optical devices, network designers invoke different optical power levels, transmission distances, data rates, and possible performance impairments in the simulation programs.

An example of such programs is the suite of software-based modeling-tool modules from VPIsystems, Inc. These design and planning tools are intended for use across all levels of network analyses, performance evaluations, and technology comparisons ranging from components, to modules, to entire networks. They are used extensively by component and system manufacturers, system integrators, network operators, and access service providers for functions such as capacity planning, comparative assessments of various technologies, optimization of transport and service networks, syntheses and analyses of wavelength division multiplexing (WDM) system and link designs, and component designs.

One particular tool from the VPIsystems suite is VPItransmissionMaker, which is a design and simulation tool for optical devices, components, subsystems, and transmission systems. This has all the tools needed to explore, design, simulate, verify, and evaluate active and passive optical components, fiber amplifiers, dense WDM transmission systems, and broadband access networks. Familiar measurement instruments offer a wide range of settable options when displaying data from multiple runs, optimizations, and multidimensional parameter sweeps. Signal processing modules allow data to be manipulated to mimic any laboratory setup.

An abbreviated version of this simulation module is available for noncommercial educational use. This version is called VPIplayer and contains predefined component and link configurations that allow interactive concept demonstrations. Among the demonstration setups are optical amplifier structures, simple single-wavelength links, and WDM links. Although the configurations are fixed, the user has the ability to interactively change the operational parameter values of components such as optical fibers, light sources, optical filters, and optical amplifiers. This can be done very simply by using the mouse to move calibrated slider controls. Results are shown in a format similar to the displays presented by laboratory instruments and show, for example, link performance degradation or improvement when various component values change.

VPIplayer can be downloaded from www.VPIphotonics.com. In addition, at www.PhotonicsComm.com there are numerous examples of optical communication components and links related to topics in this book that the reader can download and simulate. As noted above, predefined component parameters in these examples may be modified very simply via calibrated slider controls.

1.11. Summary

This chapter provides some fundamental operational concepts and definitions of terminology related to wired telecommunication systems. This "wiring" encompasses copper wires and optical fibers. Some key concepts include the following:

- Section 1.1 gives basic definitions of metric prefixes, terms used in communications (e.g., information, message, and signal), waveform characteristics, and digital signal formats.
- Section 1.9 defines the concepts of decibels and the relative power unit dBm that is used extensively in optical communication systems.
- Although most signals occurring in nature are of an analog form, in general they can be sent more easily and with a higher fidelity when changed to a digital format.
- A common digital signal configuration is the binary waveform consisting of binary 0 and binary 1 pulses called bits.
- The SONET/SDH multiplexing hierarchy defines a standard synchronous frame structure for sending multiplexed digital traffic over optical fiber transmission lines.

The designations for SONET data rates range from OC-1 (51.84 Mbps) to OC-768 (768 $\times$ OC-1 $\approx$ 40 Gbps) and beyond.

The designations for SDH data rates range from STM-1 (155.52 Mbps) to STM-256 (256 $\times$ STM-1 $\approx$ 40 Gbps) and beyond.

- For optical communications the ITU-T has defined the following six spectral bands in the 1260- to 1675-nm range:

  Original band (O-band): 1260 to 1360 nm
  Extended band (E-band): 1360 to 1460 nm
  Short band (S-band): 1460 to 1530 nm
  Conventional band (C-band): 1530 to 1565 nm
  Long band (L-band): 1565 to 1625 nm
  Ultralong band (U-band): 1625 to 1675 nm
- Computer-aided modeling and simulation software programs are essential tools to predict how an optical communication component, link, or network will function and perform. These programs are able to integrate component, link, and network functions, thereby making the design process more efficient, less expensive, and faster. One particular tool from VPIsystems, Inc. is VPItransmissionMaker, which is a design and simulation tool for optical devices, components, subsystems, and transmission systems. An abbreviated version of this simulation module is available for noncommercial educational use. This version is called VPIplayer and contains predefined component and link configurations that allow interactive concept demonstrations. VPIplayer can be downloaded from www.VPIphotonics.com. In addition, at www.PhotonicsComm.com there are numerous examples of optical communication components and links related to topics in this book that the reader can download and simulate.

Further Reading

1. A. B. Carlson, *Communication Systems*, 4th ed., McGraw-Hill, Burr Ridge, Ill., 2002. This classic book gives senior-level discussions of electrical communication systems.
2. B. A. Forouzan, *Introduction to Data Communications and Networking*, 2d ed., McGraw-Hill, Burr Ridge, Ill., 2001. This book gives intermediate-level discussions of all aspects of communication systems.
3. W. Goralski, *Optical Networking and WDM*, McGraw-Hill, New York, 2001.
4. J. Hecht, *City of Light*, Oxford University Press, New York, 1999. This book gives an excellent account of the history behind the development of optical fiber communication systems.
5. G. Keiser, *Optical Fiber Communications*, 3d ed., McGraw-Hill, Burr Ridge, Ill., 2000. This book presents more advanced discussions and theoretical analyses of optical fiber component and system performance material.
6. G. Keiser, *Local Area Networks*, 2d ed., McGraw-Hill, Burr Ridge, Ill., 2002. This book presents topics related to all aspects of local-area communications.
7. R. Ramaswami and K. N. Sivarajan, *Optical Networks*, 2d ed., Morgan Kaufmann, San Francisco, 2002. This book presents more advanced discussions and theoretical analyses of optical networking material.
8. N. Thorsen, *Fiber Optics and the Telecommunications Explosion*, Prentice Hall, New York, 1998.

Chapter

2

Optical Communication Systems Overview

Telecommunications network organizations started using optical fiber communication links just over 25 years ago. Since then, researchers have devised and developed a truly impressive collection of sophisticated passive and active optical components, transmission techniques that are unique to optical links, and software-based modeling tools for components, links, and networks. These developments enabled optical networks to deliver increasingly higher data rates over longer and longer distances for both terrestrial and transoceanic links.

The purpose of this chapter is to get an appreciation of this growth and to set the basis for understanding the material in the rest of the book. First we present the motivations for using optical communication, such as the physical differences and capacity advantages between electrical and optical signaling formats. Then we will review the evolution of optical fiber transmission systems, starting with the quest to create a viable optical fiber. This will lead us to present-day implementations. The chapter ends with an overview of the general concepts and issues of current transmission equipment, links, and networks. This includes the process of combining many independent wavelengths onto the same fiber, signal routing and switching, and standards for optical communications.

2.1. Motivations for Using Optical Fiber Systems

The motivation for developing optical fiber communication systems started with the invention of the laser in the early 1960s. The operational characteristics of this device encouraged researchers to examine the optical spectrum as an extension of the radio and microwave spectrum to provide transmission links with extremely high capacities. As research progressed, it became clear that many complex problems stood in the way of achieving such a super broadband communication system. However, it also was noted that other properties of optical fibers gave them a number of inherent cost and operational advantages over copper wires and made them highly attractive for simple on/off keyed links. Included in these advantages are the following:

- *Long transmission distance*. Optical fibers have lower transmission losses compared to copper wires. This means that data can be sent over longer distances, thereby reducing the number of intermediate repeaters needed for these spans. This reduction in equipment and components decreases system cost and complexity.
- *Large information capacity*. Optical fibers have wider bandwidths than copper wires, which means that more information can be sent over a single physical line. This property results in a decrease in the number of physical lines needed for sending a certain amount of information.
- *Small size and low weight*. The low weight and the small dimensions of fibers offer a distinct advantage over heavy, bulky wire cables in crowded underground city ducts or in ceiling-mounted cable trays. This also is of importance in aircraft, satellites, and ships where small, lightweight cables are advantageous, and in tactical military applications where large amounts of cable must be unreeled and retrieved rapidly.
- *Immunity to electrical interference*. An especially important feature of optical fibers relates to the fact that they consist of dielectric materials, which means they do not conduct electricity. This makes optical fibers immune to the electromagnetic interference effects seen in copper wires, such as inductive pickup from other adjacent signal-carrying wires or coupling of electrical noise into the line from any type of nearby equipment.
- *Enhanced safety*. Optical fibers do not have the problems of ground loops, sparks, and potentially high voltages inherent in copper lines. However, precautions with respect to laser light emissions need to be observed to prevent possible eye damage.
- *Increased signal security*. An optical fiber offers a high degree of data security, since the optical signal is well confined within the fiber and any signal emissions are absorbed by an opaque coating around the fiber. This is in contrast to copper wires where electric signals often can be tapped off easily. This makes fibers attractive in applications where information security is important, such as in financial, legal, government, and military systems.

Now let us look at a very brief history of optical communications and how this led to present-day systems. For an interesting detailed account of the development of optical communications from ancient to modern times, the reader is referred to the book *City of Light* by Jeff Hecht.

2.2. Evolution of Optical Communications

A challenge in using an optical fiber for a communications channel is to have a flexible, low-loss medium that transfers a light signal over long distances without significant attenuation and distortion. Glass is an obvious material for such applications. The earliest known glass was made around 2500 B.C., and glass already was drawn into fibers during the time of the Roman Empire. However, such glasses have very high losses and are not suitable for communication applications. One of the first known attempts of using optical fibers for communication purposes was a demonstration in 1930 by Heinrich Lamm of image transmission through a short bundle of optical fibers for potential medical imaging. However, no further work was done beyond the demonstration phase, since the technology for producing low-loss fibers with good light confinement was not yet mature.

Further work and experiments on using optical fibers for image transmission continued, and by 1960 glass-clad fibers had an attenuation of about 1 dB/m. This attenuation allowed fibers to be used for medical imaging, but it was still much too high for communications, since only 1 percent of the inserted optical power would emerge from the end of a 20-m-long fiber. Optical fibers attracted the attention of researchers at that time because they were analogous in theory to plastic dielectric waveguides used in certain microwave applications. In 1961 Elias Snitzer published a classic theoretical description of single-mode fibers with implications for information transmission use. However, to be applicable to communication systems, optical fibers would need to have a loss of no more than 10 or 20 dB/km (a power loss factor of 10 to 100).

In the early 1960s when Charles Kao was at the Standard Telecommunication Laboratories in England, he pursued the idea of using a clad glass fiber for an optical waveguide. After he and George Hockman painstakingly examined the transparency properties of various types of glass, Kao made a prediction in 1966 that losses of no more than 20 dB/km were possible in optical fibers. In July 1966, Kao and Hockman presented a detailed analysis for achieving such a loss level. Kao then went on to actively advocate and promote the prospects of fiber communications, which generated interest in laboratories around the world to reduce fiber loss. It took 4 years to reach Kao's predicted goal of 20 dB/km, and the final solution was different from what many had expected.

To understand the process of making a fiber, consider the schematic of a typical fiber structure, shown in Fig. 2.1. A fiber consists of a solid glass cylinder called the *core*. This is surrounded by a dielectric *cladding*, which has a different material property from that of the core in order to achieve light guiding in the fiber. Surrounding these two layers is a polymer buffer coating that protects the

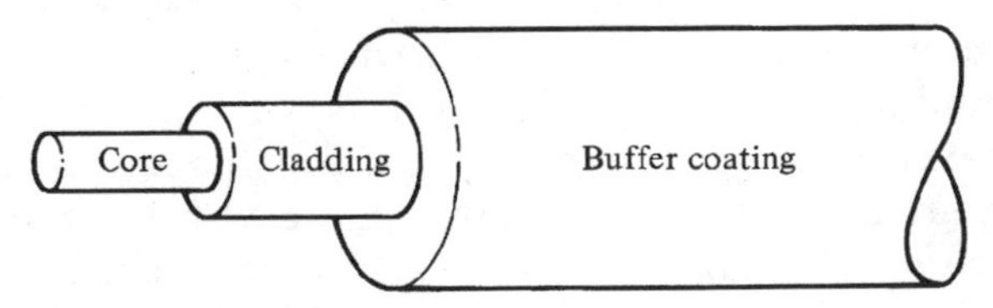

Figure 2.1. A typical fiber structure consists of a core, a cladding, and a buffer coating.

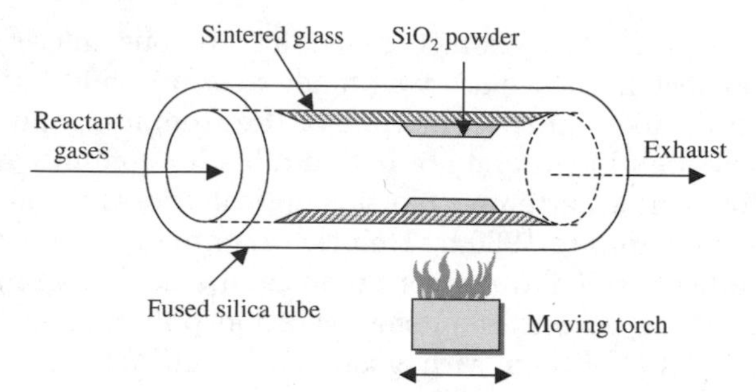

Figure 2.2. Illustration of the modified chemical vapor deposition (MCVD) process.

fiber from mechanical and environmental effects. (Chapter 4 gives the details of how a fiber guides light.)

The first step in making a fiber is to form a clear glass rod or tube called a *preform*. Currently a preform is made by one of several vapor-phase oxidation processes. In each of these processes, highly pure vapors of metal halides (e.g., $SiCl_4$ and $GeCl_4$) react with oxygen to form a white powder of SiO_2 particles. The particles are then collected on the surface of a bulk glass (such as the outside of a rod or the inside of a tube) and are *sintered* (transformed to a homogeneous glass mass by heating without melting) to form a clear glass rod or tube. Figure 2.2 shows one such process, which is known as the *modified chemical vapor deposition* (MCVD) process. Here as the SiO_2 particles are deposited, the tube is rotated and a torch travels back and forth along the tube to sinter the particles. Chapter 20 describes this and other fiber fabrication processes.

Depending on how long a fiber is desired, the preform might be a meter long and several centimeters in diameter. The preform has two distinct regions that correspond to the core and cladding of the eventual fiber. As illustrated in Fig. 2.3, fibers are made by precision feeding the preform into a circular furnace. This process softens the end of the preform to the point where it can be drawn into a long, very thin filament which becomes the optical fiber.

Prior to 1970 most researchers had tried to purify compound glasses used for standard optics, which are easy to melt and draw into fibers. A different approach was taken at the Corning Glass Works where Robert Maurer, Donald Keck, and Peter Schultz started with fused silica. This material can be made extremely pure, but has a high melting point. The Corning team made cylindrical preforms by depositing purified materials from the vapor phase. In September 1970, they announced the fabrication of single-mode fibers with an attenuation of 17 dB/km at the 633-nm helium-neon line (a loss factor of 50 over 1 km). This dramatic breakthrough was the first among the many developments that opened the door to fiber optic communications. The ensuing years saw further reductions in optical fiber attenuation. By the middle of 1972 Maurer, Keck, and Schultz had made multimode germania-doped fibers with a 4-dB/km loss and much greater strength than the earlier brittle titania-doped fibers.

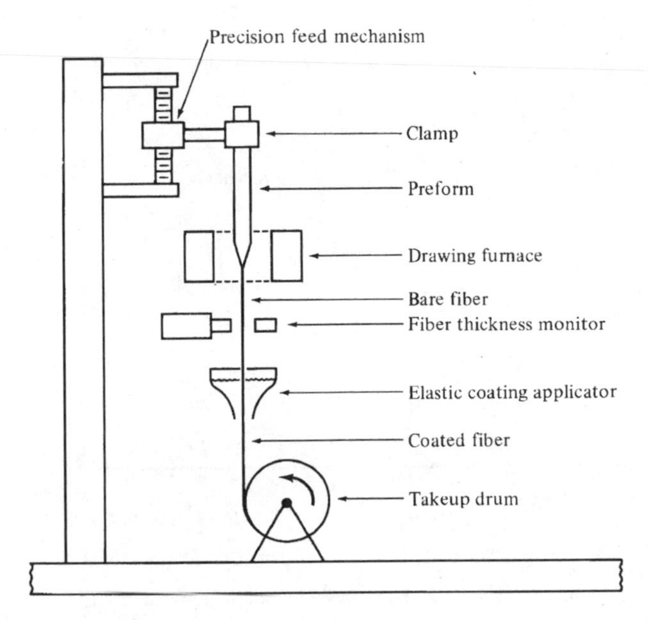

Figure 2.3. Illustration of a fiber-drawing process.

A problem with using single-mode fibers in the 1970s was that no light source existed which could couple a sufficient amount of optical power into the tiny fiber core (nominally 9 μm in diameter). Therefore, multimode fibers with larger core diameters ranging from 50 to 100 μm were used first. The main light sources were light-emitting diodes (LEDs) and laser diodes that emitted at 850 nm. The combination of early sources, multimode fibers, and operation at 850 nm limited the optical fiber links to rates of about 140 Mbps over distances of 10 km.

To overcome these limitations, around 1984 the next generation of optical systems started employing single-mode fibers and operated at 1310 nm where both the fiber attenuation and signal distortion effects are lower than at 850 nm. Figure 2.4 shows this attenuation difference in decibels per kilometer as a function of wavelength. The figure also shows that early optical fibers had three low-loss transmission windows defined by attenuation spikes due to absorption from water molecules. The *first window* ranges from 800 to 900 nm, the *second window* is centered at 1310 nm, and the *third window* ranges from 1480 to 1600 nm. In 1310-nm systems the transmission distance is limited primarily by fiber loss and not by other factors that might not allow a longer and faster transmission link. Therefore the next evolutionary step was to deploy links at 1550 nm where the attenuation was only one-half that at 1310 nm. This move started a flurry of activity in developing new fiber types, different light sources, new types of photodetectors, and a long shopping list of specialized optical components. This activity arose because transmitting in the 1550-nm region and pushing the data rate to higher and higher speeds brought about a whole series

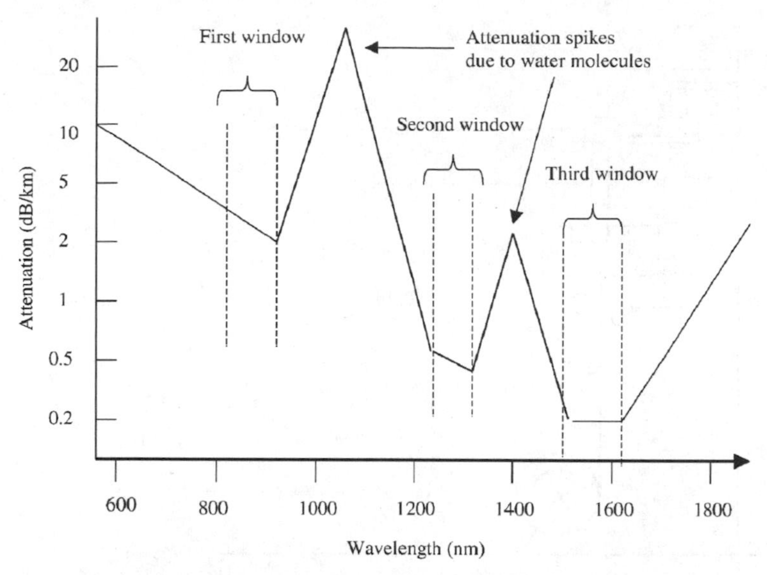

Figure 2.4. Attenuation of an optical fiber in decibels per kilometer as a function of wavelength.

of new challenging problems. The following sections give a taste of these challenges, and later chapters will elaborate on their solutions.

2.3. Elements of an Optical Link

From a simplistic point of view, the function of an optical fiber link is to transport a signal from some piece of electronic equipment (e.g., a computer, telephone, or video device) at one location to corresponding equipment at another location with a high degree of reliability and accuracy. Figure 2.5 shows the key sections of an optical fiber communications link, which are as follows:

- *Transmitter*. The transmitter consists of a light source and associated electronic circuitry. The source can be a light-emitting diode or a laser diode. The electronics are used for setting the source operating point, controlling the light output stability, and varying the optical output in proportion to an electrically formatted information input signal. Chapter 6 gives more details on sources and transmitters.
- *Optical fiber*. As Chap. 5 describes, the optical fiber is placed inside a cable that offers mechanical and environmental protection. A variety of fiber types exist, and there are many different cable configurations depending on whether the cable is to be installed inside a building, in underground pipes, outside on poles, or underwater.
- *Receiver*. Inside the receiver is a photodiode that detects the weakened and distorted optical signal emerging from the end of an optical fiber and converts

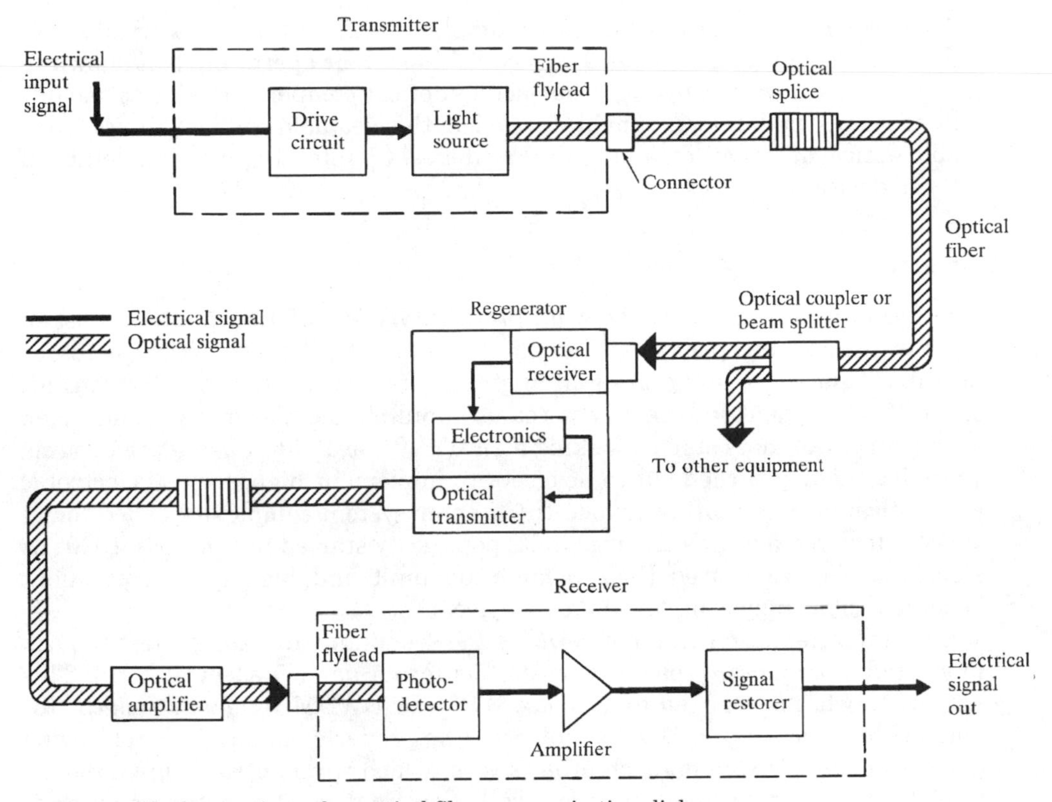

Figure 2.5. The key sections of an optical fiber communications link.

it to an electric signal. The receiver also contains amplification devices and circuitry to restore signal fidelity. Chapter 7 gives details on this topic.

- *Passive devices.* As Chap. 9 describes, passive devices are optical components that require no electronic control for their operation. Among these are optical connectors for connecting cables, splices for attaching one bare fiber to another, optical isolators that prevent unwanted light from flowing in a backward direction, optical filters that select only a narrow spectrum of desired light, and couplers used to tap off a certain percentage of light, usually for performance monitoring purposes.
- *Optical amplifiers.* After an optical signal has traveled a certain distance along a fiber, it becomes weakened due to power loss along the fiber. At that point the optical signal needs to get a power boost. Traditionally the optical signal was converted to an electric signal, amplified electrically, and then converted back to an optical signal. The invention of an optical amplifier that boosts the power level completely in the optical domain circumvented these transmission bottlenecks, as Chap. 11 describes.

- *Active components*. Lasers and optical amplifiers fall into the category of active devices, which require an electronic control for their operation. Not shown in Fig. 2.5 are a wide range of other active optical components. These include light signal modulators, tunable (wavelength-selectable) optical filters, variable optical attenuators, and optical switches. Chapter 10 gives the details of these devices.

2.4. WDM Concept

The use of *wavelength division multiplexing* (WDM) offers a further boost in fiber transmission capacity. As Fig. 2.6 illustrates, the basis of WDM is to use multiple light sources operating at slightly different wavelengths to transmit several independent information streams simultaneously over the same fiber. Although researchers started looking at WDM in the 1970s, during the ensuing years it generally turned out to be easier to implement higher-speed electronic and optical devices than to invoke the greater system complexity called for in WDM. However, a dramatic surge in its popularity started in the early 1990s as electronic devices neared their modulation limit and high-speed equipment became increasingly complex and expensive.

One implementation trend of WDM is the seemingly unending quest to pack more and more closely spaced wavelengths into a narrow spectral band. This has led to what is referred to as *dense WDM*, or DWDM. The wavelengths (or optical frequencies) in a DWDM link must be properly spaced to avoid having adjacent channels step on each other's toes, which would create signal distortion. In an optical system, interference between adjacent channels may arise from the fact that the center wavelength of laser diode sources and the spectral operating characteristics of other optical components in the link may drift with temperature and time. This may cause the signal pulses to drift or spread out spectrally. As Fig. 2.7 illustrates, if this drift or spreading is not controlled or if any guard band between wavelength channels is too small, the signal being produced at one wavelength will trespass into the spectral territory of another signal band and create interference.

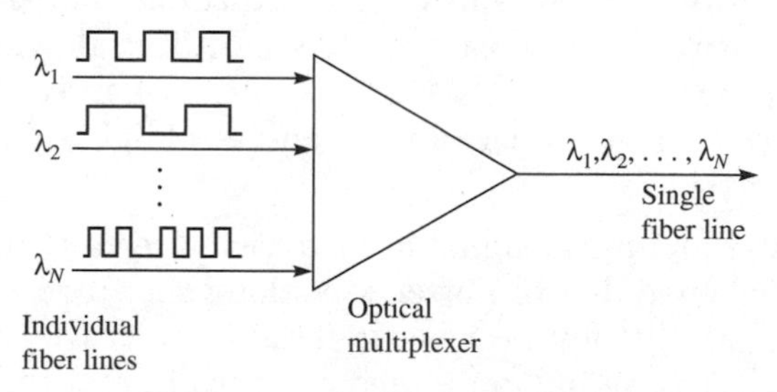

Figure 2.6. Basic concept of wavelength division multiplexing (WDM).

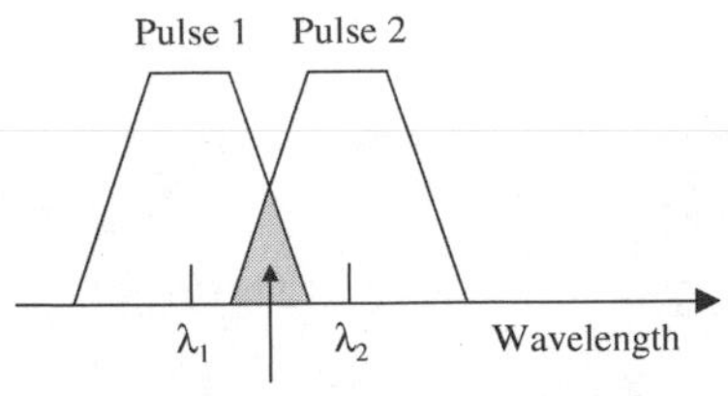

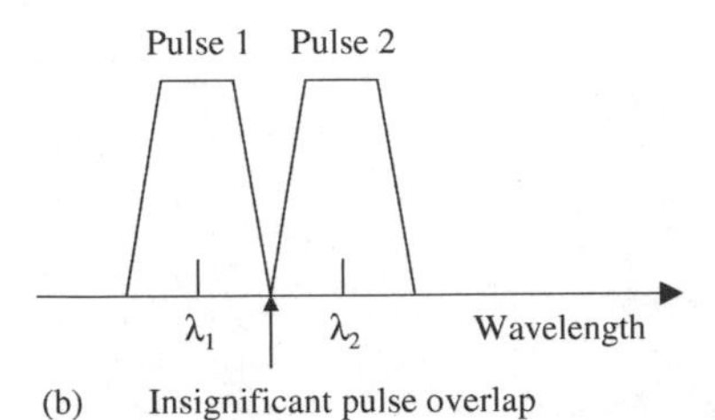

Figure 2.7. (*a*) Spectral interference between adjacent wavelength channels; (*b*) stable channels.

2.5. Applications of Optical Fiber Links

Optical fibers can be applied to interconnections ranging from localized links within an equipment rack to links that span continents or oceans. As Fig. 2.8 illustrates, networks are traditionally divided into the following three broad categories:

1. *Local-area networks* (LANs) interconnect users in a localized area such as a room, a department, a building, an office or factory complex, or a campus. Here the word *campus* refers to any group of buildings that are within reasonable walking distance of one another. For example, it could be the collocated buildings of a corporation, a large medical facility, or a university complex. LANs usually are owned, used, and operated by a single organization.
2. *Metropolitan-area networks* (MANs) span a larger area than LANs. This could be interconnections between buildings covering several blocks within a city or could encompass an entire city and the metropolitan area surrounding it. There is also some means of interconnecting the MAN resources with communication entities located in both LANs and wide-area networks. MANs are owned and operated by many organizations. When talking about MAN fiber optic applications, people tend to call them *metro applications*.
3. *Wide-area networks* (WANs) span a large geographic area. The links can range from connections between switching facilities in neighboring cities to long-haul terrestrial or undersea transmission lines running across a country or between countries. WANs invariably are owned and operated by many transmission service providers.

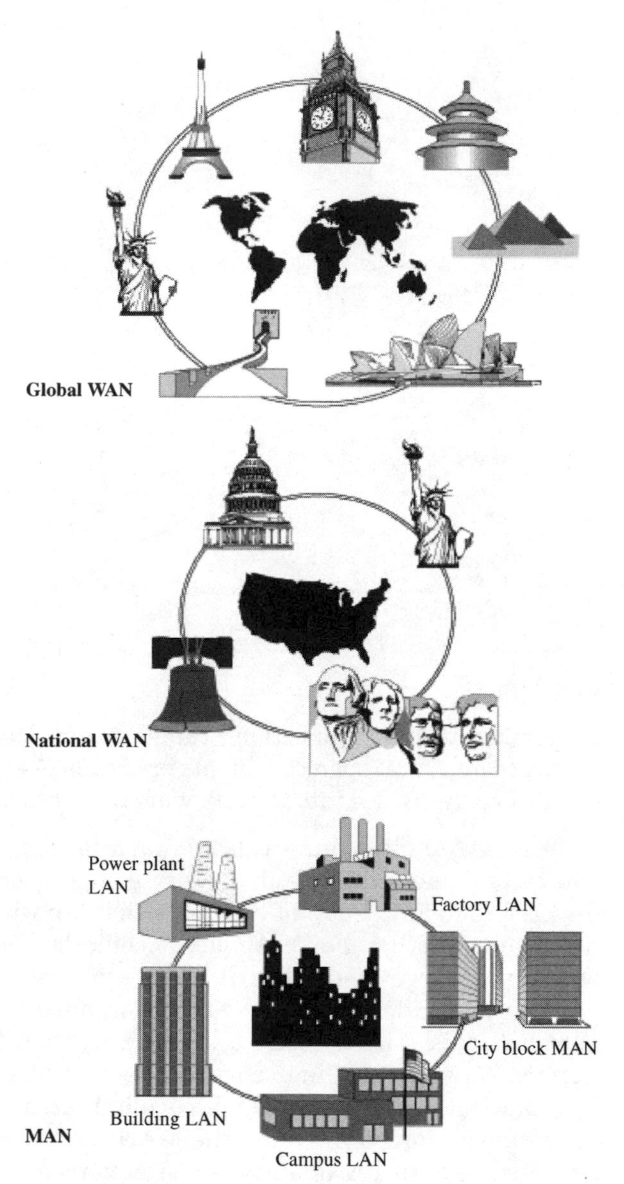

Figure 2.8. Examples of broad categories of networks.

2.6. Optical Networking and Switching

Another step toward realizing the full potential of optical fiber transmission capacity is the concept of an intelligent WDM network. In these networks the wavelength routing is done in the optical domain without going through the usual time-consuming sequence that involves an optical-to-electric signal conversion, an electric signal switching process, and then another conversion from an electrical back to an optical format. The motivation behind this concept is to extend the versatility of communication networks beyond architectures such as those provided by high-bandwidth point-to-point SONET light pipes. So far most of these types of *all-optical networks* are only concepts, since the technology still needs to mature. In particular, as described in Chap. 17, an optical cross-connect switch is one of the key elements needed to deploy agile optical networks. Developing such a component is a major challenge since it needs to switch optical signals at line rates (e.g., at 10-Gbps OC-192 or 40-Gbps OC-768 rates) without optical-to-electrical conversion, thereby providing lower switching costs and higher capacities than the currently used electrical cross-connects.

2.7. Standards for Optical Communications

When people travel from one country to another, they need to bring along an electrical adapter that will match up the voltage and plug configurations of their personal appliances to those of the other country. Even when you do this, sometimes you are greeted by sparks and black smoke as you plug something into a foreign electric socket. If the hotel personnel or clerks in stores do not speak the same language as you do, there is another interface problem. To avoid similar situations when trying to interface equipment from different manufacturers, engineers have devised many different types of standards so that diverse equipment will interface properly. However, since many people like to do things their own way, sometimes the standards in one country do not quite match those of another. Nevertheless, international standards for a wide range of component and system-level considerations have made life a lot easier.

There are three basic classes of standards for fiber optics: primary standards, component testing standards, and system standards.

Primary standards refer to measuring and characterizing fundamental physical parameters such as attenuation, bandwidth, operational characteristics of fibers, and optical power levels and spectral widths. In the United States the main organization involved in primary standards is the National Institute of Standards and Technology (NIST). This organization carries out fiber optic and laser standardization work, and it sponsors an annual conference on optical fiber measurements. Other national organizations include the National Physical Laboratory (NPL) in the United Kingdom and the Physikalisch-Technische Bundesanstalt (PTB) in Germany.

Component testing standards define relevant tests for fiber optic component performance, and they establish equipment calibration procedures. Several different organizations are involved in formulating testing standards, some very active ones being the Telecommunication Industries Association (TIA) in association with the Electronic Industries Association (EIA), the Telecommunication Standardization Sector of the International Telecommunication Union (ITU-T), and the International Electrotechnical Commission (IEC). The TIA has a list of over 120 fiber optic test standards and specifications under the general designation TIA/EIA-455-XX-YY, where XX refers to a specific measurement technique

TABLE 2.1. A Sampling of ITU-T Recommendations for Optical Links and Networks

Rec. no.	Title
G.650 to G.655	Definitions, test methods, and characteristics of various types of multimode and single-mode fibers
G.662	Generic Characteristics of Optical Amplifier Devices and Subsystems
G.671	Transmission Characteristics of Optical Components and Subsystems
G.709	Interfaces for the Optical Transport Network (OTN)
G.872	Architecture of Optical Transport Networks
G.874	Management Aspects of the Optical Transport Network Element
G.959.1	Optical Transport Network Physical Layer Interfaces
G.694.1	Spectral Grids for WDM Applications: DWDM Frequency Grid
G.694.2	Spectral Grids for WDM Applications: CWDM Wavelength Grid
G.975	Forward Error Correction for Submarine Systems

and YY refers to the publication year. These standards are also called *Fiber Optic Test Procedures* (FOTPs), so that TIA/EIA-455-XX becomes FOTP-XX. These include a wide variety of recommended methods for testing the response of fibers, cables, passive devices, and electrooptic components to environmental factors and operational conditions. For example, TIA/EIA-455-60-1997, or FOTP-60, is a method published in 1997 for measuring fiber or cable length.

System standards refer to measurement methods for links and networks. The major organizations are the American National Standards Institute (ANSI), the Institute for Electrical and Electronic Engineers (IEEE), and the ITU-T. Of particular interest for fiber optics systems are test standards and recommendations from the ITU-T. Within the G series (in the number range G.650 and higher) there are at least 44 recommendations that relate to fiber cables, optical amplifiers, wavelength multiplexing, optical transport networks (OTNs), system reliability and availability, and management and control for passive optical networks (PONs). In addition, within the same number range there are many recommendations referring to SONET and SDH. Table 2.1 lists a sampling of these ITU-T recommendations, which aim at all aspects of optical networking.

2.8. Summary

The dielectric properties of optical fibers give them a number of inherent cost and operational advantages over copper wires. Among these are lower weight, smaller size, greater information capacity, and immunity to signal interference. On the other hand, this comes with some increased complexity with respect to handling and connecting the hair-thin fibers.

Optical fiber communications has rapidly become a mature technology and now is ubiquitous in the telecommunications infrastructure. As is the case with any other technology, the challenges to improve performance are never-ending. Researchers are working to pack more and more wavelengths closer together in a given spectral band, to increase the data rates per wavelength, and to go longer distances by developing new types of optical amplifiers.

As a further step toward realizing the full potential of optical fiber transmission capacity, researchers are considering the concept of an intelligent WDM network. The major activity in this area is the development of an optical cross-connect (OXC) that will switch optical signals at line rates (e.g., at 10-Gbps OC-192 or 40-Gbps OC-768 rates) without optical-to-electrical conversion. The eventual creation of such a component will provide lower switching costs and higher capacities than the currently used electrical cross-connects.

A key ingredient for the widespread implementation of optical fiber technology is an extensive body of test, interface, and system design standards. For example, the TIA has published over 120 fiber optic test standards and specifications for testing the response of fibers, cables, passive devices, and electrooptic components to environmental factors and operational conditions. Furthermore, within the ITU-T G series there are at least 44 recommendations that relate to performance specifications for fiber cables, optical amplifiers, wavelength multiplexing, optical transport networks, system reliability and availability, and management and control for passive optical networks. In addition, within this G series there are many recommendations referring to SONET and SDH. Table 2.1 lists a sampling of ITU-T recommendations, which aim at all aspects of optical networking.

Further Reading

1. J. Hecht, *City of Light*, Oxford University Press, New York, 1999.
2. K. C. Kao and G. A. Hockman, "Dielectric-fibre surface waveguides for optical frequencies," *Proceedings IEE*, vol. 113, pp. 1151–1158, July 1966.
3. E. Snitzer, "Cylindrical dielectric waveguide modes," *J. Opt. Soc. Amer.*, vol. 51, pp. 491–498, May 1961.
4. G. Keiser, "A review of WDM technology and applications," *Opt. Fiber Technol.*, vol. 5, no. 1, pp. 3–39, 1999.
5. National Institute of Standards and Technology (NIST), 325 Broadway, Boulder, CO 80303 (http://www.nist.gov).
6. National Physical Laboratory (NPL), Teddington, Middlesex, United Kingdom (http://www.npl.co.uk).
7. Physikalisch-Technische Bundesanstalt (PTB), Braunschweig, Germany (http://www.ptb.de).
8. Telecommunication Industries Association (TIA) (http://www.tiaonline.org).
9. Electronic Industries Association (EIA), 2001 Eye Street, Washington, D.C. 20006.
10. Telecommunication Standardization Sector of the International Telecommunication Union (ITU-T), Geneva, Switzerland (http://www.itu.int).
11. International Electrotechnical Commission (IEC), Geneva, Switzerland (http://www.iec.ch).
12. American National Standards Institute (ANSI), New York (http://www.ansi.org).
13. Institute of Electrical and Electronic Engineers (IEEE), New York (http://www.ieee.org).
14. A. McGuire and P. A. Bonenfant, "Standards: The blueprints for optical networks," *IEEE Commun. Mag*, vol. 36, pp. 68–78, February 1998.

Chapter

3

The Behavior of Light

The concepts of how light travels along an optical fiber and how it interacts with matter are essential to understanding why certain components are needed and what their functions are in an optical fiber communication system. In this chapter discussions on the properties of light cover the dual wave-particle nature of light, the speed of light in different materials, reflection, refraction, and polarization. These concepts relate to optical phenomena that we see every day, such as light traveling through a solid (e.g., glass), reflection, and refraction. Obviously these factors also play a major role in optical fiber communications. So, let's get "enlightened" with the following discussions.

3.1. The Dual Wave-Particle Nature of Light

The fundamental behavior of light is somewhat mysterious since some phenomena can be explained by using a wave theory whereas in other cases light behaves as though it is composed of miniature particles. This results in a *dual wave-particle nature of light*. The wave nature is necessary to explain how light travels through an optical fiber, but the particle theory is needed to explain how optical sources generate signals and how photodetectors change these optical signals to electric signals.

Light particles are known as *photons*, which have a certain energy associated with them. As described in Sec. 3.3, the most common measure of photon energy is the *electron volt* (eV), which is the energy a photon gains when moving through a 1-V electric field. Photons travel in straight lines called *rays* and are used to explain certain light phenomena using the so-called *ray theory* or *geometric optics* approach. This approach is valid when the object with which the light interacts is much larger than the wavelength of the light. This theory explains large-scale optical effects such as reflection and refraction (which are described in Sec. 3.4) and describes how devices such as light sources, photodetectors, and optical amplifiers function.

As noted in Chap. 1, all types of waves including light waves can interfere with one another. Thus if two light waves line up with each other (or are in phase), they produce a bright spot. However, when two light waves are 180° out of phase, then the peaks of one wave are aligned with the troughs of the other wave. In this case the two waves will interfere destructively, thereby canceling each other out. To explain effects such as these, we need to turn to the *electromagnetic wave theory* or *physical optics* viewpoint of light. The concepts involved here are important when we examine the behavior of devices such as wavelength-sensitive optical couplers.

Whereas the geometric optics approach deals with light rays, the physical optics viewpoint uses the concept of electromagnetic field distributions called *modes*. We will examine the concept of modes in greater detail in Chap. 4 when discussing optical fibers. Basically the discussion in Chap. 4 shows that modes are certain allowable distributions of light power in an optical fiber. Later chapters describe other specific physical aspects of the wave theory as they relate to optical components.

3.2. The Speed of Light

One of the earliest recorded discussions of the speed of light is that by Aristotle (384 to 347 B.C.), when he quoted Empedocles of Acragas (495 to 435 B.C.) as saying the light from the sun must take some time to reach the earth. However, Aristotle himself disagreed with the concept that light has a finite speed and thought that it traveled instantaneously. Galileo disagreed with Aristotle and tried to measure the speed of light with a shuttered lantern experiment, but was unsuccessful. Finally, about 600 years later in the 1670s, the Danish astronomer Ole Roemer measured the speed of light while making detailed observations of the movements of Jupiter's moon Io.

In free space a light wave travels at a speed $c = 3 \times 10^8$ m/s (300,000,000 m/s), which is known as the *speed of light*. Actually this is a convenient and fairly accurate estimate. To be exact, $c = 299{,}792{,}458$ m/s in a vacuum, which is equivalent to 186,287.490 mi/s, if you prefer those units. The speed of light is related to the wavelength λ (Greek lambda) and the wave frequency ν (Greek nu) through the equation $c = \lambda\nu$.

3.3. Measuring Properties of Light

The physical property of the radiation in different parts of the spectrum can be measured in several interrelated ways (see the "Measurements in the EM Spectrum" discussion below). These are the length of one period of the wave, the energy of a photon, or the oscillating frequency of the wave. Whereas electric signal transmission tends to use frequency to designate the signal operating bands, optical communications generally uses *wavelength* to designate the spectral operating region and *photon energy* or *optical power* when discussing topics such as signal strength or electrooptical component performance.

Measurements in the EM Spectrum As can be seen from Fig. 1.7, there are three different ways to measure various regions in the EM spectrum. These measurement units are related by some simple equations. First, the speed of light c is equal to the wavelength λ times the frequency ν, so that $c = \lambda\nu$. Rearranging this equation gives the relationship between wavelength and frequency. For example, if the frequency is known and we want to find the wavelength, then we use

$$\lambda = \frac{c}{\nu} = \frac{3\times 10^8\,\text{m/s}}{\nu} \tag{3.1}$$

where the frequency ν is measured in cycles per second or *hertz* (Hz). Conversely, if the wavelength is known and we want to find the frequency, then we use the relationship $\nu = c/\lambda$.

The relationship between the energy of a photon and its frequency (or wavelength) is determined by the equation known as *Planck's law*

$$E = h\nu \tag{3.2}$$

where the parameter $h = 6.63 \times 10^{-34}$ J·s = 4.14 eV·s is called *Planck's constant*. The unit J means joules, and the unit eV stands for *electron volts*. In terms of wavelength (measured in units of micrometers), the energy in electron volts is given by

$$E\ \text{eV} = \frac{1.2406}{\lambda\ \mu\text{m}} \tag{3.3}$$

3.4. Refractive Index

A fundamental optical parameter of a material relates to how fast light travels in it. Upon entering a dielectric or nonconducting medium, a light wave slows down and now travels at a speed s, which is characteristic of the material and is less than c. The ratio of the speed of light in a vacuum to that in matter is known as the *refractive index* or *index of refraction* n of the material and is given by

$$n = \frac{c}{s} \tag{3.4}$$

Typical values of n to two decimal places are 1.00 for air, 1.33 for water, 1.45 for silica glass, and 2.42 for diamond. Note that if we have two different materials, then the one with the larger value of n is said to be *optically denser* than the material with the lower value of n. For example, glass is optically denser than air.

Number Accuracy People who design optical test and measurement equipment often must know the precise value of the refractive index for air, and they need to take into account its variation with wavelength, temperature, pressure, and gas composition. The wavelength dependence of the index of refraction n_{air} of standard dry air at a pressure of 760 torr and 15°C is

$$n_{air} = 1 + 10^{-8}\left(8342.13 + \frac{2{,}406{,}030}{130 - 1/\lambda^2} + \frac{15{,}997}{38.9 - 1/\lambda^2}\right) \tag{3.5}$$

where the wavelength λ of the light is measured in micrometers (10^{-6} m). Using this equation, we find that for a wavelength of 1550 nm = 1.550 μm used in fiber optic communications, the refractive index for air is n_{air} = 1.00027325, which yields a speed of light c_{air} = 299,710,562 m/s.

3.5. Reflection and Refraction

The concepts of reflection and refraction can be understood most easily by using light rays. When a light ray encounters a boundary separating two materials that have different refractive indices, part of the ray is reflected to the first medium and the remainder is bent (or *refracted*) as it enters the second material. This is shown in Fig. 3.1 where $n_1 > n_2$. The bending or refraction of the light ray at the interface is a result of the difference in the speed of light in two materials with different refractive indices.

Snell's Law The relationship describing refraction at the interface between two different light-transmitting materials is known as *Snell's law* and is given by

$$n_1 \sin \phi_1 = n_2 \sin \phi_2 \tag{3.6}$$

or equivalently as

$$n_1 \cos \theta_1 = n_2 \cos \theta_2 \tag{3.7}$$

where the angles are defined in Fig. 3.1. The angle ϕ_1 between the incident ray and the normal to the surface is known as the *angle of incidence*.

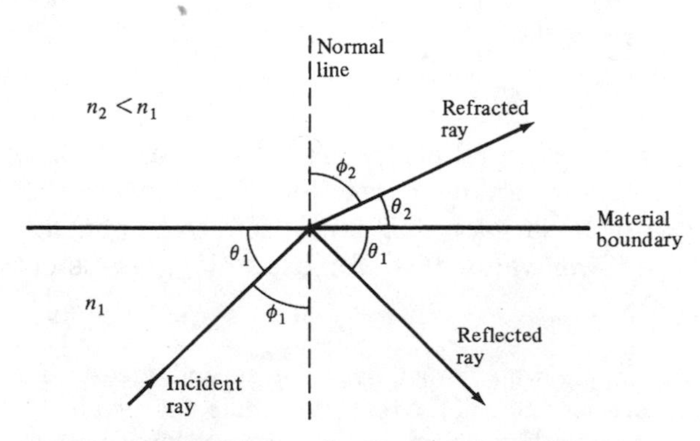

Figure 3.1. Refraction and reflection of a light ray at a material boundary.

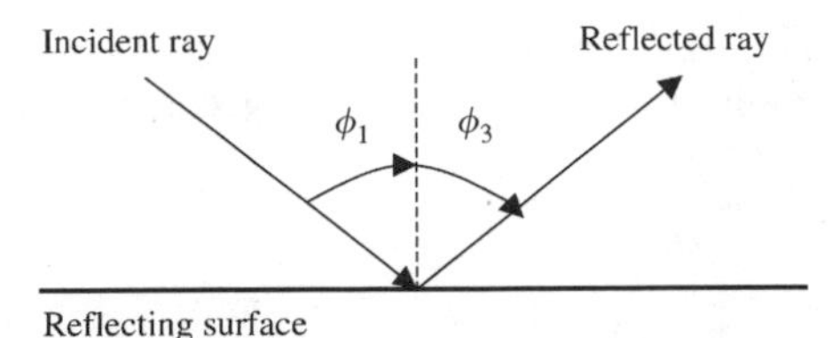

Figure 3.2. Illustration of the law of reflection.

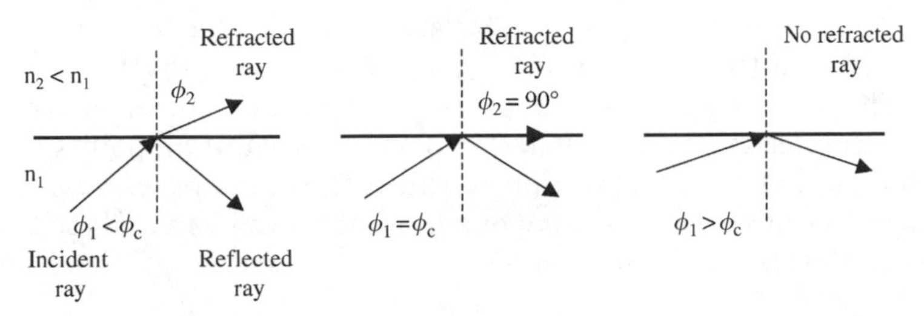

Figure 3.3. Representation of the critical angle and total internal reflection at a glass-air interface.

According to the law of reflection, as illustrated in Fig. 3.2, the angle ϕ_1 at which the incident ray strikes the interface is exactly equal to the angle ϕ_3 that the reflected ray makes with the same interface. In addition, the incident ray, the normal to the interface, and the reflected ray all lie in the same plane, which is perpendicular to the interface plane between the two materials. This is called the *plane of incidence*.

When light traveling in a certain medium is reflected off an optically denser material (one with a higher refractive index), the process is referred to as *external reflection*. Conversely, the reflection of light off a less optically dense material (such as light traveling in glass being reflected at a glass-to-air interface) is called *internal reflection*.

As the angle of incidence ϕ_1 in an optically denser material becomes larger, the refracted angle ϕ_2 approaches $\pi/2$. Beyond this point no refraction into the adjoining material is possible, and the light rays become *totally internally reflected*. The conditions required for total internal reflection can be determined by using Snell's law [see Eq. (3.6)]. Consider Fig. 3.3, which shows a glass surface in air. A light ray gets bent toward the glass surface as it leaves the glass in accordance with Snell's law. If the angle of incidence ϕ_1 is increased, a point will eventually be reached where the light ray in air is parallel to the glass surface. This point is known as the *critical angle of incidence* ϕ_c. When ϕ_1 is greater than ϕ_c, the condition for total internal reflection is satisfied; that is, the light is totally reflected back into the glass with no light escaping from the glass surface.

Example If we look at the glass-air interface in Fig. 3.3, when the refracted light ray is parallel to the glass surface, then $\phi_2 = 90°$ so that $\sin \phi_2 = 1$. Thus $\sin \phi_c = n_2/n_1$.

Using $n_1 = 1.48$ for glass and $n_2 = 1.00$ for air, we get $\phi_c = 42°$. This means that any light in the glass incident on the interface at an angle ϕ_1 greater than 42° is totally reflected back into the glass.

3.6. Polarization

Light is composed of one or more *transverse electromagnetic waves* that have both an electric field (called an E field) and a magnetic field (called an H field) component. As shown in Fig. 3.4, in a transverse wave the directions of the vibrating electric and magnetic fields are perpendicular to each other and are at right angles to the direction of propagation (denoted by the vector **k**) of the wave. The wave shown in Fig. 3.4 is *plane-polarized*. This means that the vibrations in the electric field are parallel to one another at all points in the wave, so that the electric field forms a plane called the *plane of vibration*. Likewise all points in the magnetic field component of the wave lie in a plane that is at right angles to the electric field plane.

3.6.1. Unpolarized light

An ordinary light wave is made up of many transverse waves that vibrate in a variety of directions (i.e., in more than one plane) and is referred to as *unpolarized light*. However, any arbitrary direction of vibration can be represented as a combination of a parallel vibration and a perpendicular vibration, as shown in Fig. 3.5. Therefore, unpolarized light can be viewed as consisting of two orthogonal plane polarization components, one that lies in the plane of incidence (the plane containing the incident and reflected rays) and the other that lies in a plane perpendicular to the plane of incidence. These are denoted as the *parallel polarization* and the *perpendicular polarization* components,

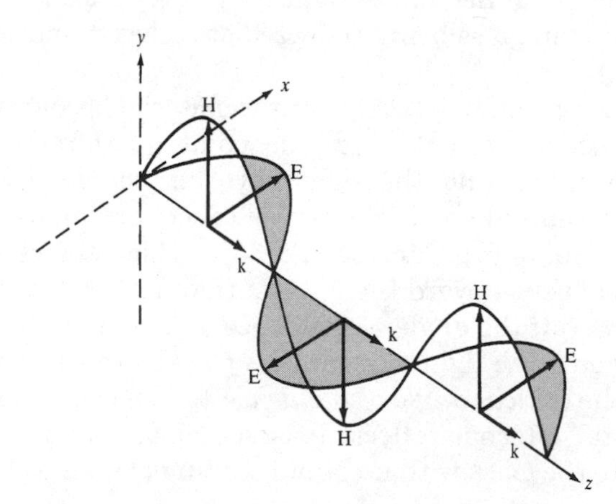

Figure 3.4. Electric and magnetic field distributions in a train of plane electromagnetic waves at a given instant in time.

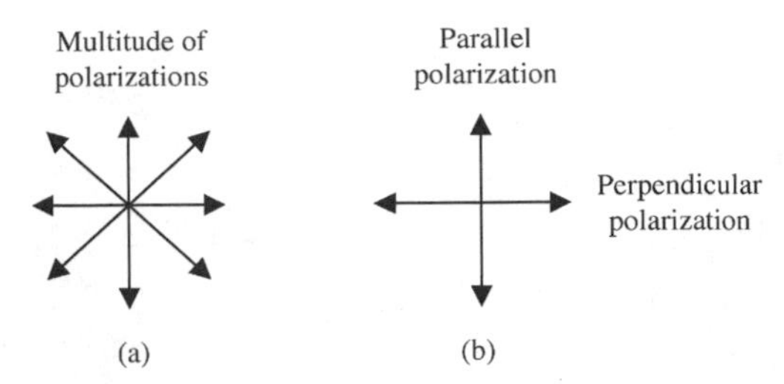

Figure 3.5. Polarization represented as a combination of a parallel vibration and a perpendicular vibration.

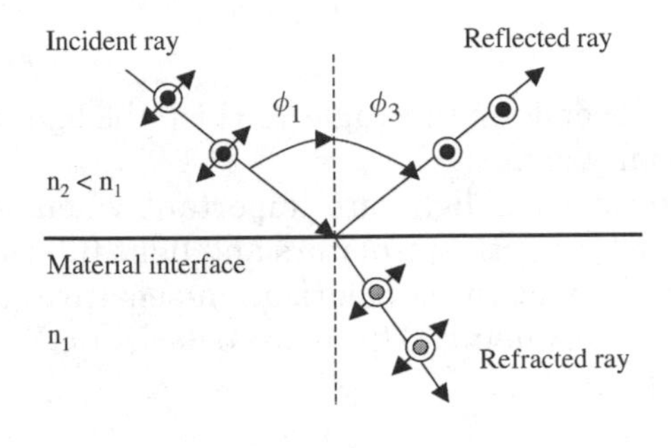

Figure 3.6. Behavior of an unpolarized light beam at the interface between air and a nonmetallic surface.

respectively. In the case when all the electric field planes of the different transverse waves are aligned parallel to one another, then the light wave is *linearly polarized*. This is the simplest type of polarization.

Unpolarized light can be split into separate polarization components either by reflection off a nonmetallic surface or by refraction when the light passes from one material to another. As noted earlier in Fig. 3.1, when an unpolarized light beam traveling in air impinges on a nonmetallic surface such as glass, part of the beam is reflected and part is refracted into the glass. A circled dot and an arrow designate the parallel and perpendicular polarization components, respectively, in Fig. 3.6. The reflected beam is partially polarized and at a specific angle (known as *Brewster's angle*) is completely perpendicularly polarized. A familiar example of this is the use of polarizing sunglasses to reduce the glare of partially polarized sunlight reflections from road or water surfaces. The parallel component of the refracted beam is entirely transmitted into the glass, whereas the perpendicular component is only partially refracted. How much of

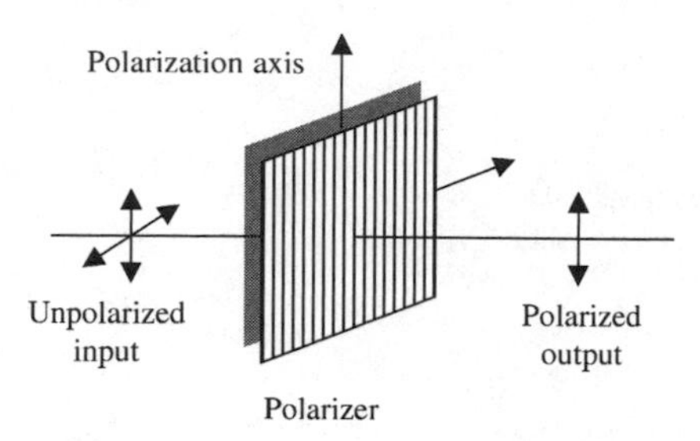

Figure 3.7. Only the vertical polarization component passes through this polarizer.

the refracted light is polarized depends on the angle at which the light approaches the surface and on the material itself.

These polarization characteristics of light are important when we examine the behavior of components such as optical isolators and light filters, which are described in Chap. 9. Here we look at three polarization-sensitive materials or devices that are used in such components. These are polarizers, Faraday rotators, and birefringent crystals.

3.6.2. Polarizers

A *polarizer* is a material or device that transmits only one polarization component and blocks the other. For example, when unpolarized light enters a polarizer that has a vertical transmission axis as shown in Fig. 3.7, only the vertical polarization component passes through the device. As noted earlier, a familiar application is the use of polarizing sunglasses. To see the polarization property of the sunglasses you are wearing, tilt your head sideways. A number of glare spots will then appear. The sunglasses block out the polarized light from these spots when you hold your head normally.

3.6.3. Faraday rotators

A *Faraday rotator* is a device that rotates the *state of polarization* (SOP) of light passing through it by a specific amount. For example, a popular device rotates the SOP clockwise by 45° or one-quarter wavelength, as shown in Fig. 3.8. This rotation is independent of the SOP of input light, but the rotation angle is different depending on the direction in which the light passes through the device. That is, the rotation is not reciprocal. In addition, the SOP of the input light is maintained after the rotation; for example, if the input light to a 45° Faraday rotator is linearly polarized in a vertical direction, then the rotated light exiting the crystal also is linearly polarized at a 45° angle. The material is usually some type of asymmetric crystal such as yttrium iron garnet (YIG), and the degree of angular rotation is proportional to the thickness of the device.

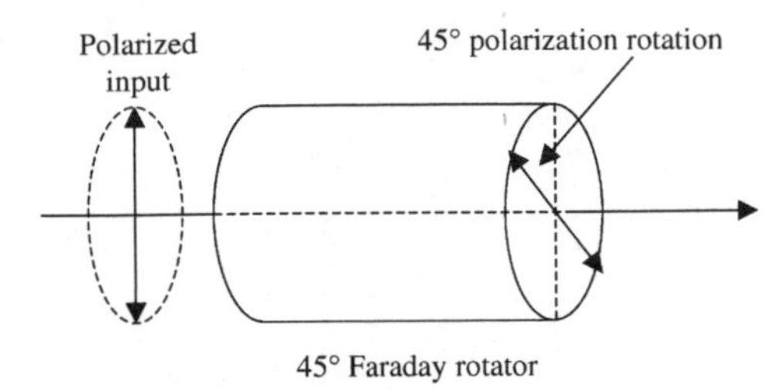

Figure 3.8. A Faraday rotator is a device that rotates the state of polarization clockwise by 45° or one-quarter wavelength.

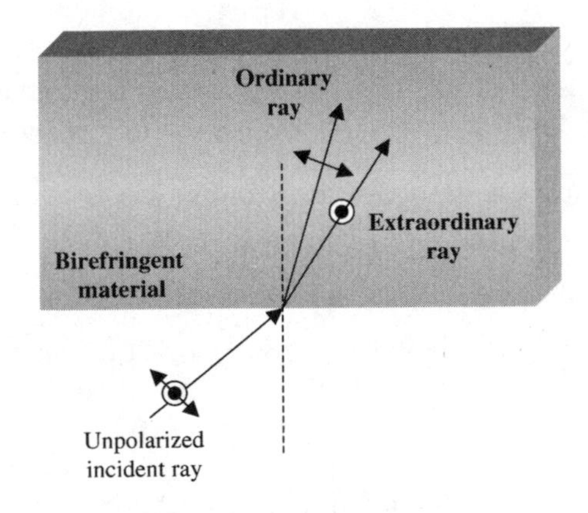

Figure 3.9. A birefringent crystal splits the light signal entering it into two perpendicularly polarized beams.

3.6.4. Double-refractive crystals

Certain crystalline materials have a property called *double refraction* or *birefringence*. This means that the indices of refraction are slightly different along two perpendicular axes of the crystal, as shown in Fig. 3.9. A device made from such materials is known as a *spatial walk-off polarizer* (SWP). The SWP splits the light signal entering it into two orthogonally (perpendicularly) polarized beams. One of the beams is called an *ordinary ray* or o ray, since it obeys Snell's law of refraction at the crystal surface. The second beam is called the *extraordinary ray* or e ray, since it refracts at an angle that deviates from the prediction of the standard form of Snell's law. Each of the two orthogonal polarization components thus is refracted at a different angle, as shown in Fig. 3.9. For example, if the incident unpolarized light arrives at a perpendicular angle to the surface of the device, the o ray can pass straight through the device whereas

TABLE 3.1. Some Common Birefringent Crystals and Their Ordinary and Extraordinary Indices of Refraction

Crystal name	Symbol	n_o	n_e
Calcite	$CaCO_3$	1.658	1.486
Lithium niobate	$LiNbO_3$	2.286	2.200
Rutile	TiO_2	2.616	2.903
Yttrium vanadate	YVO_4	1.945	2.149

the e ray component is deflected at a slight angle so it follows a different path through the material.

Table 3.1 lists the ordinary index n_o and the extraordinary index n_e of some common birefringent crystals that are used in optical communication components. As will be described in later chapters, they have the following applications:

- Calcite is used for polarization control and in beam splitters.
- Lithium niobate is used for light signal modulation.
- Rutile is used in optical isolators and circulators.
- Yttrium vanadate is used in optical isolators, circulators, and beam displacers.

3.7. Summary

Some optical phenomena can be explained using a wave theory whereas in other cases light behaves as though it is composed of miniature particles called *photons*. The wave nature explains how light travels through an optical fiber and how it can be coupled between two adjacent fibers, but the particle theory is needed to explain how optical sources generate light and how photodetectors change an optical signal to an electric signal.

In free space a light wave travels at a speed $c = 3 \times 10^8$ m/s, but it slows down by a factor $n > 1$ when entering a material, where the parameter n is the index of refraction (or refractive index) of the material. Values of the refractive index for materials related to optical communications are 1.00 for air and between 1.45 and 1.50 for various glass compounds. Thus light travels at about 2×10^8 m/s in a glass optical fiber.

When a light ray encounters a boundary separating two media that have different refractive indices, part of the ray is reflected back into the first medium and the remainder is bent (or *refracted*) as it enters the second material. As will be discussed in later chapters, these concepts play a major role in describing the amount of optical power that can be injected into a fiber and how light waves travel along a fiber.

The polarization characteristics of light waves are important in examining the behavior of components such as optical isolators and filters. Polarization-sensitive devices include light signal modulators, polarization filters, Faraday rotators, beam splitters, and beam displacers. Birefringent crystals such as calcite, lithium niobate, rutile, and yttrium vanadate are polarization-sensitive materials used in such components.

Further Reading

1. B. E. A. Saleh and M. C. Teich, *Fundamentals of Photonics*, Wiley, New York, 1991. This book presents an advanced encyclopedic treatment of all aspects of the fundamentals of photonics.
2. E. Hecht, *Optics*, 4th ed., Addison-Wesley, Reading, Mass., 2002. A classical undergraduate-level university textbook on optics.
3. J. L. Miller and E. Friedman, *Photonics Rules of Thumb: Optics, Electro-Optics, Fiber Optics, and Lasers*, McGraw-Hill, New York, 1996.

Chapter

4

Optical Fibers

The optical fiber is a key part of a lightwave communication system. An optical fiber is nominally a cylindrical dielectric waveguide that confines and guides light waves along its axis. Except for certain specialty fibers, basically all fibers used for telecommunication purposes have the same physical structure. The variations in the material and the size of this structure dictate how a light signal is transmitted along different types of fiber and also influence how the fiber responds to environmental perturbations, such as stress, bending, and temperature variations. This chapter describes various fiber structures, physical characteristics, operational properties, and applications.

4.1. Light Propagation in Fibers

Figure 4.1 shows the end-face cross section and a longitudinal cross section of a standard optical fiber, which consists of a cylindrical glass core surrounded by a glass cladding. The *core* has a refractive index n_1, and the *cladding* has a refractive index n_2. Surrounding these two layers is a polymer *buffer coating* that protects the fiber from mechanical and environmental effects. Traditionally the core radius is designated by the letter a. In almost all cases, for telecommunication fibers the core and cladding are made of silica glass (SiO_2).

The refractive index of pure silica varies with wavelength, ranging from 1.453 at 850 nm to 1.445 at 1550 nm. By adding certain impurities such as germanium or boron to the silica during the fiber manufacturing process, the index can be changed slightly, usually as an increase in the core index. This is done so that the refractive index n_2 of the cladding is slightly smaller than the index of the core (that is, $n_2 < n_1$), which is the condition required for light traveling in the core to be totally internally reflected at the boundary with the cladding. The difference in the core and cladding indices also determines how light signals behave as they travel along a fiber. Typically the index differences range from 0.2 to 3.0 percent depending on the desired behavior of the resulting fiber.

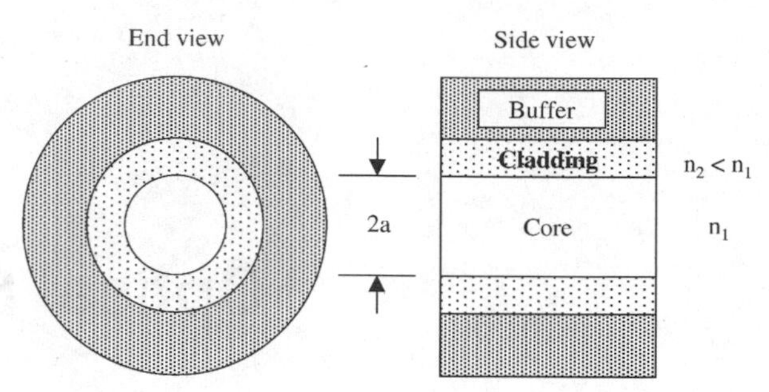

Figure 4.1. End-face cross section and a longitudinal cross section of a standard optical fiber.

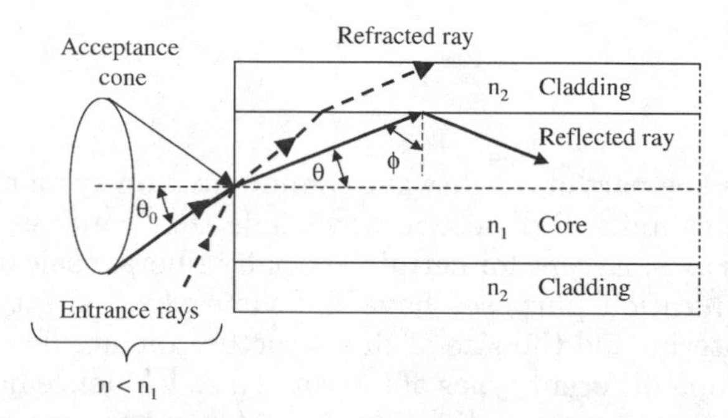

Figure 4.2. Ray optics representation of the propagation mechanism in an ideal step-index optical waveguide.

To get an understanding of how light travels along a fiber, let us first examine the case when the core diameter is much larger than the wavelength of the light. As discussed in Chap. 3, for such a case we can consider a simple geometric optics approach using the concept of light rays. Figure 4.2 shows a light ray entering the fiber core from a medium of refractive index n, which is less than the index n_1 of the core. The ray meets the core end face at an angle θ_0 with respect to the fiber axis and is refracted into the core. Inside the core the ray strikes the core-cladding interface at a normal angle ϕ. If the light ray strikes this interface at such an angle that it is totally internally reflected, then the ray follows a zigzag path along the fiber core.

Now suppose that the angle θ_0 is the largest entrance angle for which total internal reflection can occur at the core-cladding interface. Then rays outside of the acceptance cone shown in Fig. 4.2, such as the ray given by the dashed line, will refract out of the core and be lost in the cladding. This condition defines a *critical angle* ϕ_c, which is the smallest angle ϕ that supports total internal reflection at the core-cladding interface.

Critical Angle Referring to Fig. 4.2, from Snell's law the minimum angle $\phi = \phi_{\min}$ that supports total internal reflection is given by $\phi_c = \phi_{\min} = \arcsin(n_2/n_1)$. Rays striking the core-cladding interface at angles less than $\phi_{\min}$ will refract out of the core and be lost in the cladding. Now suppose the medium outside of the fiber is air for which $n = 1.00$. By applying Snell's law to the air-fiber interface boundary, the condition for the critical angle can be related to the maximum entrance angle $\theta_{0,\max}$ through the relationship

$$\sin\theta_{0,\max} = n_1 \sin\theta_c = \left(n_1^2 - n_2^2\right)^{1/2}$$

where $\theta_c = \pi/2 - \phi_c$. Thus those rays having entrance angles θ_0 less than $\theta_{0,\max}$ will be totally internally reflected at the core-cladding interface.

Example

1. Suppose the core index $n_1 = 1.480$ and the cladding index $n_2 = 1.465$. Then the critical angle is $\phi_c = \arcsin(1.465/1.480) = 82°$, so that $\theta_c = \pi/2 - \phi_c = 8°$.
2. With this critical angle, the maximum entrance angle is

$$\theta_{0,\max} = \arcsin(n_1 \sin\theta_c) = \arcsin(1.480 \sin 8°) = 11.9°$$

4.2. Optical Fiber Modes

Although it is not directly obvious from the ray picture shown in Fig. 4.2, only a finite set of rays at certain discrete angles greater than or equal to the critical angle ϕ_c is capable of propagating along a fiber. These angles are related to a set of electromagnetic wave patterns or field distributions called *modes* that can propagate along a fiber. When the fiber core diameter is on the order of 8 to 10 μm, which is only a few times the value of the wavelength, then only the one single *fundamental ray* that travels straight along the axis is allowed to propagate in a fiber. Such a fiber is referred to as a *single-mode fiber*. The operational characteristics of single-mode fibers cannot be explained by a ray picture, but instead need to be analyzed in terms of the *fundamental mode* by using the electromagnetic wave theory. Fibers with larger core diameters (e.g., greater than or equal to 50 μm) support many propagating rays or modes and are known as *multimode fibers*. A number of performance characteristics of multimode fibers can be explained by ray theory whereas other attributes (such as the optical coupling concept presented in Chap. 8) need to be described by wave theory.

Figure 4.3 shows the field patterns of the three lowest-order *transverse electric* (TE) modes as seen in a cross-sectional view of an optical fiber. They are the TE_0, TE_1, and TE_2 modes and illustrate three of many possible power distribution patterns in the fiber core. The subscript refers to the *order* of the mode, which is equal to the number of zero crossings within the guide. In single-mode fibers only the lowest-order or *fundamental mode* (TE_0) will be guided along the fiber core. Its $1/e^2$ width is called the *mode field diameter*.

As the plots in Fig. 4.3 show, the power distributions are not confined completely to the core, but instead extend partially into the cladding. The fields vary harmonically within the core guiding region of index n_1 and decay exponentially outside of this region (in the cladding). For low-order modes the

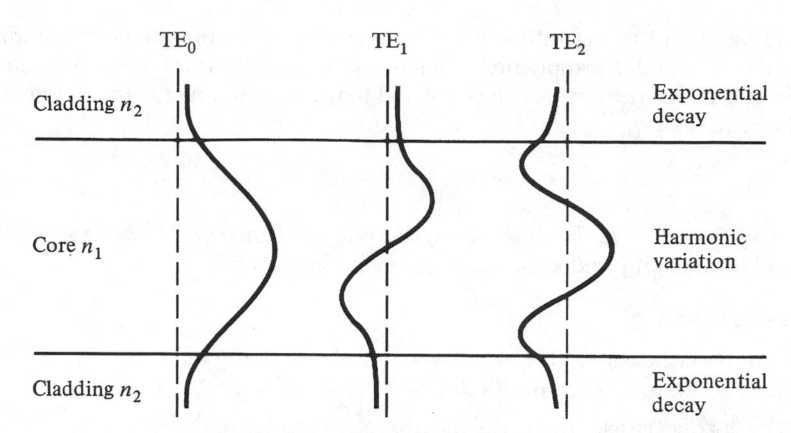

Figure 4.3. Electric field patterns of the three lowest-order guided modes as seen in a cross-sectional view of an optical fiber.

fields are concentrated tightly near the axis of the fiber with little penetration into the cladding. On the other hand, for higher-order modes the fields are distributed more toward the edges of the core and penetrate farther into the cladding region. The importance of the characteristic that the power of a mode extends partially into the cladding will be seen in later chapters which discuss applications such as coupling of power from one fiber to another.

4.3. Variations of Fiber Types

Variations in the material composition of the core and the cladding give rise to the two basic fiber types shown in Fig. 4.4*a*. In the first case, the refractive index of the core is uniform throughout and undergoes an abrupt change (or step) at the cladding boundary. This is called a *step-index fiber*. In the second case, the core refractive index varies as a function of the radial distance from the center of the fiber. This defines a *graded-index fiber*. Section 4.6 describes the advantages of graded-index fibers over step-index fibers for high-speed data transfer when using multimode fibers. More complex structures of the cladding index profile allow fiber designers to tailor the signal dispersion characteristics of the fiber (see Sec. 4.6). Figure 4.4*b* shows two of many different possible configurations.

Table 4.1 lists typical core, cladding, and buffer coating sizes of optical fibers for use in telecommunications, in a metropolitan-area network (MAN), or in a local-area network (LAN). The outer diameter of the buffer coating can be either 250 or 500 μm. Single-mode fibers are used for long-distance communication and for transmissions at very high data rates. The larger-core multimode fibers typically are used for local-area network applications in a campus environment, particularly for gigabit or 10-Gbit rate Ethernet links, which are known popularly as GigE and 10GigE, respectively. Here the word *campus* refers to any group of buildings that are within reasonable walking distance of one another (see Sec. 2.5).

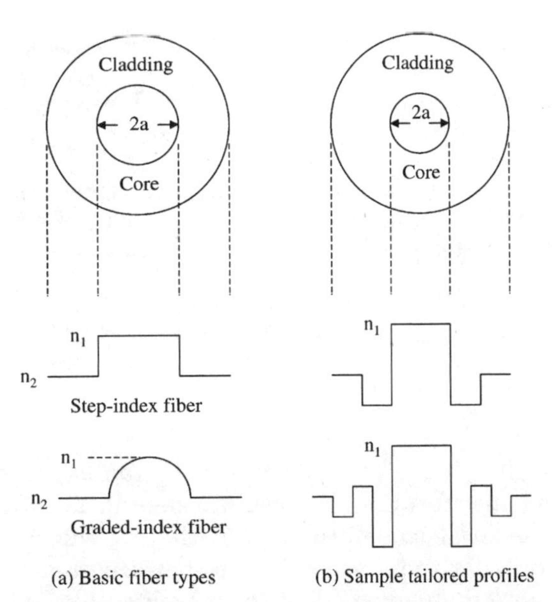

Figure 4.4. Variations in the material composition of the core and cladding yield different fiber types. (*a*) Simple profiles define step- and graded-index fibers; (*b*) complex cladding-index profiles tailor the signal dispersion characteristics of a fiber.

TABLE 4.1. Typical Core, Cladding, and Buffer Coating Sizes of Optical Fibers

Fiber type	Core diameter, μm	Cladding outer diameter, μm	Buffer outer diameter, μm	Application
Single-mode	7–10	125	250 or 500	Telecommunications
Multimode	50.0	125	250 or 500	LAN or MAN
Multimode	62.5	125	250 or 500	LAN
Multimode	85	125	250 or 500	Older LAN fiber
Multimode	100	140	250 or 500	Older fiber type

The critical angle also defines a parameter called the *numerical aperture* (NA), which is used to describe the light acceptance or gathering capability of fibers that have a core size much larger than a wavelength. This parameter defines the size of the acceptance cone shown in Fig. 4.2. The numerical aperture is a dimensionless quantity which is less than unity, with values ranging from 0.14 to 0.50.

Numerical Aperture The critical angle condition on the entrance angle defines the *numerical aperture* (NA) of a step-index fiber. This is given by

$$\mathrm{NA} = n \sin \theta_{0,\max} = n_1 \sin \theta_c = \left(n_1^2 - n_2^2\right)^{1/2} \approx n_1 \sqrt{2\Delta}$$

where the parameter Δ is called the *core-cladding index difference* or simply the *index difference*. It is defined through the equation $n_2 = n_1(1-\Delta)$. Typical values of Δ range from 1 to 3 percent for multimode fibers and from 0.2 to 1.0 percent for single-mode fibers. Thus, since Δ is much less than 1, the approximation on the right-hand side of the above equation is valid.

Since the numerical aperture is related to the maximum acceptance angle, it is used commonly to describe the light acceptance or gathering capability of a multimode fiber and to calculate the source-to-fiber optical power coupling efficiencies.

Example A multimode step-index fiber has a core index $n_1 = 1.480$ and an index difference $\Delta = 0.01$. The numerical aperture for this fiber is $\mathrm{NA} = 1.480\sqrt{0.02} = 0.21$.

4.4. Single-Mode Fibers

An important parameter for single-mode fibers is the *cutoff wavelength*. This is designated by λ_{cutoff} and specifies the smallest wavelength for which all fiber modes except the fundamental mode are cut off; that is, the fiber transmits light in a single mode only for those wavelengths that are greater than λ_{cutoff}. The fiber can support more than one mode if the wavelength of the light is less than the cutoff. Thus if a fiber is single-mode at 1310 nm, it is also single-mode at 1550 nm, but not necessarily at 850 nm.

When a fiber is fabricated for single-mode use, the cutoff wavelength usually is chosen to be much less than the desired operating wavelength. For example, a fiber for single-mode use at 1310 nm may have a cutoff wavelength of 1275 nm.

Cutoff Wavelength For a fiber to start supporting only a single mode at a wavelength λ_{cutoff}, the following condition (derived from solutions to Maxwell's equations for a circular waveguide) needs to be satisfied:

$$\lambda_{\text{cutoff}} = \frac{2\pi a}{2.405}\left(n_1^2 - n_2^2\right)^{1/2}$$

where a is the radius of the fiber core, n_1 is the core index, and n_2 is the cladding index.

Example Suppose we have a fiber with $a = 4.2\,\mu\text{m}$, $n_1 = 1.480$, and $n_2 = n_1(1 - 0.0034) = 1.475$. Its cutoff wavelength then is

$$\lambda_{\text{cutoff}} = \frac{2\pi(4.2\,\mu\text{m})}{2.405}\left[(1.480)^2 - (1.475)^2\right]^{1/2} = 1334\,\text{nm}$$

4.5. Optical Fiber Attenuation

Light traveling in a fiber loses power over distance, mainly because of absorption and scattering mechanisms in the fiber. The fiber loss is referred to as *signal attenuation* or simply *attenuation*. Attenuation is an important property of an optical fiber because, together with signal distortion mechanisms, it determines the maximum transmission distance possible between a transmitter and a receiver (or an amplifier) before the signal power needs to be boosted to an appropriate level above the signal noise for high-fidelity reception. The degree of the attenuation depends on the wavelength of the light and on the fiber material.

Figure 4.5 shows a typical attenuation versus wavelength curve for a silica fiber. The loss of power is measured in decibels (see Chap. 1), and the loss within a cable is described in terms of decibels per kilometer (dB/km).

Example Suppose a fiber has an attenuation of 0.4 dB/km at a wavelength of 1310 nm. Then after it travels 50 km, the optical power loss in the fiber is 20 dB (a factor of 100).

Figure 4.5 also shows that early optical fibers had a large attenuation spike between 900 and 1200 nm due to the fourth-order absorption peak from water molecules. Another spike from the third-order water absorption occurs between 1350 and 1480 nm for commonly fabricated fibers. Because of such absorption peaks, three transmission windows were defined initially. The *first window* ranges from 800 to 900 nm, the *second window* is centered at 1310 nm, and the *third window* ranges from 1480 to 1600 nm. Since the attenuation of low-water-peak fibers makes the designation of these windows obsolete, the concept of operational spectral bands arose for the 1260- to 1675-nm region, as described in Chap. 1. Figure 4.6 shows the attenuation as a function of wavelength for a low-water-peak fiber in the region covered by the six operational bands.

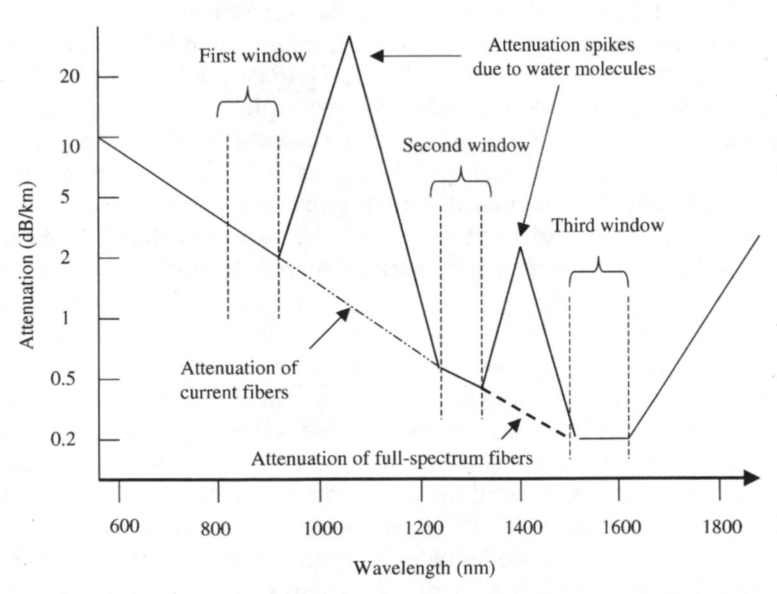

Figure 4.5. A typical attenuation versus wavelength curve for a silica fiber. Early fibers had a high loss spike around 1100 nm. Full-spectrum (low-water-content) fibers allow transmission in all spectral bands.

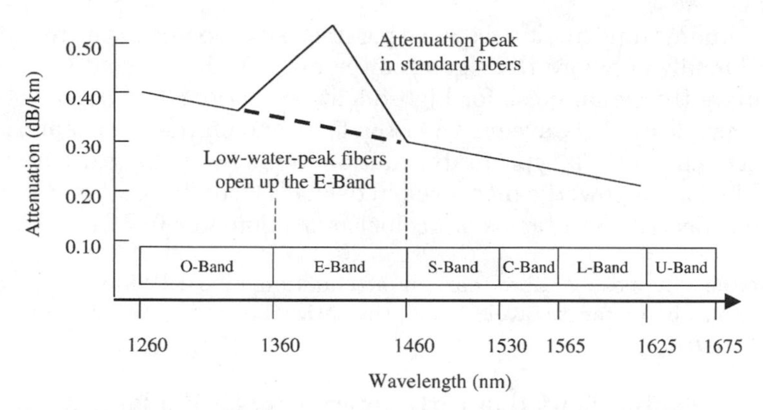

Figure 4.6. Attenuation versus wavelength for a low-water-peak fiber in the six operational spectral bands.

TABLE 4.2. Typical Losses in Standard 9-μm Fiber for Three Fiber Turns on a Specific Mandrel

Mandrel radius	Loss at 1310 nm	Loss at 1550 nm
1.15 cm	2.6 dB	23.6 dB
1.80 cm	0.1 dB	2.6 dB

In addition to the intrinsic absorption and scattering loss mechanisms in a fiber, light power can be lost as a result of fiber bending. Fibers can be subject to two types of bends: (1) *macroscopic bends* that have radii which are large compared with the fiber diameter, for example, those that occur when a fiber cable turns a corner, and (2) random *microscopic bends* of the fiber axis that can arise when fibers are incorporated into cables. Since the microscopic bending loss is determined in the manufacturing process, the user has little control over the degree of loss resulting from them. In general cable fabrication processes keep these values to a very low value, which is included in published cable loss specifications.

For slight bends, the excess optical power loss due to macroscopic bending is extremely small and is essentially unobservable. As the radius of curvature decreases, the loss increases exponentially until at a certain critical bend radius the curvature loss becomes observable. If the bend radius is made a bit smaller once this threshold has been reached, the losses suddenly become extremely large. Bending losses depend on wavelength and are measured by winding several loops of fiber on a rod of a specific diameter. Table 4.2 gives typical bending loss values when three loops of a standard 9-μm core-diameter single-mode fiber are wound on rods with radii of 1.15 and 1.80 cm. Note the large difference in losses between operation at 1310 and 1550 nm. As a rule of thumb, it is best not to make the bend radius of such a fiber be less than 2.5 cm.

Since often fibers need to be bent into very tight loops within component packages, special fibers that are immune to bending losses have been developed for such applications. This class of specialty fibers is described in Sec. 4.7.

4.6. Fiber Information Capacity

The information-carrying capacity of the fiber is limited by various distortion mechanisms in the fiber, such as signal dispersion factors and nonlinear effects. The three main dispersion categories are modal, chromatic, and polarization mode dispersions. These distortion mechanisms cause optical signal pulses to broaden as they travel along a fiber. As Fig. 4.7 shows, if optical pulses travel sufficiently far in a fiber, they will eventually overlap with neighboring pulses, thereby creating errors in the output since they become indistinguishable to the receiver. Nonlinear effects occur when there are high power densities (optical power per cross-sectional area) in a fiber. Their impact on signal fidelity includes shifting of power between wavelength channels, appearances of spurious signals at other wavelengths, and decreases in signal strength. Chapter 15 discusses these nonlinear-induced degradations in greater detail.

Modal dispersion arises from the different path lengths associated with various modes (as represented by light rays at different angles). It appears only in multimode fibers, since in a single-mode fiber there is only one mode. By looking at Fig. 4.8 it can be deduced that rays bouncing off the core-cladding interface follow a longer path compared to the fundamental ray that travels straight down the fiber axis. For example, since ray 2 makes a steeper angle than ray 1,

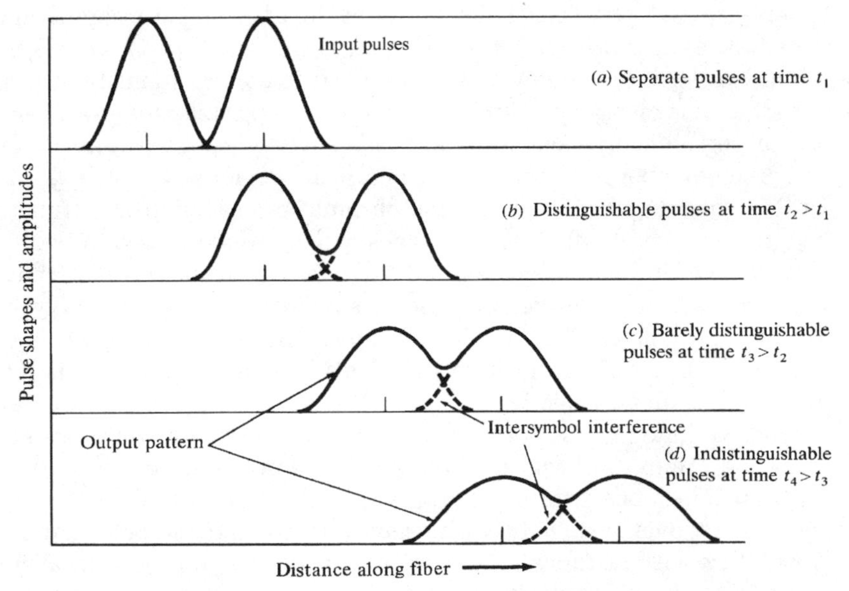

Figure 4.7. Broadening and attenuation of two adjacent pulses as they travel along a fiber.

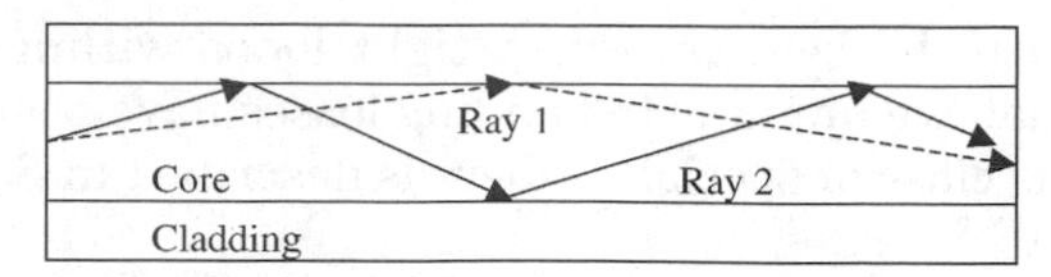

Figure 4.8. Rays that have steeper angles have longer path lengths.

ray 2 has a longer path length from the beginning to the end of a fiber. If all the rays are launched into a fiber at the same time in a given light pulse, then they will arrive at the fiber end at slightly different times. This causes the pulse to spread out and is the basis of modal dispersion.

In a graded-index fiber, the index of refraction is lower near the core-cladding interface than at the center of the core. Therefore, in such a fiber the rays that strike this interface at a steeper angle will travel slightly faster as they approach the cladding than those rays arriving at a smaller angle. For example, this means that the light power in ray 2 shown in Fig. 4.8 will travel faster than that in ray 1. Thereby the various rays tend to keep up with one another to some degree. Consequently the graded-index fiber exhibits less pulse spreading than a step-index fiber where all rays travel at the same speed.

The index of refraction of silica varies with wavelength; for example, it ranges from 1.453 at 850 nm to 1.445 at 1550 nm. In addition, as described in Chap. 6, a light pulse from an optical source contains a certain slice of wavelength spectrum. For example, a laser diode source may emit pulses that have a 1-nm spectral width. Consequently, different wavelengths within an optical pulse travel at slightly different speeds through the fiber (recall from Chap. 3 that $s = c/n$). Therefore each wavelength will arrive at the fiber end at a slightly different time, which leads to pulse spreading. This factor is called *chromatic dispersion*, which often is referred to simply as *dispersion*. It is a fixed quantity at a specific wavelength and is measured in units of picoseconds per kilometer of fiber per nanometer of optical source spectral width, abbreviated as ps/(km·nm). For example, a single-mode fiber might have a chromatic dispersion value of $D_{CD} = 2$ ps/(km·nm) at 1550 nm. Figure 4.9 shows the chromatic dispersion as a function of wavelength for several different fiber types, which are described in Sec. 4.8.

Polarization mode dispersion (PMD) results from the fact that light-signal energy at a given wavelength in a single-mode fiber actually occupies two orthogonal polarization states or modes. Figure 4.10 shows this condition. At the start of the fiber the two polarization states are aligned. However, fiber material is not perfectly uniform throughout its length. In particular, the refractive index is not perfectly uniform across any given cross-sectional area. This condition is known as the *birefringence* of the material. Consequently, each polarization mode will encounter a slightly different refractive index, so that each will travel at a slightly different velocity and the polarization orientation will rotate with distance. The resulting difference in propagation times between the two orthogonal polarization modes will result in pulse spreading. This is the basis of polarization mode dispersion. PMD is not a fixed quantity but fluctuates

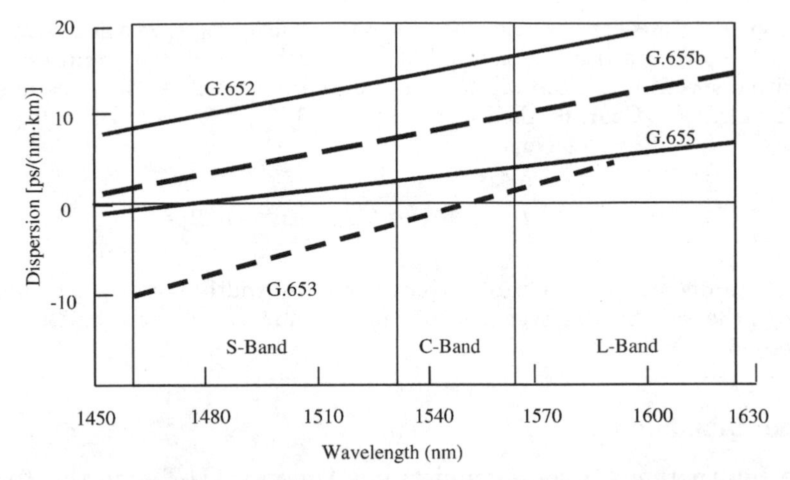

Figure 4.9. Chromatic dispersion as a function of wavelength in various spectral bands for several different fiber types.

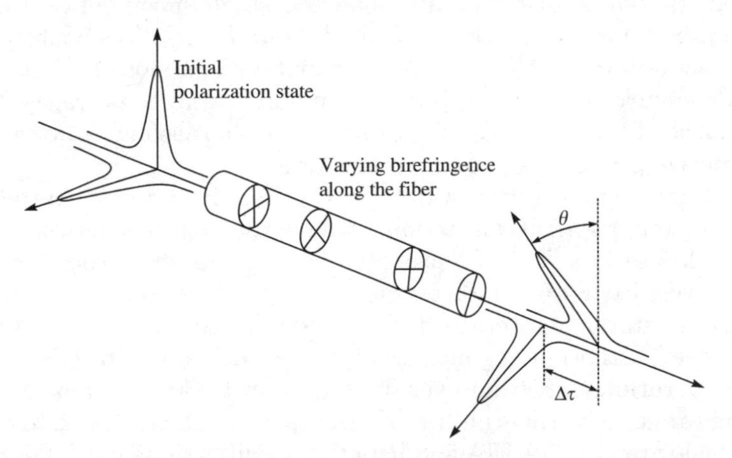

Figure 4.10. Variation in the polarization states of an optical pulse as it passes through a fiber that has varying birefringence along its length.

with time due to factors such as temperature variations and stress changes on the fiber. It varies as the square root of distance and thus is specified as a mean value in units of $\text{ps}/\sqrt{\text{km}}$. A typical value is $D_{\text{PMD}} = 0.05\,\text{ps}/\sqrt{\text{km}}$.

Dispersion Calculation If t_{mod}, t_{CD}, and t_{PMD} are the modal, chromatic, and polarization mode dispersion times, respectively, then the total dispersion t_T can be calculated by the relationship

$$t_T = \sqrt{(t_{\text{mod}})^2 + (t_{\text{CD}})^2 + (t_{\text{PMD}})^2}$$

Note that $t_{\text{mod}} = 0$ for single-mode fibers. As a rule of thumb, the information-carrying capacity over a certain length of fiber then is determined by specifying that the pulse spreading not be more than 10 percent of the pulse width at a designated data rate.

Example Consider a single-mode fiber for which $D_{CD} = 2\,\text{ps/(km}\cdot\text{nm)}$ and $D_{PMD} = 0.1\,\text{ps}/\sqrt{\text{km}}$. If a transmission link has a length $L = 500\,\text{km}$ and uses a laser source with a spectral emission width of $\Delta\lambda = 0.01\,\text{nm}$ (we will see more details about what this means in Chap. 6), then we have $t_{mod} = 0$, $t_{CD} = D_{CD} \times L \times \Delta\lambda = 10\,\text{ps}$, and $t_{PMD} = D_{PMD} \times \sqrt{L} = 2.24\,\text{ps}$. Thus

$$t_T = \sqrt{(10\,\text{ps})^2 + (2.24\,\text{ps})^2} = 10.2\,\text{ps}$$

If t_T can be no more than 10 percent of a pulse width, then the maximum data rate R_{max} that can be sent over this 500-km link is $R_{max} = 0.1/t_T = 9.8\,\text{Gbps}$ (gigabits per second).

4.7. Optical Fiber Standards

The International Telecommunications Union (ITU-T) and the Telecommunications Industry Association (TIA/EIA) are the main organizations that have published standards for both multimode and single-mode optical fibers used in telecommunications. The recommended bounds on fiber parameters (e.g., attenuation, cutoff wavelength, and chromatic dispersion) designated in these standards ensure the users of product capability and consistency. In addition, the standards allow fiber manufacturers to have a reasonable degree of flexibility to improve products and develop new ones.

Multimode fibers are used widely in LAN environments, storage area networks, and central-office connections, where the distance between buildings is typically 2 km or less. The two principal multimode fiber types for these applications have either 50- or 62.5-μm core diameters, and both have 125-μm cladding diameters. To meet the demands for short-reach low-cost transmission of high-speed Ethernet signals, a 50-μm multimode fiber is available for 10-Gbps operation at 850 nm over distances up to 300 m. Table 4.3 shows the operating ranges of various multimode fibers for applications up to 10GigE. The standards document TIA/EIA-568 lists the specifications for 10GigE fiber. The ITU-T recommendation G.651 describes other multimode fiber specifications for LAN applications using 850-nm optical sources.

The ITU-T has published a series of recommendations for single-mode fibers. The characteristics of these fibers are given in the following listing. They are summarized in Table 4.4.

TABLE 4.3. Operating Ranges of Various Multimode Fibers for Applications up to 10GigE

Parameter	62.5-μm fiber		50-μm fiber			Unit
Bandwidth	160	200	400	500	2000	MHz-km
Range	26	33	66	82	300	m
Attenuation	2.5	2.5	2.5	2.5	2.5	dB/km

TABLE 4.4. ITU-T Recommendations for Single-Mode Fibers

ITU-T recommendation no.	Description
G.651	Multimode fiber for use at 850 nm in a LAN
G.652	Standard single-mode fiber (1310-nm optimized)
G.652.C	Low-water-peak fiber for CWDM applications
G.653	Dispersion-shifted fiber (made obsolete by NZDSF)
G.654	Submarine applications (1500-nm cutoff wavelength)
G.655	Nonzero dispersion-shifted fiber (NZDSF)
G.655b	Advanced nonzero dispersion-shifted fiber (A-NZDSF)

ITU-T G.652. This recommendation deals with the single-mode fiber that was installed widely in telecommunication networks in the 1990s. It has a Ge-doped silica core which has a diameter between 5 and 8 μm. Since early applications used 1310-nm laser sources, this fiber was optimized to have a zero-dispersion value at 1310 nm. Thus it is referred to as a *1310-nm optimized fiber*. With the trend toward operation in the lower-loss 1550-nm spectral region, the installation of this fiber has decreased dramatically. However, the huge base of G.652 fiber that is installed worldwide will still be in service for many years. If network operators want to use installed G.652 fiber at 1550 nm, complex dispersion compensation techniques are needed, as described in Chap. 15.

ITU-T G.652.C. *Low-water-peak fiber* for CWDM applications is created by reducing the water ion concentration in order to eliminate the attenuation spike in the 1360- to 1460-nm E-band. The fibers have core diameters ranging from 8.6 to 9.5 μm and an attenuation of less than 0.4 dB/km. The main use of this fiber is for low-cost short-reach CWDM (coarse WDM) applications in the E-band. In CWDM the wavelength channels are sufficiently spaced that minimum wavelength stability control is needed for the optical sources, as described in Chap. 13.

ITU-T G.653. *Dispersion-shifted fiber* (DSF) was developed for use with 1550-nm lasers. In this fiber type the zero-dispersion point is shifted to 1550 nm where the fiber attenuation is about one-half that at 1310 nm. Although this fiber allows a high-speed data stream of a single-wavelength channel to maintain its fidelity over long distances, it presents dispersion-related problems in DWDM applications where many wavelengths are packed into one or more of the operational bands. As a result, this fiber type became obsolete with the introduction of G.655 NZDSF.

ITU-T G.654. This specification deals with *cutoff-wavelength-shifted fiber* that is designed for long-distance high-power signal transmission. Since it has a high cutoff wavelength of 1500 nm, this fiber is restricted to operation at 1550 nm. It typically is used only in submarine applications.

ITU-T G.655. Nonzero dispersion-shifted fiber (NZDSF) was introduced in the mid-1990s for WDM applications. Its principal characteristic is that it has a nonzero dispersion value over the entire C-band, which is the spectral operating region for erbium-doped optical fiber amplifiers (see Chap. 11). This is in contrast to G.653 fibers in which the dispersion varies from negative values through zero to positive values in this spectral range.

ITU-T G.655b. Advanced nonzero dispersion-shifted fiber (A-NZDSF) was introduced in October 2000 to extend WDM applications into the S-band. Its principal characteristic is that it has a nonzero dispersion value over the entire S-band and the C-band. This is in contrast to G.655 fibers in which the dispersion varies from negative values through zero to positive values in the S-band.

4.8. Specialty Fibers

Whereas telecommunication fibers, such as those described above, are designed to transmit light over long distances with minimal change in the signal, *specialty fibers* are used to manipulate the light signal. Specialty fibers interact with light and are custom-designed for specific applications such as optical signal amplification, wavelength selection, wavelength conversion, and sensing of physical parameters.

A number of both passive and active optical devices use specialty fibers to direct, modify, or strengthen an optical signal as it travels through the device. Among these optical devices are light transmitters, optical signal modulators, optical receivers, wavelength multiplexers, couplers, splitters, optical amplifiers, optical switches, wavelength add/drop modules, and light attenuators. Table 4.5 gives a summary of some specialty fibers and their applications. Later chapters describe the applications of each of these devices in greater detail.

4.8.1. Erbium-doped fiber

Erbium-doped optical fibers have small amounts of erbium ions added to the silica material and are used as a basic building block for optical fiber amplifiers. As described in Chap. 11, a length of Er-doped fiber ranging from 10 to 30 m is used as a gain medium for amplifying optical signals in the C-band (1530 to 1560 nm). There are many variations on the doping level, cutoff wavelength,

TABLE 4.5. Summary of Some Specialty Fibers and Their Applications

Specialty fiber type	Application
Erbium-doped fiber	Gain medium for optical fiber amplifiers
Photosensitive fibers	Fabrication of fiber Bragg gratings
Bend-insensitive fibers	Tightly looped connections in device packages
High-loss attenuating fiber	Termination of open optical fiber ends
Polarization-preserving fibers	Pump lasers, polarization-sensitive devices, sensors
High-index fibers	Fused couplers, short-λ sources, DWDM devices
Holey (photonic crystal) fibers	Switches; dispersion compensation

TABLE 4.6. Generic Parameter Values of an Erbium-Doped Fiber for Use in the C-Band

Parameter	Specification
Peak absorption at 1530 nm	5 to 10 dB/m
Effective numerical aperture	0.14 to 0.31
Cutoff wavelength	900 ± 50 nm; or 1300 nm
Mode field diameter at 1550 nm	5.0 to 7.3
Cladding diameter	125 μm standard; 80 μm for tight coils
Coating material	UV-cured acrylic

mode field diameter, numerical aperture, and cladding diameter for these fibers. Higher erbium concentrations allow the use of shorter fiber lengths, smaller claddings are useful for compact packages, and a higher numerical aperture allows for the fiber to be coiled tighter in small packages. Table 4.6 lists some generic parameter values of an erbium-doped fiber for use in the C-band.

Photosensitive fiber. A photosensitive fiber is designed so that its refractive index changes when it is exposed to ultraviolet light. This sensitivity may be provided by doping the fiber material with germanium and boron ions. The main application is to create a fiber Bragg grating, which is a periodic variation of the refractive index along the fiber axis (Chap. 9 has details on this). Applications of fiber Bragg gratings include light-coupling mechanisms for pump lasers used in optical amplifiers, wavelength add/drop modules, optical filters, and chromatic dispersion compensation modules.

Bend-insensitive fiber. A bend-insensitive fiber has a moderately higher numerical aperture (NA) than that in a standard single-mode telecommunication fiber. The numerical aperture can be varied to adjust the mode field diameter. Increasing the NA reduces the sensitivity of the fiber to bending loss by confining optical power more tightly within the core than in conventional single-mode fibers. Bend-insensitive fibers are available commercially in a range of core diameters to provide optimum performance at specific operating wavelengths, such as 820, 1310, or 1550 nm. These fibers are offered with either an 80-μm or a 125-μm cladding diameter as standard products. The 80-μm *reduced-cladding fiber* results in a much smaller volume compared with a 125-μm cladding diameter when a fiber length is coiled up within a device package. Whereas Table 4.2 shows there is a high bending loss for tightly wound conventional single-mode fibers, the induced attenuation at the specified operating wavelength due to 100 turns of bend-insensitive fiber on a 10-mm-radius mandrel is less than 0.5 dB.

Attenuating fiber. These fibers have a uniform attenuation in the 1250- to 1620-nm band. This makes an attenuating fiber useful for WDM applications to lower the power level at the input of receivers or at the output of an EDFA. The fibers are offered commercially with attenuation levels available from 0.4 dB/cm to

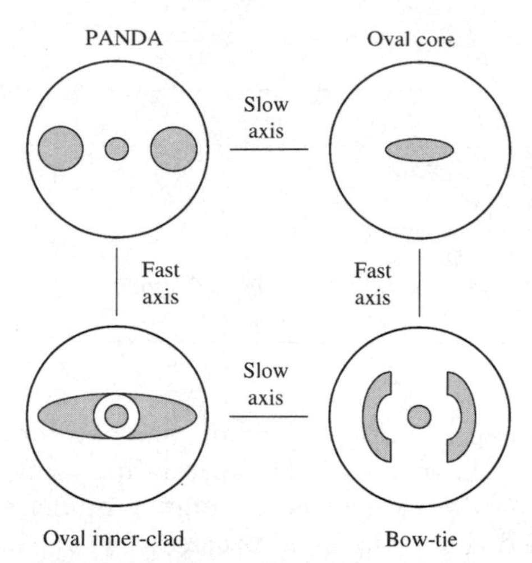

Figure 4.11. Cross-sectional geometry of four different polarization-maintaining fibers.

greater than 15 dB/cm. Fibers with an attenuation of 15 dB/cm (a loss factor of 32 within 1 cm) may be used to terminate the end of a fiber optic link so that there are no return reflections, or as a high-level plug-type attenuator.

Polarization-preserving fiber. In contrast to standard optical fibers in which the state of polarization fluctuates as a light signal propagates through the fiber, polarization-preserving fibers have a special core design that maintains the polarization. Applications of these fibers include light signal modulators fabricated from lithium niobate, optical amplifiers for polarization multiplexing, light-coupling fibers for pump lasers, and polarization-mode dispersion compensators. Figure 4.11 illustrates the cross-sectional geometry of four different polarization-maintaining fibers. The light circles represent the cladding, and the dark areas are the core configurations. The goal in each design is to introduce a deliberate birefringence into the core so that the two polarization modes become decoupled within a very short distance, which leads to preservation of the individual polarization states.

High-index fiber. These fiber types have a higher core refractive index, which results in a larger numerical aperture. Consequently, since a higher NA enables optical power to be coupled more efficiently into a core, a short (nominally 1-m) length of such a fiber may be attached directly to an optical source. Such a fiber section is referred to as a *pigtail* or a *flylead*. The fibers can be designed specifically for short-wavelength or long-wavelength optical sources (see Chap. 6). In addition, they have applications in fused-fiber couplers (Chap. 8) and in wavelength division multiplexing (Chaps. 12 and 13).

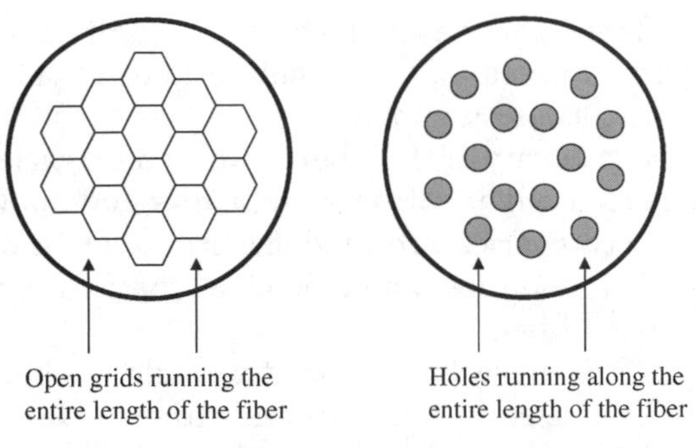

Figure 4.12. End-face patterns of two possible holey fiber structures.

Holey fiber. A holey or photonic crystal fiber typically consists of a silica material which contains numerous air-filled microscopic holes. Figure 4.12 shows the end-face patterns of two possible holey fiber structures. The tubular holes run along the entire length of the fiber parallel to the fiber axis. The size, position, and number of holes provide the fiber with specific waveguide properties. This technology is under development. Potential applications of holey fibers in telecommunications include dispersion compensation, wavelength conversion, optical switching, and high-power optical amplification.

4.9. Summary

An optical fiber is nominally a cylindrical dielectric waveguide that confines and guides light waves along its axis. Basically all fibers used for telecommunication purposes have the same physical structure, which consists of a cylindrical glass core surrounded by a glass cladding. The difference in the core and cladding indices determines how light signals travel along a fiber.

An important physical concept is that only a finite set of light rays that impinge on the core walls at specific angles may propagate along a fiber. These angles are related to a set of electromagnetic wave patterns called *modes*. For a *single-mode fiber,* the core diameter is around 8 to 10 μm (several wavelengths), and only the *fundamental ray* is allowed to propagate. *Multimode fibers* have larger core diameters (e.g., around 50 μm) and support many modes. The ray theory can explain a number of fiber performance characteristics, but other attributes require the wave theory. The power distribution of modes is not confined completely to the core, but extends partially into the cladding. This concept is important when we examine concepts such as optical power coupling.

Light traveling in a fiber loses power over distance, mainly because of absorption and scattering mechanisms in the fiber. This attenuation is an important property of an optical fiber because, together with signal distortion mechanisms, it determines the maximum transmission distance possible. The degree of the attenuation depends on the wavelength of the light and on the fiber material. The loss of power is measured in decibels, and the loss within a cable is described in terms of decibels per kilometer.

The information-carrying capacity of the fiber is limited by various distortion mechanisms in the fiber, such as signal dispersion factors and nonlinear effects. The three main dispersion categories are modal, chromatic, and polarization mode dispersions. These distortion mechanisms cause optical signal pulses to broaden as they travel along a fiber.

The ITU-T and the TIA/EIA have published standards for both multimode and single-mode optical fibers used in telecommunications. The recommended bounds on fiber parameters (e.g., attenuation, cutoff wavelength, and chromatic dispersion) designated in these standards ensure the users of product capability and consistency. Multimode fibers are used in LAN environments, storage area networks, and central-office connections, where the distance between buildings is typically 2 km or less. The two principal multimode fiber types have either 50- or 62.5-μm core diameters, and both have 125-μm cladding diameters. For short-reach, low-cost transmission of high-speed Ethernet signals, a 50-μm multimode fiber is available for 10-Gbps operation at 850 nm over distances up to 300 m.

The ITU-T also has published a series of recommendations for single-mode fibers. Of these, two key ones for DWDM use are the ITU-T G.655 (nonzero dispersion-shifted fiber, or NZDSF) and the ITU-T G.655b (advanced nonzero dispersion-shifted fiber, or A-NZDSF). The G.655 fibers are designed for the C-band, and the G.655b fibers allow DWDM operation over the entire S-band and the C-band. In addition, the G.652.C recommendation describes fibers for CWDM applications.

Whereas telecommunication fibers, such as those described above, are designed to transmit light over long distances with minimal change in the signal, *specialty fibers* are used to manipulate the light signal. Specialty fibers interact with light and are custom-designed for specific applications such as optical signal amplification, wavelength selection, wavelength conversion, and sensing of physical parameters.

Further Reading

1. B. Comycz, *Fiber Optic Installer's Field Manual*, McGraw-Hill, New York, 2000.
2. G. Keiser, *Optical Fiber Communications*, 3d ed., McGraw-Hill, Burr Ridge, Ill., 2000.
3. T. R. Jordal, "How to test laser-based premises networks," *Lightwave*, vol. 18, pp. 126, 128, 156, February 2001 (www.lightwaveonline.com).
4. R. E. Kristiansen, "Holey fibers," *SPIE OE Magazine*, vol. 2, pp. 25–27, June 2002 (www.oemagazine.com).
5. TIA/EIA-568-B.3, *Optical Fiber Cabling Components Standard*, April 2000.
6. ITU-T Recommendation G.651, *Characteristics of a 50/125-μm Multimode Graded-Index Optical Fiber Cable*, February 1998.
7. ITU-T Recommendation G.652, *Characteristics of a Single-Mode Optical Fiber Cable*, October 2000.
8. ITU-T Recommendation G.653, *Characteristics of a Dispersion-Shifted Optical Fiber Cable*, October 2000.

9. ITU-T Recommendation G.654, *Characteristics of a Cutoff-Shifted Single-Mode Optical Fiber Cable*, October 2000.
10. ITU-T Recommendation G.655, *Characteristics of a Nonzero Dispersion-Shifted Single-Mode Optical Fiber Cable*, October 2000.
11. J. Ryan, "Next-generation NZ-DSF fibers will balance performance characteristics," *Lightwave*, vol. 18, pp. 146–152, March 2000 (www.light-wave.com).
12. F. Aranda, "Asymmetry maintains polarization," *Laser Focus World*, vol. 38, pp. 187–190, May 2002.

Chapter

5

Optical Fiber Cables

Optical cables are essential elements of an optical communications link. In addition to providing protection to the optical fibers contained within the cable, the construction of an optical cable determines whether it can withstand the environments in which it will be used. If engineers select the wrong cable configuration, the cost of retrofitting installed cable can be prohibitively high (and afterward the engineers may be told to seek employment elsewhere). When a circuit card fails or an equipment rack needs replacement, it is a fairly straightforward process to enter an equipment room to remove the items and put in new ones. Optical cables, on the other hand, cannot be replaced easily without undergoing major disruptive digging or having huge piles or spools of cables lying in the way in hallways or on sidewalks. This chapter addresses cabling design considerations, cable types for diverse applications, and installation methods in different environments.

5.1. Fiber-Related Design Issues

Cabling of optical fibers involves enclosing them within some type of protective structure. The cable structure will vary greatly depending on whether the cable is to be pulled or blown into underground or intrabuilding tubes (called *ducts*), buried directly in the ground, installed on outdoor poles, or placed underwater. Different cable configurations are required for each type of application, but certain fundamental cable design principles will apply in every case. The objectives of cable manufacturers have been that the optical fiber cables should be installable with the same type of equipment, installation techniques, and precautions as those used for conventional wire cables. This requires special cable designs because of the unique properties of optical fibers such as their strength, dielectric (nonmetallic) nature, small size, and low weight. Let's take a closer look at these properties.

5.1.1. Fiber strength

Glass is quite a strong material and typically withstands large pulling forces. However, it does have its strength limits, as any other material. Thus, one important mechanical property is the maximum allowable *axial load* on the cable, that is, how hard one can pull on the cable before something snaps. In copper-based cables the wires themselves generally are the principal load-bearing members of the cable, and elongations of more than 20 percent are possible. On the other hand, typical high-quality optical fibers break after stretching around 0.5 to 1.0 percent. Since damage occurs very quickly at axial stress levels above 40 percent of the permissible stretching point and very slowly below 20 percent, fiber elongations during cable installation and afterward when it is in operation should be limited to 0.1 to 0.2 percent.

To prevent excessive stretching, the cabling process usually includes the incorporation of *strength members* into the cable design. This is especially important in the design of aerial cables that can experience severe stresses due to factors such as wind forces or ice loading. Of course if a tree falls on an aerial transmission line during a storm or if someone with a backhoe or a bulldozer accidentally encounters a buried cable, then the chances of cable survivability are extremely small! Common strength members are strong nylon yarns, steel wires, and fiberglass rods. Some examples of strength members are described in Sec. 5.2 in the discussions on cable materials and structures.

5.1.2. Dielectric nature of fibers

Although a variety of cable designs use steel strength members, in certain applications it is advantageous to have a completely nonmetallic cable. Such cables have a low weight, are immune to ground-loop problems, are resistant to electromagnetic coupling arising from adjacent electronic equipment or nearby lightning strikes, and can be run through explosive environments where electric sparks would not be very welcome.

In some cases signal amplification equipment located within the optical link requires electric power. In that case copper wires may be integrated into the optical cable, since the fibers cannot carry electric power to run the equipment.

5.1.3. Small size and low weight

Cable designs must take into account the small size of optical fibers both from the perspective of handling ease and from the desire to have strong, low-weight cables. Their size and weight characteristics make optical cables smaller and easier to handle than their copper counterparts, which tend to be heavier and bulkier. The cabling process itself also *color-codes* each fiber by means of different jacket colors and arranges the small fibers systematically within the cable (see Sec. 5.5). This allows them to be seen and identified easily when one is attaching connectors or splicing fibers. In addition, low weight and small size are advantageous for applications such as tactical military communication links where long lengths of cable need to be unrolled and retrieved rapidly, LAN links where fiber cables are placed in cable trays above suspended ceilings, and interbuilding links where cables often need to be pulled through crowded cable ducts.

5.2. Cable Materials and Structures

In most cases there are some common fundamental materials that are used in the cabling process. To see what they are, let us examine the generic cable configuration shown in Fig. 5.1. Here there is a central *strength member*, which can be strong nylon yarns, steel wires, or fiberglass rods. A commonly used yarn is Kevlar, which is a soft but tough yellow synthetic nylon material belonging to a generic yarn family known as *aramids*. Note that aramid yarns also are used for making bulletproof vests, trampolines, and tennis rackets. Individual fibers or modules consisting of fiber groupings are wound loosely around the central member. Optionally, a cable wrapping tape and another strength member then encapsulate these fiber groupings.

Surrounding all this is a tough polymer *jacket* that provides crush resistance and handles any tensile stresses applied to the cable so that the fibers inside are not damaged. The jacket also protects the fibers inside against abrasion, moisture, oil, solvents, and other contaminants. The jacket type defines the application characteristics; for example, heavy-duty cables for direct-burial and aerial use have thicker, tougher jackets than light-duty cables for indoor use.

An important factor for using a cable in a building is the *flammability rating*. The National Electrical Code (NEC) in the United States establishes flame ratings for cables, while on a global scale the Underwriters Laboratories (UL) has developed cable test procedures. For example, the NEC requires that all cables which run through plenums (the air-handling space between walls, under floors, and above drop ceilings) must either be placed in fireproof conduits or be constructed of low-smoke and fire-retardant materials. Table 5.1 lists some popular jacket materials and their properties.

One of two basic structures is used to house individual fibers for any type of fiber optic cable design. The structures are the *tight-buffered fiber cable* design and the *loose-tube cable* configuration. Cables with tight-buffered fibers nominally

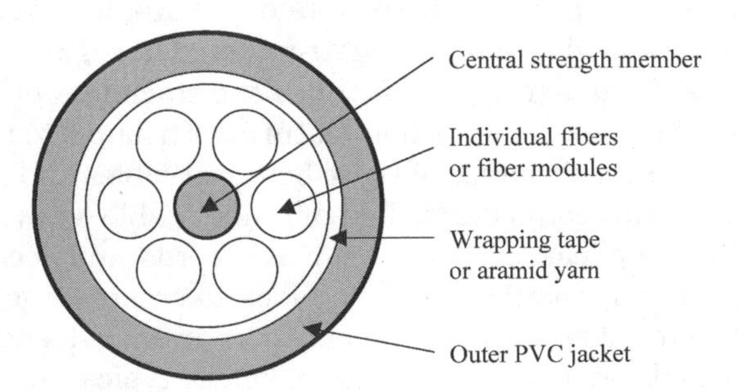

Figure 5.1. Cross section of a generic optical fiber cable.

TABLE 5.1. Optical Cable Jacketing Materials and Their Applications

Jacket material	Material properties
Polyvinyl chloride (PVC)	Flame-retardant; good mechanical protection and flexibility. Widely used for indoor cables. Used outdoors with UV light inhibitors added
Polyethylene (PE)	Excellent resistance to UV light, water, abrasion, and scrapes. Excellent low-temperature flexibility. Since it burns, it is not suitable for indoor use
Flame-retardant polyethylene (FRPE)	Adding a flame-retardant to PE results in a durable, highly abrasion-resistant jacketing. Widely used for indoor and outdoor cables
Polyurethane (PU)	Excellent resistance to abrasion, resistant to UV light, and flexible at extremely low temperatures because of its natural pliability. Widely used for military field communication cables and outdoor deployable cables
ETFE (Tefzel)	Resistant to high temperatures (150°C), abrasion, and flames. Stiffer and more expensive than PVC so used only where its properties are required
Low-smoke zero-halogen polyolefin	Will not emit toxic fumes, smoke, or acidic gases in case of a fire. Has mechanical performance comparable to that of a PVC or FRPE jacket
Cross-linked polyolefin	This material is widely used for highly demanding indoor/outdoor applications due to its high resistance to abrasion, cuts, solvents, ozone, and stress cracking

are used indoors whereas the loose-tube structure is intended for long-haul outdoor applications. A *ribbon cable* is an extension of the tight-buffered cable. In all cases the fibers themselves consist of the normally manufactured glass core and cladding which are surrounded by a protective 250-μm-diameter coating. Let's take a closer look at these structures.

5.2.1. Tight-buffered fiber cable

As shown in Fig. 5.2 for a *simplex* one-fiber cable, in the *tight-buffered* design each fiber is individually encapsulated within its own 900-μm-diameter plastic buffer structure, hence the designation *tight-buffered design*. The 900-μm buffer is nearly 4 times the diameter and 5 times the thickness of the 250-μm protective coating material. This construction feature contributes to the excellent moisture and temperature performance of tight-buffered cables and also permits their direct termination with connectors. The one-fiber cable shown in Fig. 5.2 is for light-duty indoor applications such as patch cords and local-area networks, or it can be used as a modular building block to create larger cables.

Surrounding the 900-μm fiber structure is a layer of aramid strength material which in turn is encapsulated in a PVC outer jacket. A typical outer diameter might be 2.4 mm. Compared to other cable designs, tight-buffered structures

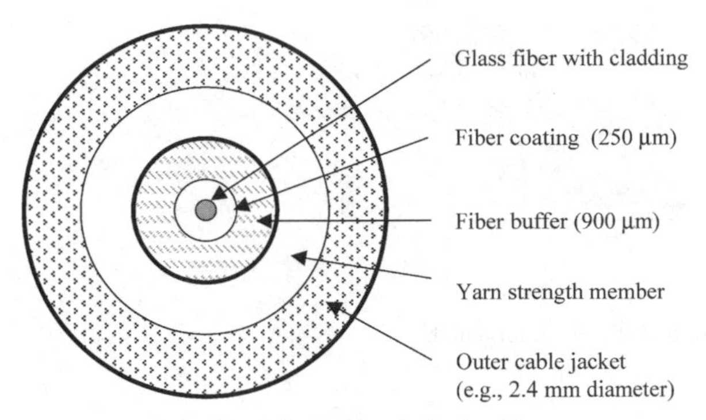

Figure 5.2. Simplex tight-buffered fiber cable.

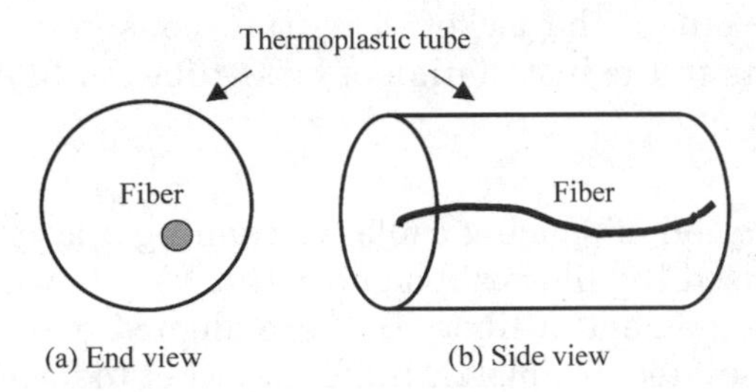

Figure 5.3. Loose-tube concept.

require less care to avoid damaging fibers when stripping back the outer plastic buffer that surrounds the fibers. This fiber preparation process needs to be done when one is attaching connectors or splicing one fiber to another.

5.2.2. Loose-tube cable configuration

In the *loose-tube* cable configuration, one or more standard-coated fibers are enclosed in a thermoplastic tube that has an inner diameter which is much larger than the fiber diameter, as shown in Fig. 5.3. The fibers in the tube are slightly longer than the cable itself. These two conditions isolate the fibers from the cable and allow them to move freely in the tube. This allows the cable to stretch under tensile loads without applying stress on the fibers. Invariably the tube is filled with a *gel* that acts as a buffer, permits the fibers to move freely within the tube, and prevents moisture from entering the tube. This type of configuration thus is known as a *gel-filled cable*. The purpose of this construction is to isolate the fiber from external stresses on the surrounding cable structure caused by factors such as temperature changes. Historically, loose-tube gel-filled cable has been used for outdoor long-haul routes, in which the cables are hung on poles, installed in ducts, or buried directly in the ground.

Fibers within the loose-tube gel-filled cables typically have a 250-μm coating, so they are more fragile than the larger tight-buffered fibers. Thus, greater care

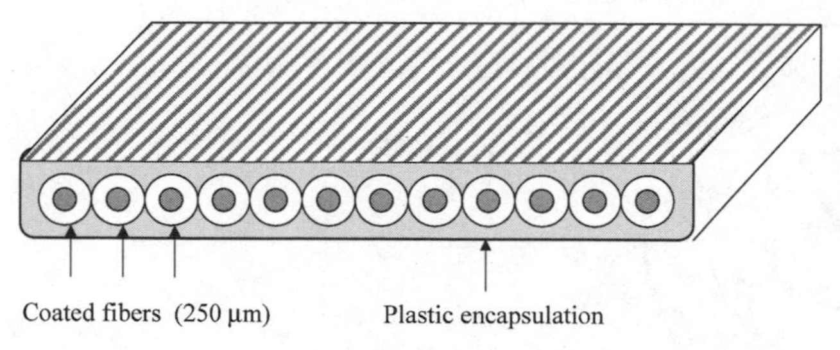

Figure 5.4. Example of a 12-fiber ribbon cable.

and effort are needed when attaching connectors or splicing fibers in gel-filled cables. Each fiber must be individually cleaned to remove the gel, and the breakout point of the main cable must be blocked by some method to prevent the cable gel from oozing out of the jacket. This time-consuming and labor-intensive process adds costs to the installation of loose-tube gel-filled cable.

5.2.3. Ribbon cable

To facilitate the field operation of splicing cables containing a large number of fibers, cable designers devised the fiber-ribbon structure. As shown in Fig. 5.4, the *ribbon cable* is an arrangement of fibers that are aligned precisely next to each other and then encapsulated in a plastic buffer or jacket to form a long continuous ribbon. The number of fibers in a ribbon typically ranges from 4 to 12. These ribbons then can be stacked on top of one another to form a densely packed arrangement of many fibers (say, 144 fibers) within a cable structure.

5.3. Indoor Cables

Just as an interior decorator can think of innumerable ways to rearrange living quarters, there are many different ways in which to arrange fibers inside a cable. The particular arrangement of fibers and the cable design itself need to take into account issues such as the physical environment, the services that the optical link will provide, and any anticipated maintenance and repair that may be needed.

5.3.1. Indoor cable designs

Indoor cables can be used for interconnecting instruments, for distributing signals among office users, for connections to printers or servers, and for short patch cords in telecommunication equipment racks. The three main types are described here.

- *Interconnect cable*. Interconnect cables are designed for light-duty low-fiber-count indoor applications such as fiber-to-the-desk links, patch cords, and point-to-point runs in conduits and trays. The cable is flexible, compact, and lightweight with a tight-buffered construction. By using the simplex tight-buffered unit shown in Fig. 5.2, a number of different indoor cable types can

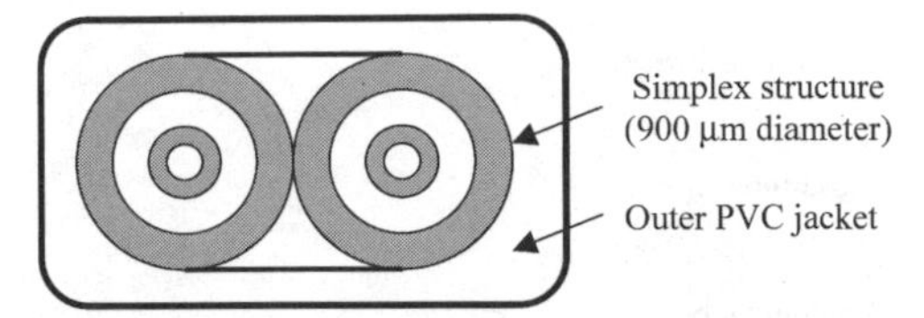

Figure 5.5. Duplex tight-buffered interconnect cable.

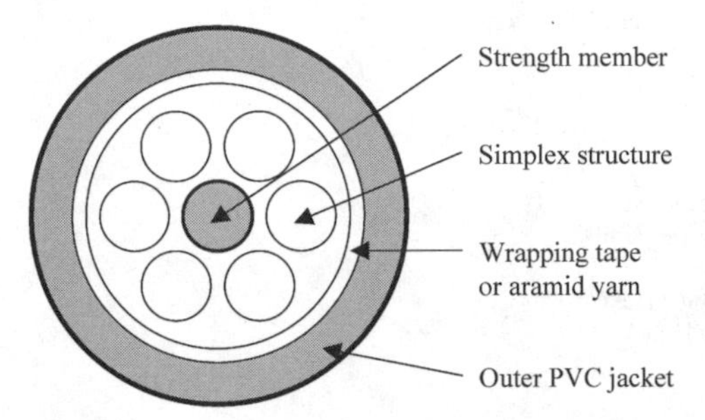

Figure 5.6. Example of an indoor six-fiber breakout cable.

be created. For example, two of these modules can be encapsulated in an outer PVC jacket to form a popular *duplex cable*, as shown in Fig. 5.5. Fiber optic *patch cords*, also known as *jumper cables*, are short lengths (usually less than 2 m) of simplex or duplex cable with connectors on both ends. They are used to connect lightwave test equipment to a fiber patch panel or to interconnect optical transmission modules within an equipment rack.

- *Breakout or fanout cable*. Up to 12 tight-buffered fibers can be stranded around a central strength member to form what is called a *breakout* or *fanout* cable. Breakout cables are designed specifically for low- to medium-fiber-count applications where it is necessary to protect individual jacketed fibers. Figure 5.6 illustrates a six-fiber cable for local-area network applications. The breakout cable facilitates easy installation of fiber optic connectors. All that needs be done to prepare the ends of the cable for connector attachment is to remove the outer jacket, thereby exposing what are essentially individual single-fiber cables. An independent connector then can be attached to each fiber. With such a cable configuration, breaking the individually terminated fibers out to separate pieces of equipment can be achieved easily.
- *Distribution cable*. As shown in Fig. 5.7, individual or small groupings of tight-buffered fibers can be stranded around a nonconducting central strength member to form what is called a *distribution cable*. This cable can be used for a wide range of intrabuilding and interbuilding network applications for sending data, voice, and video signals. If groupings of fibers are desired, they can be wound around a smaller strength member and held together with a cable wrapping tape, or they can be in a loose-tube structure. Distribution cables

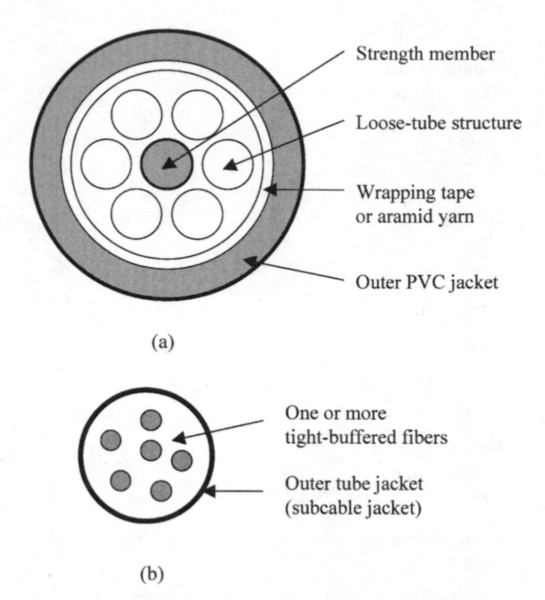

Figure 5.7. Example of an indoor distribution cable. (*a*) Distribution cable cross section; (*b*) loose-tube cross section (one or more fibers).

are designed for use in intrabuilding trays, conduits, backbone premise pathways, and dropped ceilings. A main feature is that they enable groupings of fibers within the cable to be branched (distributed) to various locations.

5.3.2. Indoor cable applications

For indoor cables the National Electrical Code (NEC) in the United States requires that the cables be marked correctly and installed properly in accordance with their intended use. The NEC identifies three indoor cable types for different building regions: plenum, riser, and general-purpose areas. Table 5.2 lists some common indoor cable applications and their markings. A basic difference in the cable types is the material used for the outer protective cable jacket. NEC Article 770 addresses the flammability and smoke emission requirements of indoor fiber optic cables. These requirements vary according to the particular application, as described here:

- *Plenum cables*. A *plenum* is the empty space between walls, under floors, or above drop ceilings used for airflow; or it can form part of an air distribution system used for heating or air conditioning. Plenum-rated cables are UL-certified by the UL-910 *plenum fire test method* as having adequate fire resistance and low smoke-producing characteristics for installations in these spaces without the use of a conduit. As noted in Table 5.2, these cables are termed OFNP (*optical fiber nonconductive plenum*) for all-dielectric cables or OFCP (*optical fiber conductive plenum*) when they contain metallic components.

TABLE 5.2. Cable Markings Based on the U.S. National Electric Code and UL Test Specifications

Marking	Cable name	UL test	Substitute
OFNP	Optical fiber nonconductive plenum	UL 910	None
OFCP	Optical fiber conductive plenum	UL 910	None
OFNR	Optical fiber nonconductive riser	UL 1666	OFNP
OFCR	Optical fiber conductive riser	UL 1666	OFCP
OFN	Optical fiber nonconductive	UL 1581	OFNP or OFNR
OFC	Optical fiber conductive	UL 1581	OFCP or OFCR

- *Riser cables*. A *riser* is an opening, shaft, or duct that runs vertically between one or more floors. Riser cables can be used in these vertical passages. Riser-rated cables are UL-certified by the UL-1666 *riser fire test method* as having adequate fire resistance for installation without conduit in areas such as elevator shafts and wiring closets. As noted in Table 5.2, these cables are termed OFNR (*optical fiber nonconductive riser*) or OFCR (*optical fiber conductive riser*). Note that plenum cables may be substituted for riser cables, but not vice versa.
- *General-purpose cables*. A general-purpose area refers to all other regions on the same floor that are not plenum or riser spaces. General-purpose cables can be installed in horizontal, single-floor connections, for example, to connect from a wall jack to a computer. However, they cannot be used in riser or plenum applications without being placed in fireproof conduits. To qualify as a general-purpose cable, it must pass the UL 1581 *vertical-tray fire test*. As shown in Table 5.2, these cables are rated OFN (*optical fiber nonconductive*) or OFC (*optical fiber conductive*). Note that plenum or riser cables may be substituted for general-purpose cables, but not vice versa.

5.4. Outdoor Cables

Outdoor cable installations include aerial, duct, direct-burial, underwater, and tactical military applications. Invariably these cables consist of a loose-tube structure. Many different designs and sizes of outdoor cables are available depending on the physical environment in which the cable will be used and the particular application. Some important designs are described here.

5.4.1. Aerial cable

An *aerial cable* is intended for mounting outside between buildings or on poles or towers. The two main designs that are being used are the self-supporting and the facility-supporting cable structures. The *self-supporting cable* contains an internal strength member that allows the cable to be strung between poles without implementing any additional support to the cable. For the *facility-supporting cable*, first a separate wire or strength member is strung between the poles, and then the cable is lashed or clipped to this member. Three common self-supporting aerial cable structures known as OPGW, ADSS, and figure 8 are described below.

In addition to housing the optical fibers, the *optical ground wire* (OPGW) cable structure contains a steel or aluminum tube that is designed to carry the ground current of an electrical system. The metal structure acts as the strength member of the cable. OPGW cables with up to 144 fibers are available.

The *all-dielectric self-supporting* (ADSS) cable uses only dielectric materials, such as aramid yarns and glass-reinforced polymers, for strength and protection of the fibers. An ADSS cable typically contains 288 fibers in a loose-tube stranded-cable-core structure.

Figure 5.8 illustrates a popular aerial cable known as a *figure 8* cable because of its shape. A key feature is the factory-attached *messenger*, which is a support member used in aerial installations. The built-in messenger runs along the entire length of the cable and is an all-dielectric material or a high-tension steel cable with a diameter between 0.25 and 0.625 in (0.64 and 1.6 cm). This configuration results in a self-supporting structure that allows the cable to be installed easily and quickly on low-voltage utility or railway poles. In some cases the steel messenger is placed at the center of a self-supporting aerial cable to reduce stresses on the cable from wind or ice loading. This configuration usually is for short distances between poles or for short distances between adjacent buildings.

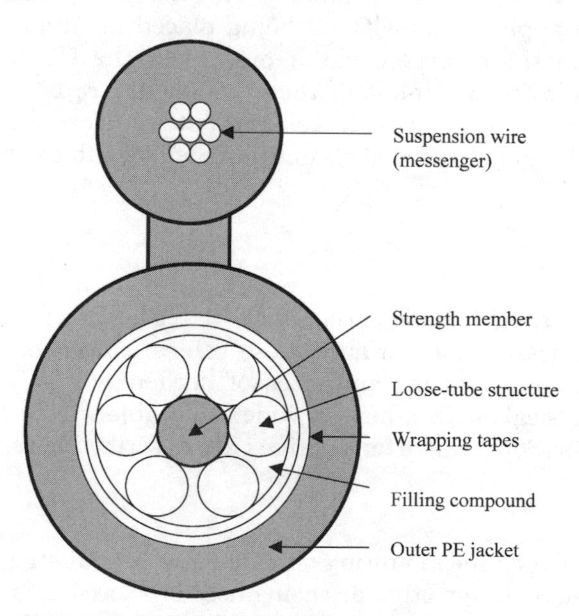

Figure 5.8. Figure 8 aerial cable.

Alternatively, a cable for an aerial application does not contain a built-in messenger. Instead, first a separate steel messenger is strung between poles, and then the optical cable is lashed to this messenger. This lashing method supports the cable at short intervals between poles instead of just at the poles themselves, thereby reducing stress along the length of the cable.

5.4.2. Armored cable

An *armored cable* for direct-burial or underground duct applications has one or more layers of steel-wire or steel-sheath protective armoring below a layer of polyethylene (PE) jacketing, as shown in Fig. 5.9. This not only provides additional strength to the cable but also protects it from gnawing animals such as squirrels or burrowing rodents. Since burrowing gophers do not drink (not even water), they often cause damage to underground cables, which to them look like tasty roots that contain food and moisture. For example, in the United States the plains pocket gopher (*Geomys busarius*) will destroy unprotected cable that is buried less than 6 ft deep.

5.4.3. Underwater cable

Underwater cable, also known as *submarine cable*, can be used in rivers, lakes, and ocean environments. Since such cables normally are exposed to high water pressures, they have much more stringent requirements than underground cables. For example, as shown in Fig. 5.10 for a cable that can be used in rivers and lakes, they have various water-blocking layers and a heavier armor jacket. Cables that run under the ocean have further layers of armoring and contain copper wires to provide electric power for submersed optical amplifiers or regenerators. In addition, if such a cable is damaged, the ruptured portion needs to be lifted to the surface for repair.

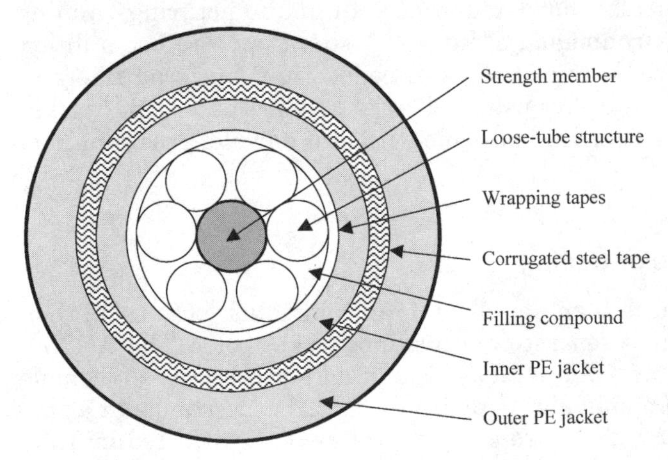

Figure 5.9. Example of an armored outdoor cable.

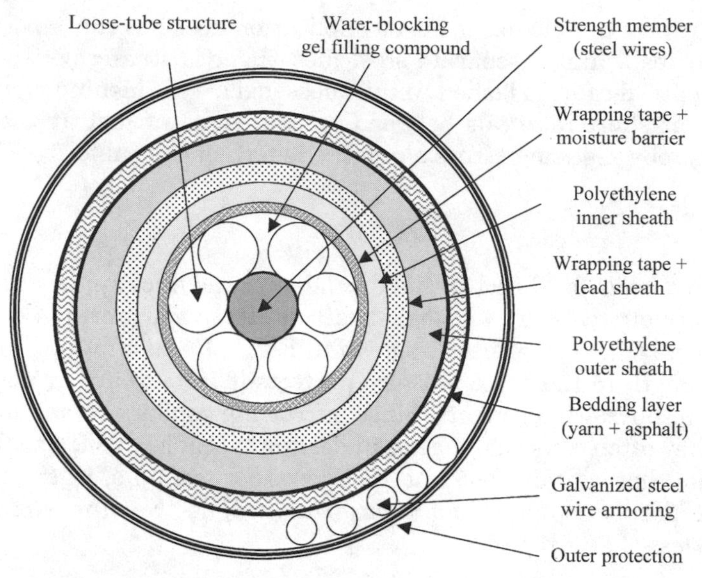

Figure 5.10. Fiber optic underwater cable.

5.4.4. Military cable

Extremely strong, lightweight, rugged, survivable tight-buffered cables have been designed for *military tactical field use*. That means they need to be crush-resistant and resilient so they can withstand being run over by military vehicles, including tanks, and they need to function in a wide range of harsh environments. In addition, since often they are deployed in the field from reels attached to the back of a rapidly moving jeep, they must survive hard pulls. As a result of being developed for such hostile environments, these cables also have found use in manufacturing, mining, and petrochemical environments. Figure 5.11 shows a two-fiber military distribution cable. This consists of two color-coded tight-buffered fibers surrounded by aramid yarn and encapsulated in a polyurethane (PU) jacket. Cable sizes with up to 24 fibers are possible with standard lengths ranging from 300 m to 2 km.

5.5. Fiber and Jacket Color Coding

If there is more than one fiber in an individual loose tube, then each fiber is designated by a separate and distinct jacket color. The ANSI/TIA/EIA-598-A standard, *Optical Fiber Cable Color Coding*, prescribes a common set of fiber colors. Since nominally there are up to 12 fiber strands in a single loose tube, strands 1 through 12 are uniquely color-coded, as listed in Table 5.3. If there are more than 12 fibers within an individual loose tube, then strands 13

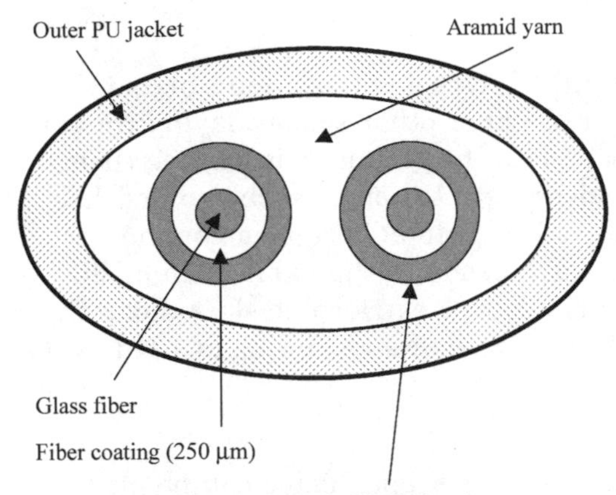

Figure 5.11. Military two-fiber distribution cable.

TABLE 5.3. Standard Optical Fiber Jacket and Loose-Tube Color Identifications

Fiber number	Color	Fiber number	Color
1	Blue	13	Blue/black tracer
2	Orange	14	Orange/yellow tracer
3	Green	15	Green/black tracer
4	Brown	16	Brown/black tracer
5	Slate (gray)	17	Slate/black tracer
6	White	18	White/black tracer
7	Red	19	Red/black tracer
8	Black	20	Black/yellow tracer
9	Yellow	21	Yellow/black tracer
10	Violet	22	Violet/black tracer
11	Rose (pink)	23	Rose/black tracer
12	Aqua	24	Aqua/black tracer

through 24 repeat the same fundamental color code used for strands 1 through 12, with the addition of a black or yellow dashed or solid tracer line, as noted in Table 5.3. For cables having more than one loose tube, the tubes also are color-coded in the same manner as the fibers; that is, tube 1 is blue, tube 2 is orange, and so on.

Ribbon cables follow the same color-coding scheme. Thus, one of the outside fibers would have a blue jacket, the next fiber would be orange, and so on until the other outer edge is reached where, for a 12-fiber ribbon, the fiber would be aqua (light blue).

5.6. Installation Methods

Workers can install optical fiber cables by pulling or blowing them through ducts (both indoor and outdoor) or other spaces, laying them in a trench outside, plowing them directly into the ground, suspending them on poles, or laying or plowing them underwater. Although each method has its own special handing procedures, they all need to adhere to a common set of precautions. These include avoiding sharp bends of the cable, minimizing stresses on the installed cable, periodically allowing extra cable slack along the cable route for unexpected repairs, and avoiding excessive pulling or hard yanks on the cable.

5.6.1. Direct-burial installations

For *direct-burial installations* a fiber optic cable can be plowed directly underground or placed in a trench which is filled in later. Figure 5.12 illustrates a plowing operation that may be carried out in nonurban areas. The cables are mounted on large reels on the plowing vehicle and are fed directly into the ground by means of the plow mechanism. Since a plowing operation normally is not feasible in an urban environment, a *trenching* method must be used.

Figure 5.12. Plowing operation for direct burial of optical fiber cables. (*Photo courtesy of Vermeer Manufacturing Company; www.vermeer.com.*)

Trenching is more time-consuming than direct plowing since it requires a trench to be dug by hand or by machine to some specified depth. However, trenching allows the installation to be more controlled than in plowing. For example, in direct plowing it is not known if a sharp rock is left pressing against the installed cable or if the cable was damaged in some way that may cause it to fail later.

Usually a combination of the two methods is used, with plowing being done in isolated open areas and trenching being done where plowing is not possible, such as in urban areas. In addition, another technique called *directional boring* or *horizontal drilling* may be needed in areas where the surface cannot be disturbed, for example, a multiple-lane highway. These machines come in at least a dozen different sizes depending on the depth and distance that holes need to be bored. For example, the horizontal drilling machine illustrated in Fig. 5.13 can bore a 3- to 8-cm-diameter hole below the surface for distances of over 100 m.

During direct-burial installations, a bright (usually orange) *warning tape* normally is placed a short distance (typically 18 in) above the cable to alert future digging operators to the presence of a cable. The tape may contain metallic strips so that it can be located from aboveground with a metal detector. In addition, a warning post or a cable marker that is flush with the ground may be used to indicate where a cable is buried. Figure 5.14 illustrates some typical cable indication methods. Besides indicating to repair crews where a cable is located, these precautions are intended to minimize the occurrence of what is known popularly in the telecommunications world as *backhoe fade* (the rupture of a cable by an errant backhoe).

Figure 5.13. Directional boring machine in use. (*Photo courtesy of Vermeer Manufacturing Company; www.vermeer.com.*)

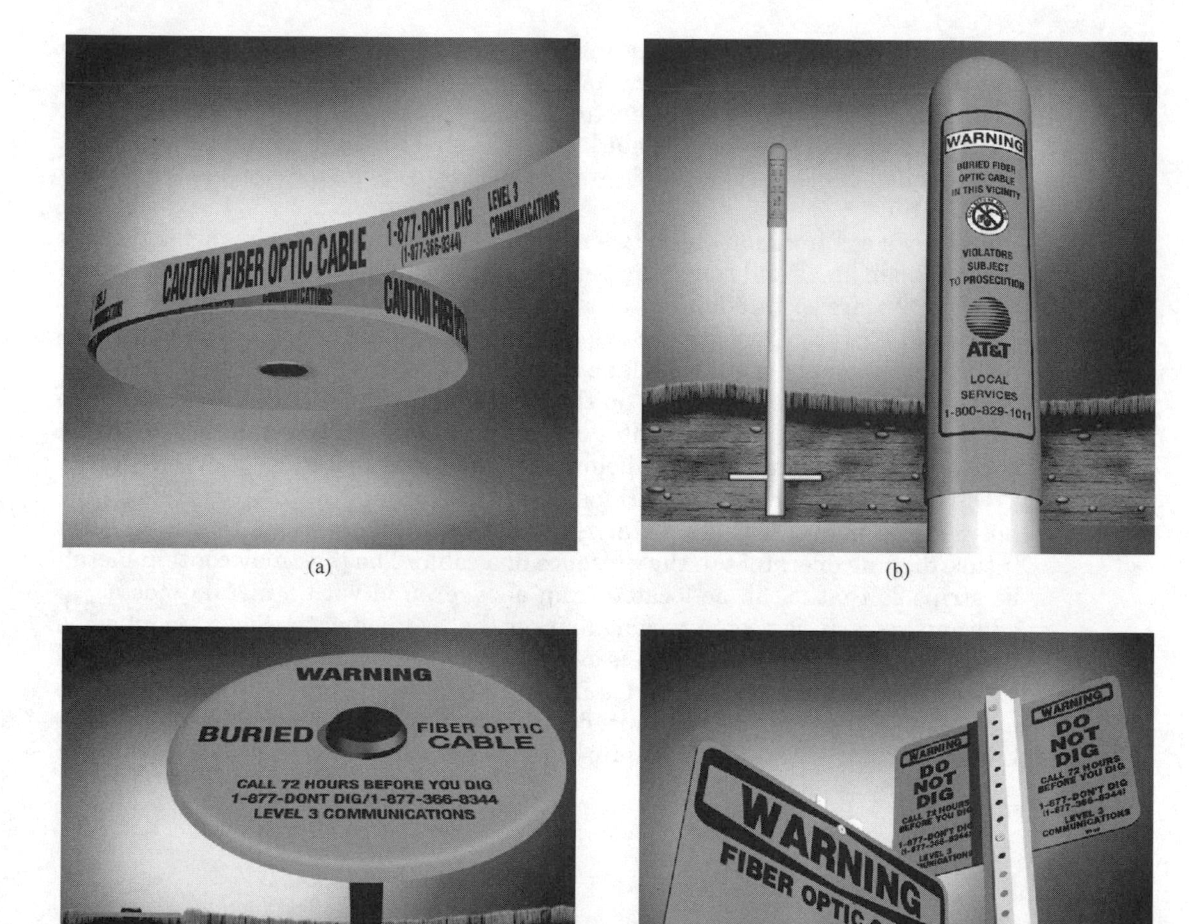

Figure 5.14. Examples of tapes, posts, and ground-flush markers to indicate that a cable is buried in that location. (*Photos courtesy of William Frick & Company; www.fricknet.com.*)

5.6.2. Pulling into ducts

Most *ducts* are constructed of a high-density polyurethane, PVC, or an epoxy fiberglass compound. To reduce pulling tensions during cable installation, the inside walls can have longitudinal or corrugated ribs, or they may have been lubricated at the factory. Alternatively, a variety of pulling lubricants are available that may be applied to the cable itself as it is pulled into a long duct or into one that has numerous bends. A duct also can contain a pulling tape running along its length that was installed by the duct manufacturer. This is a flat tape similar to a measuring tape that is marked every meter for easy identification of distance. If the duct does not already contain a pulling tape, the tape can be fished through or blown into a duct length. After the fiber optic cable is installed in a duct, end plugs can be added to prevent water and debris from entering the duct. Similar to direct-burial installations, a warning tape may be placed underground above the duct, or warning posts or markers may be placed aboveground to alert future digging operators to its presence.

5.6.3. Air-assisted installation

Using forced air to blow a fiber cable into a duct is an alternative method to a pulling procedure. The cable installation scheme of utilizing the friction of the air moving over the cable jacket is referred to as either a *cable jetting* or a *high-airspeed blown* (HASB) method. Cable jetting must overcome the same frictional forces to move cable as in a pulling operation, but it does this differently and with much less stress on the cable. As shown in Fig. 5.15, the force in the cable jetting method comes, first, from a mechanical device which pushes the cable into the duct and, second, from the force of moving air on the cable jacket. The advantage of cable jetting is that the cable moves freely around bends, whereas the pulling method puts a high lateral stress on the cable when it is passed through bends in a duct. Figure 5.16 shows an example of how a cable is fed into a truck-mounted cable jetting machine at the beginning of a large duct. The cable jetting machine is at the far left in the photo.

As a variation to this method, the force of air can be on a piston or carrier attached to the front end of the cable. Figure 5.15*b* illustrates this method. The force of air on this piston then pulls the cable through the duct. The air-blowing method utilizing such an air-capturing device is referred to as the *push/pull installation* method. However, in this variation the cable experiences the same lateral stresses as in a standard pulling method when traversing bends in a duct.

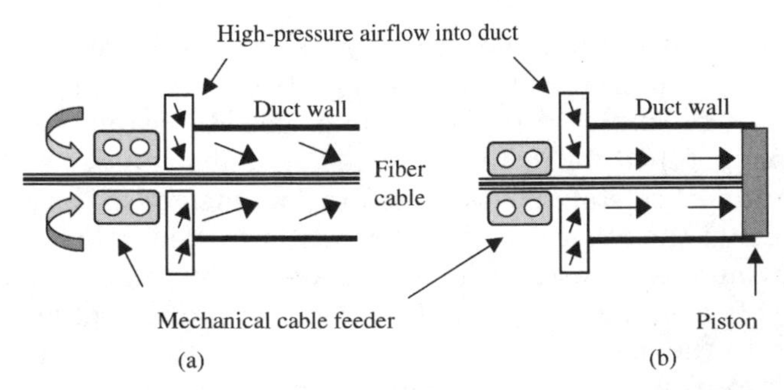

Figure 5.15. Two air-assisted cable installation methods. (*a*) Force of air is on the fiber; (*b*) force of air is on the end piston.

Figure 5.16. Microducts being fed into the mechanical drive of a cable jetting machine. (*Photo courtesy of Sherman & Reilly, Inc.; www.sherman-reilly.com.*)

5.6.4. Aerial installation

Cable crews can install an aerial cable either by lashing it onto an existing steel messenger wire or by directly suspending it between poles, if it is a self-supporting cable. Several different methods can be used to install the fiber optic cables. The primary method for installing self-supporting cable is a *stationary reel technique*. This method stations the payoff reel at one end of the cable route and the take-up reel at the other end. A pull rope is attached to the cable and is threaded through pulleys on each pole. The take-up reel gradually pulls the cable from the payoff reel, the pulleys guide it into position along the route, and it is then attached to the poles.

If a *messenger wire* is used, first this wire is installed between poles with an appropriate tension and sag calculated to support the fiber optic cable. The messenger wire must be grounded properly and should be kept on one pole side along the route whenever possible. One of at least three techniques then can be used to attach the fiber optic cable to the messenger wire. Each of these methods uses a special lashing machine that hangs on the messenger and attaches the cable as it moves along the messenger.

5.6.5. Submarine installation

Over a million kilometers of submarine cable already are submerged in oceans around the world, which is enough to circle the globe 30 times. Specially designed cable-laying ships are used to install an undersea cable. The ships have several large circular containers inside of them called *cable tanks*. In modern cable ships these tanks together can hold up to 8000 km of underwater cable. Such a length of cable is assembled onshore in a factory environment along with underwater signal amplifiers that need to be located every 60 km or so. The amplifiers are housed in large beryllium-copper tubes that are about a meter long and 50 cm in diameter. After being assembled, this cable unit is coiled by hand into the cable holding tanks on the ship at a rate of around 80 km/day. During installation, near the shore a sea plow buries the cable to a depth of about 1 m under the ocean floor to protect it from fishing nets and other factors that might damage the cable. In the middle of the ocean the cable simply lays exposed on the ocean floor.

5.7. Summary

Optical cables come in many sizes, styles, and configurations. Typically, in addition to optical fibers they contain aramid yarn or steel strength members and are encapsulated in one of the jacketing materials listed in Table 5.1. Fibers within the cable or within a ribbon may be identified individually by means of the standard color coding method listed in Table 5.3. The three fundamental cable structures are

- *Tight-buffered fiber cable* where each fiber is individually encapsulated within its own 900-μm-diameter plastic buffer structure
- *Loose-tube cable* in which one or more standard coated free-moving fibers are enclosed in a tube that has an inner diameter which is much larger than the fiber diameter
- *Ribbon cable* where up to 12 fibers are aligned precisely next to one another and encapsulated in a plastic jacket to form a ribbon

Indoor cables can be used in plenum, riser, and general-purpose areas as listed in Table 5.2. The different indoor cable types include

- *Interconnect cables* for light-duty low-fiber-count indoor applications such as fiber-to-the-desk links, patch cords, and point-to-point runs in conduits and trays.

- *Breakout or fanout cables* for low- to medium-fiber-count applications where it is necessary to protect individual jacketed fibers. One application is the termination of the individual fibers of a breakout cable in a protective patch-panel box to permit connections to equipment.
- *Distribution cables* for use in intrabuilding trays, conduit, backbone premise pathways, and dropped ceilings. A main feature is that they enable groupings of fibers within the cable to be branched (distributed) to various locations.

Outdoor cables come in a wide range of configurations. Some important ones are

- *Aerial cable*, which is intended for mounting outside between buildings or on poles or towers.
- *Armored cable*, which is designed for direct-burial or underground duct applications.
- *Underwater cable*, also known as *submarine cable*, that can be used in rivers, lakes, and ocean environments.
- *Military cable*, which is extremely strong, lightweight, rugged, survivable tight-buffered cable designed for military tactical field use. These cables also have found use in manufacturing, mining, and petrochemical environments.

Workers can install optical fiber cables by pulling or blowing them through ducts (both indoor and outdoor) or other spaces, laying them in a trench outside, plowing them directly into the ground, suspending them on poles, or laying or plowing them underwater. Although each method has its own special handling procedures, they all need to adhere to a common set of precautions. These include avoiding sharp bends of the cable, minimizing stresses on the installed cable, periodically allowing extra cable slack along the cable route for unexpected repairs, and avoiding excessive pulling or hard yanks on the cable.

Further Reading

1. *NFPA 70—National Electrical Code*, Article 770, "Optical Fiber Cables and Raceways," National Fire Protection Association, Quincy, Mass., 2002.
2. *Canadian Electrical Code*, "Optical Fiber Cables," Canadian Standards Association, 2002.
3. Underwriters Laboratories (http://www.UL.com).
4. ANSI/TIA/EIA-598-A, *Optical Fiber Cable Color Coding*, 1995.
5. Bob Chomycz, *Fiber Optic Installer's Field Manual*, McGraw-Hill, New York, 2000.
6. BiCSI, *Telecommunications Cabling Installation*, McGraw-Hill, New York, 2001.
7. For an article by W. F. DeWitt, "Looking up: Selection criteria for fiber optic cables in the aerial plant," *Outside Plant Magazine*, June 2000, see http://www.ospmag.com/features/2000/lookingup.htm.
8. Special issue on "Undersea Communications Technology," *AT&T Technical Journal*, vol. 74, no. 1, January/February 1995.

Chapter

6

Light Sources and Transmitters

When you go to a large hardware store, you will see dozens of different lightbulbs ranging in size and power from flashlight to floodlight applications. A similar situation holds for light sources used in optical communications. In this case the sources are much smaller, but they also range from simple, inexpensive light-emitting diodes (LEDs) to costly, high-power laser diodes with complex semiconductor structures.

Normally a user does not obtain an isolated light source, but purchases it as part of an optical transmitter. However, it is important to know the characteristics of the source in order to choose the transmitter properly. Thus this chapter first presents the characteristics of LED and laser diode light sources. Following this we will see how the sources are incorporated into transmitters. The transmitter can range from a simple inexpensive, LED-based package that fits on a circuit board for short-distance links, to expensive, laser diode-based modules that also contain sophisticated electronics for controlling the temperature, wavelength stability, and optical output power level of the laser diode.

6.1. General Source Characteristics

To understand the differences between LEDs and laser diodes, let us first look at some general characteristics. These include the operating wavelengths of various device materials, the source output spectral width (the range of wavelengths that are contained in an optical output pulse), and the device modulation capability (how fast a device can be turned on and off).

6.1.1. Materials

Semiconductor-based light sources are about the size of a grain of salt. This size allows efficient coupling of their light output into the small diameters of fibers. In addition, their semiconductor structure and low-power dissipation characteristics make them compatible with integrated-circuit electronics. Let us first

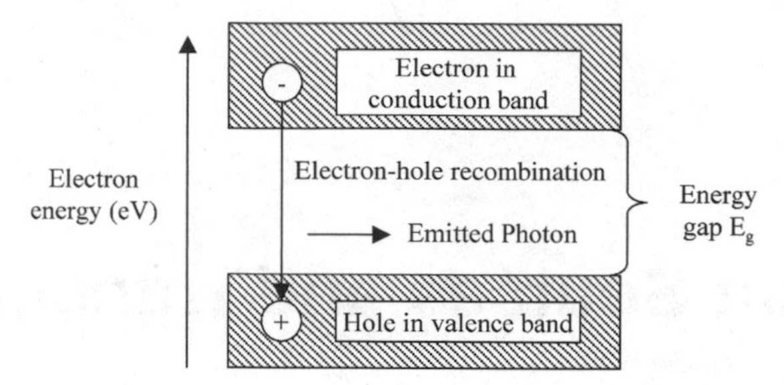

Figure 6.1. Electrons in semiconductor materials can reside in only two specific energy bands separated by an energy gap.

look how these materials behave. The electrons in semiconductor materials are allowed to reside in only two specific energy bands, as shown in Fig. 6.1. The two allowed bands are separated by a forbidden region, called an *energy gap*, in which electrons cannot reside. The energy difference between the top and bottom bands is referred to as the *bandgap energy*. In the upper band, called the *conduction band*, electrons are not bound to individual atoms and are free to move around in the material. The lower band is called the *valence band*. Here *holes* (which are vacancies in an atom that are not occupied by an electron) are free to move. The mobile electrons and holes set up a current flow when an external electric field is applied.

The conduction of electrons and holes in a material can be increased greatly by adding trace amounts of impurity atoms to a material. For example, suppose an element that has five electrons in the outer shell replaces a Si atom that has four outer-shell electrons. The fifth electron is loosely bound and thus is available for conduction. Since in this type of material there is an excess of negatively charged electrons, the material is called *n-type material*. Similarly replacing a Si atom with an element that has three electrons in the outer shell results in an excess of mobile holes in the valence band. This is called *p-type material* because conduction is a result of (positive) hole flow.

An electron sitting in the conduction band can drop down into a hole in the valence band, thereby returning an atom to its neutral state. This process is called *recombination* (or *electron-hole pair recombination*), since an electron recombines with a hole. This recombination process releases energy in the form of a photon and is the basis by which a source emits light. As Chap. 3 describes, the energy E emitted during such a recombination is related to a specific wavelength of light λ through the relationship $E = 1.240/\lambda$, where λ is given in micrometers and E is specified in electron volts. Since each type of material has a unique bandgap energy, electron-hole recombination in different materials results in different wavelengths being emitted.

To create a light-emitting device for use in the spectral transmission bands of optical fibers, material engineers fabricate layered structures consisting of different alloy mixtures. Table 6.1 lists some LED and laser diode material

TABLE 6.1. Some LED and Laser Diode Material Mixtures and Their Characteristics

Material	Wavelength range, nm	Bandgap energies, eV
GaAs	900	1.4
GaAlAs	800–900	1.4–1.55
InGaAs	1000–1300	0.95–1.24
InGaAsP	900–1700	0.73–1.35

mixtures together with their operating wavelength range and approximate bandgap energies. Alloys consisting of three elements are called *ternary compounds*, and four-element alloys are known as *quaternary compounds*. A specific operating wavelength can be selected for AlGaAs, InGaAs, and InGaAsP devices by varying the proportions of the constituent atoms. Thus devices can be tailored to emit at a selected wavelength in the 780- to 850-nm band or in any of the other transmission bands ranging from 1280 to 1675 nm for glass fibers.

6.1.2. Spectral output width

A major difference between light sources is the spectral width of their light output. This is an important factor when one is choosing an optical source since, as Chap. 4 notes, signal spreading in an optical fiber due to chromatic dispersion is directly proportional to the wavelength band over which a source emits light. Recall that chromatic dispersion occurs since each wavelength in a light signal travels at a slightly different velocity. This effect progressively smears out an optical signal as it travels along a fiber. Consequently in order to send a high-speed signal (consisting of very narrow light pulses) over long distances, the source needs to emit light within as narrow a spectral width as possible. This can be done only with laser diodes. An LED has spectral widths ranging from 30 nm at a central wavelength of 850 nm to around 120 nm at a 1550-nm central wavelength. Laser diodes, on the other hand, can have spectral widths of a few picometers (10^{-3} nm) at 1550 nm. For a laser diode the spectral width is referred to as the *linewidth*.

Example Consider a fiber that has a chromatic dispersion $D = 2\,\text{ps/(km}\cdot\text{nm)}$. Suppose we want to send a 10-Gbps signal over this fiber. At 10 Gbps the pulse widths are $1/(10\,\text{Gbps}) = 100\,\text{ps}$. If 20 percent pulse spreading is tolerable, then this data rate has an allowable spread of 20 ps before adjacent pulses overlap too much. Then if we use an inexpensive laser source with a 5-nm linewidth, this signal can travel only 2 km before the pulse overlap limit is reached. For a high-quality laser with a 0.05-nm linewidth, the chromatic dispersion distance limit is 2000 km.

6.1.3. Modulation speed

Direct modulation is the process of using a varying electric signal to change the optical output level of a device. The term *modulation speed* refers to how fast a device can be turned on and off by an electric signal to produce a corresponding optical output pattern. As the next two sections describe, laser diodes can be modulated significantly faster than LEDs. However, there is a speed limit beyond which even the laser does not respond fast enough to the changes in an electric signal. Beyond this point a steady light output stream from the laser diode is fed into an external device, which can change the intensity of the light that passes through it very rapidly. This process is known as *external modulation*. Section 6.5 describes external modulators in greater detail.

6.2. LEDs

If we look around us, we notice LEDs everywhere. They can be seen glowing green, yellow, or red in vehicles, computer equipment, kitchen appliances, telephones, cameras, and every imaginable piece of electronic equipment. They are inexpensive and highly reliable light sources. The LEDs used in optical communications are much smaller and emit in the infrared region, but compared to the other telecommunication light sources used, they are much less expensive and easier to use in transmitter designs. However, because of their relatively low power output, broad emission pattern, and slow turn-on time, their use is limited to low-speed (less than 200-Mbps), short-distance (up to a few kilometers) applications using multimode fibers.

6.2.1. Principles of operation

To create a material structure for an LED, *n*-type and *p*-type semiconductor materials are joined together. The boundary between the two joined materials is called a *pn junction*, as shown in Fig. 6.2. To create a supply of electrons and holes that may flow across the *pn* junction to recombine and thereby emit light, one applies a voltage across the junction. This is called a *bias voltage*. Variations in the applied voltage, or the driving current, then correspond to a varying optical output from the device. Figure 6.3 shows a typical relationship between the optical power generated by an LED and the drive current. Nominally LEDs

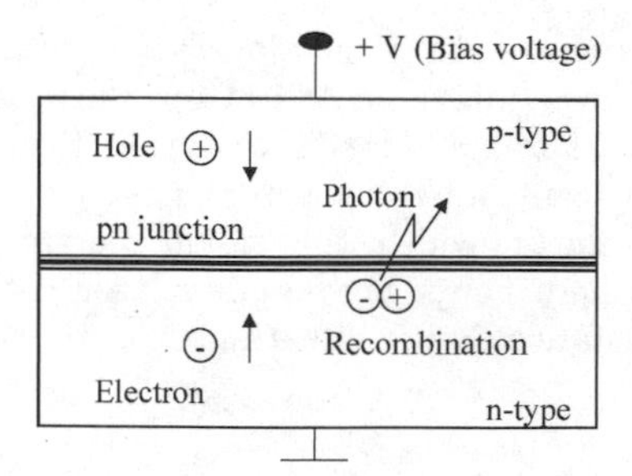

Figure 6.2. Electrons and holes recombine at a *pn* junction and thereby emit light.

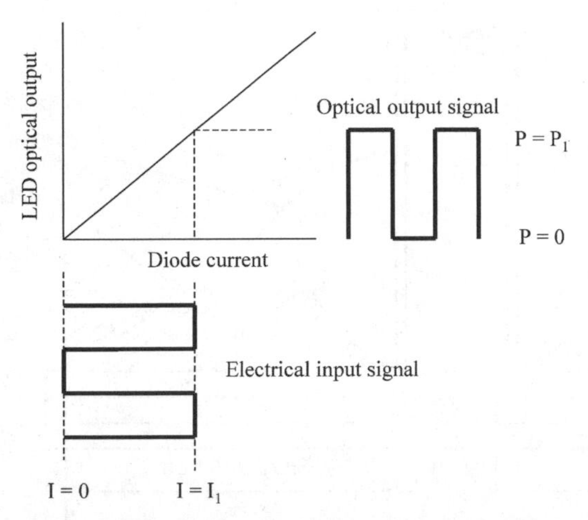

Figure 6.3. Typical relationship between the optical signal generated by an LED and the drive current.

operate around 50- to 100-mA drive currents and require a bias voltage of around 1.5 V. As shown in Fig. 6.3, to send a digital signal, the drive current can simply turn the LED on during a 1 pulse (current $I = I_1$ produces power level P_1) and off during a 0 pulse (current $I = 0$).

The light that is emitted during electron-hole recombination can go in all directions. Therefore engineers have varied the device geometry and internal structure in different ways to confine and guide the light so it can be coupled efficiently into a fiber (see the text below on photon and carrier confinement for details). Even so, LEDs have fairly broad emission patterns. Because of this they are used mainly with multimode fibers, since they couple too little light into single-mode fibers. Two LED configurations are the surface-emitting and edge-emitting structures.

6.2.2. Surface emitters

In the *surface emitter* a circular metal contact defines the active region in which light is generated, as Fig. 6.4 shows. The contact is nominally 50 μm in diameter. To couple the light into a fiber, first a well is etched into the substrate on which the device is fabricated. A multimode fiber that has a core diameter of 50 μm or larger then can be brought close to the region where the light is generated and can be cemented in place. The light emission pattern is *isotropic* (also called *lambertian*), which means it is equally bright when viewed from any direction. Since the emission area and the fiber core area are the same, most of the emitted light will shine on the fiber core end face. However, not all the light will be coupled into the fiber since it has a limited numerical aperture (i.e., a limited light acceptance cone), as Fig. 6.5 shows. Chapter 7 presents more details on power coupling factors.

Figure 6.4. Schematic (not to scale) of a high-radiance surface-emitting LED. The active region is a circular section that has an area compatible with the fiber-core end face.

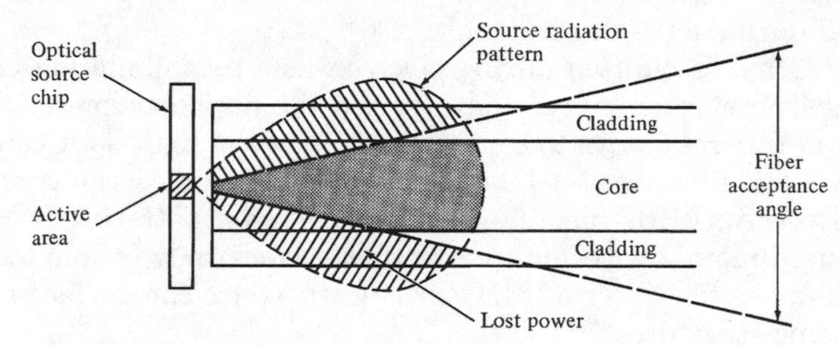

Figure 6.5. Schematic of an optical source coupled to a fiber. Light outside of the acceptance angle is lost.

6.2.3. Edge emitters

The *edge emitter* shown in Fig. 6.6 consists of an active *pn* junction region where the photons are generated and two light-guiding layers that function in the same manner as an optical fiber. This structure forms a waveguide channel that directs the optical radiation toward the edge of the device, where it can be coupled into an optical fiber. The emission pattern of the edge emitter is more directional than that of a surface emitter, which allows a greater percentage of the emitted light to be coupled into a fiber.

Photon and Carrier Confinement To be useful in fiber optic transmission applications, an LED must have a high radiance output and a high quantum efficiency. Its *radiance* (or *brightness*) is a measure, in watts (W), of how much optical power radiates

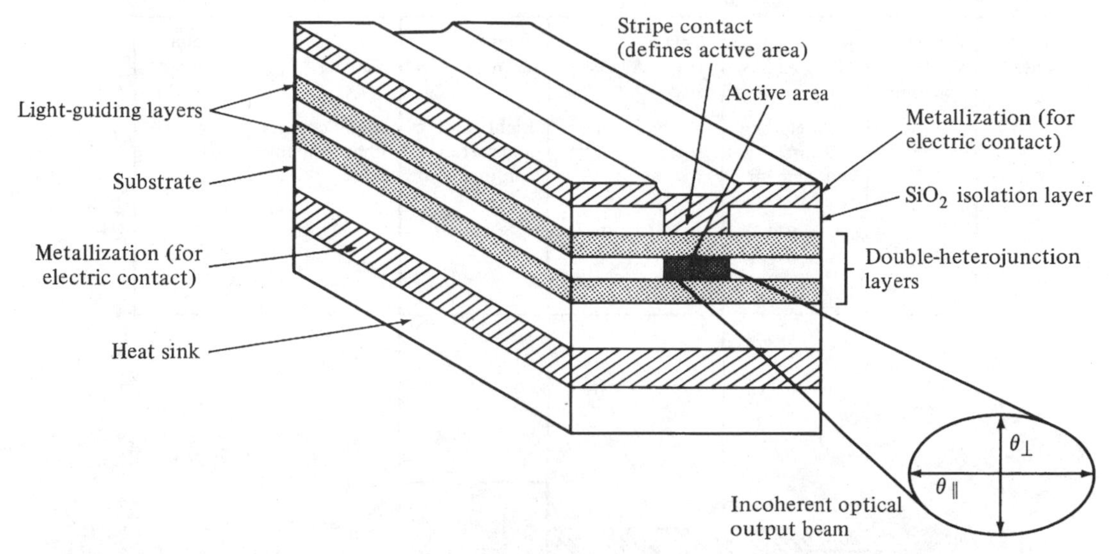

Figure 6.6. Schematic (not to scale) of an edge-emitting LED. The output beam is lambertian in the plane of the *pn* junction ($\theta_{\parallel} = 120°$) and highly directional perpendicular to the *pn* junction ($\theta_{\perp} \approx 30°$).

into a unit solid angle per unit of emitting surface area. A high radiance is needed to couple sufficiently high optical power levels into a fiber core. The *quantum efficiency* describes the fraction of injected electron-hole pairs that emit photons when they recombine at a *pn* junction (not every recombination results in light being emitted). Conventionally, the electron-hole pairs are called *charge carriers* or simply *carriers*.

To achieve a high radiance and a high quantum efficiency, the LED structure must provide a means of confining the carriers to the active region of the *pn* junction where radiative recombination takes place. In addition, the structure needs to confine the light so that the emitted photons are guided to the fiber end and are not absorbed by the material surrounding the *pn* junction.

An effective structure for achieving carrier and optical confinements is the sandwich configuration shown in Fig. 6.7 for an LED operating around 850 nm. This is referred to as a *double-heterostructure* (or *heterojunction*) device because of the two different alloy layers on each side of the active region. The bandgap differences of adjacent layers confine the charge carriers, while the differences in the indices of refraction of the adjoining layers confine the emitted photons to the central active layer. Light sources at longer wavelengths consist of similar structures but different materials.

6.3. Laser Diodes

Semiconductor-based *laser diodes* are the most widely used optical sources in fiber communication systems. The four main laser types are the Fabry-Perot (FP) laser, the distributed-feedback (DFB) laser, tunable lasers, and the vertical cavity surface-emitting laser (VCSEL). Key properties of these lasers include high optical output powers (greater than 1 mW), narrow linewidths (a fraction of a nanometer, except for the FP laser), and highly directional output beams for efficient coupling of light into fiber cores.

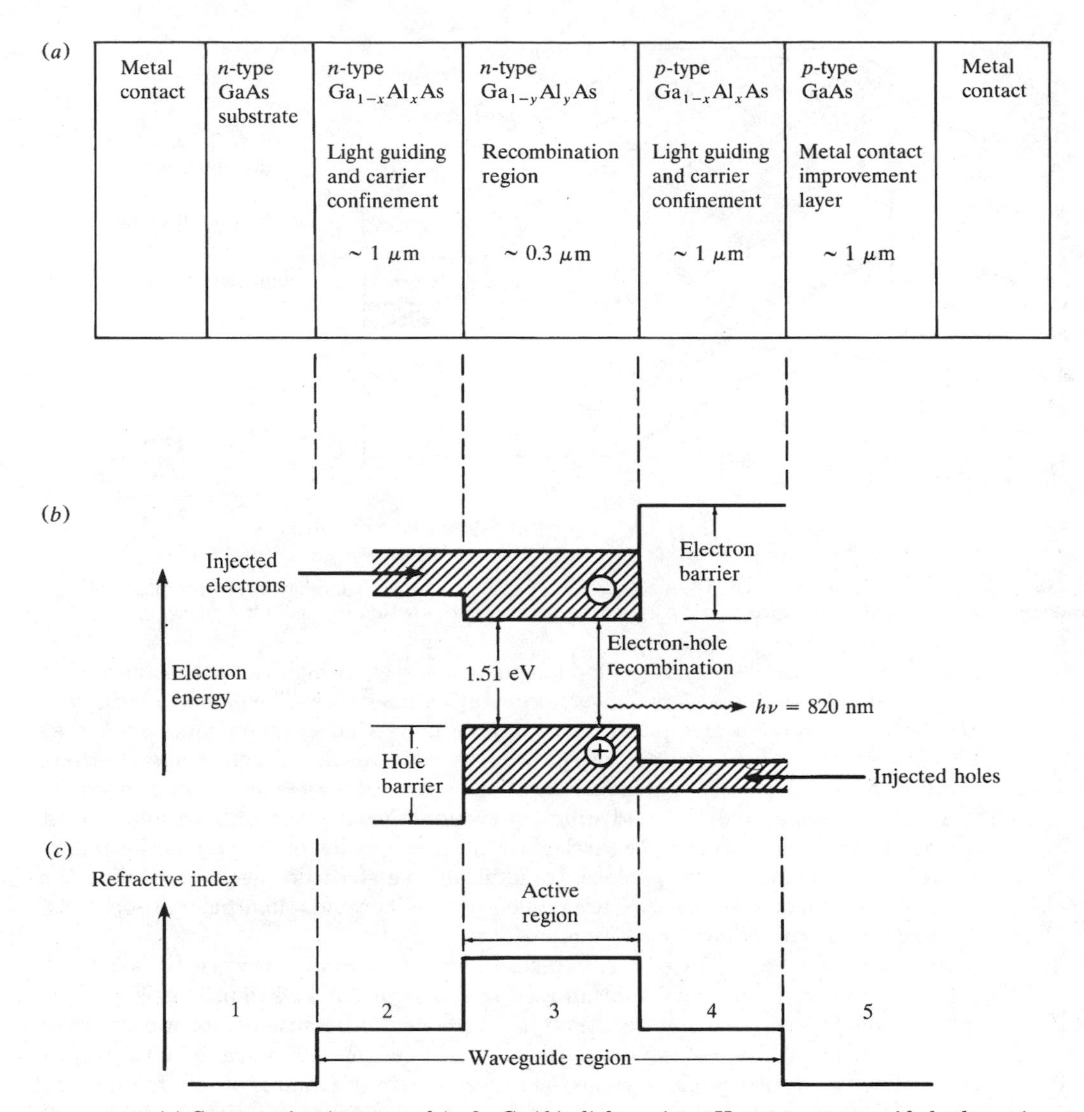

Figure 6.7. (*a*) Cross-section (not to scale) of a GaAlAs light emitter. Here $x > y$ to provide both carrier confinement and optical guiding. (*b*) Energy band diagram showing the active region and the electron and hole confinement barriers. (*c*) Variations in the refractive index of the layers.

Laser Action The term *laser* actually is an acronym for the phrase "*l*ight *a*mplification by *s*timulated *e*mission of *r*adiation." So what is stimulated emission of radiation and how does it result in light amplification? First, let us look at the possible photon emission and absorption processes shown in Fig. 6.8 for a two-level atomic system. An electron can move from an energy level E_1 in the valence band (called the *ground state*) to a higher state (called an *excited state*) at energy level E_2 in the conduction band either by being pumped externally to that level or by absorbing the energy $h\nu_{12}$ from a passing photon. The latter process is called *stimulated absorption*. In either case, an excited electron then can return to the ground state either spontaneously or

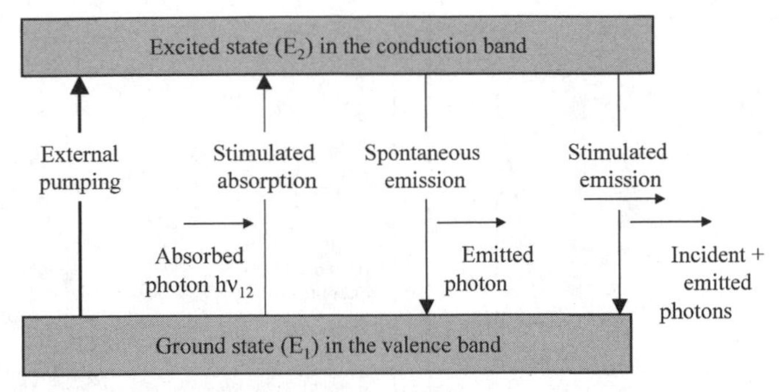

Figure 6.8. Possible photon emission and absorption processes for a two-level atomic system.

by stimulation, thereby emitting a photon. The corresponding photon generation processes are called *spontaneous emission* and *stimulated emission*, respectively. Spontaneous emission occurs randomly "at the will" of the excited electron. Consequently, spontaneously generated photons have random phases and frequencies (or equivalently, random wavelengths), since the electrons can return to the ground state from any level in the conduction band. Therefore, this type of light has a broad spectral width and is called *incoherent*.

Stimulated emission occurs when some external stimulant (such as an incident photon) causes an excited electron to drop to the ground state. The photon emitted in this process has the same energy (i.e., the same wavelength) as the incident photon and is in phase with it. Recall from Chap. 2 that this means their amplitudes add to produce a brighter light. Thus this type of light is called *coherent*. Under normal conditions the number of excited electrons is very small, so that stimulated emission is essentially negligible. For stimulated emission to occur, there must be a *population inversion* of carriers. This fancy term simply means that there are more electrons in an excited state than in the ground state. Since this is not a normal condition, population inversion is achieved by supplying additional external energy to pump electrons to a higher energy level. The "pumping" techniques can be optical or electrical. For example, the bias voltage from a power supply provides the external energy in a semiconductor device.

Laser action normally takes place within a region called the *gain medium* or *laser cavity*. To achieve lasing action within this region, the photon density needs to be built up so that the stimulated emission rate becomes higher than the rate at which photons are absorbed by the semiconductor material. A variety of mechanisms can be used either at the ends or within the cavity to reflect most of the photons back and forth through the gain medium. With each pass through the cavity, the photons stimulate more excited electrons to drop to the ground state, thereby emitting more photons of the same wavelength. This process thus builds up the photon density in the gain region.

If the gain is sufficient to overcome the losses in the cavity, the device will start to oscillate at a particular optical frequency. The point where this oscillation occurs is called the *lasing threshold*. Below this point both the spectral range and the lateral beam width of the light output are broad, like that from an LED. Beyond the lasing threshold the device behaves as a laser, and the light output increases sharply with bias voltage, as shown in Fig. 6.9. As the lasing transition point is approached, the spectral width and the beam pattern both narrow dramatically with increasing drive current.

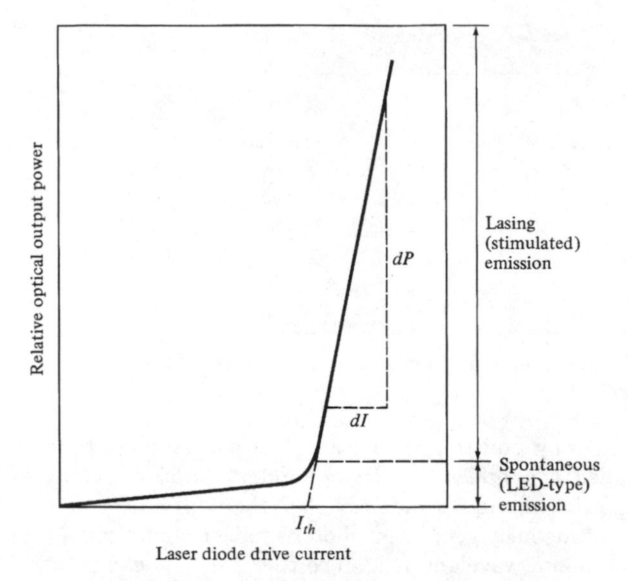

Figure 6.9. Relationship between light output and laser drive current. Below the lasing threshold the optical output is a spontaneous LED-type emission.

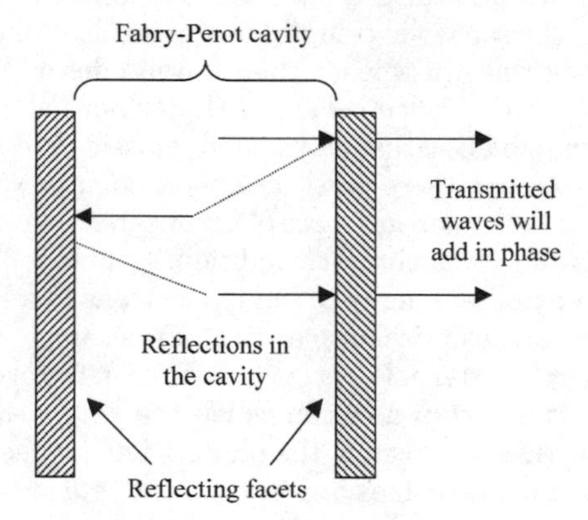

Figure 6.10. Two parallel light-reflecting mirrored surfaces define a Fabry-Perot cavity or an etalon.

6.3.1. Fabry-Perot laser

In the *Fabry-Perot* (FP) laser, the lasing cavity is defined by the two end faces of the cavity, as shown in Fig. 6.10. These are called *facets* and act as light-reflecting mirrors. This structure is called a Fabry-Perot cavity or an *etalon* (see Sec. 9.2 for more details on etalon theory). Since this cavity is fairly long, the laser will oscillate (or *resonate*) simultaneously in several modes or frequencies. When these resonant frequencies are transmitted through the right-hand facet, they add in phase. This results in a greatly increased amplitude for the resonant wavelengths compared to other wavelengths, and thus defines the lasing modes of the FP laser. This effect produces a broad output spectrum that has a width of several nanometers at its half-maximum point. Although this broad spectral output does not make the FP laser feasible for high-speed, long-haul transmissions, it is popular for short- and intermediate-distance (up to 15 km) applications running at data rates of up to 622 Mbps. These applications use the small-form-factor (SFF) transceiver described in Sec. 6.4.4.

6.3.2. Distributed-feedback laser

In a *distributed-feedback* (DFB) laser, a series of closely spaced reflectors provide light feedback in a distributed fashion throughout the cavity, as shown in Fig. 6.11. Through a suitable design of these reflectors, which normally are some type of grating, the device can be made to oscillate in only a single mode with a very narrow linewidth. This means that it emits at a fairly well-defined wavelength. The particular operating wavelength can be selected at the time of device fabrication by an appropriate choice of the reflector spacing. Single-mode DFB lasers are used extensively in high-speed transmission systems. Table 6.2 lists some typical performance parameter values of commercially available DFB lasers used for data transmission. Note that a 2-MHz linewidth is equivalent to 10^{-5} pm at 1550 nm. As described in Chap. 12, uncooled lasers are cost-effective sources for coarse WDM networks, whereas the more expensive cooled lasers are needed for dense WDM applications.

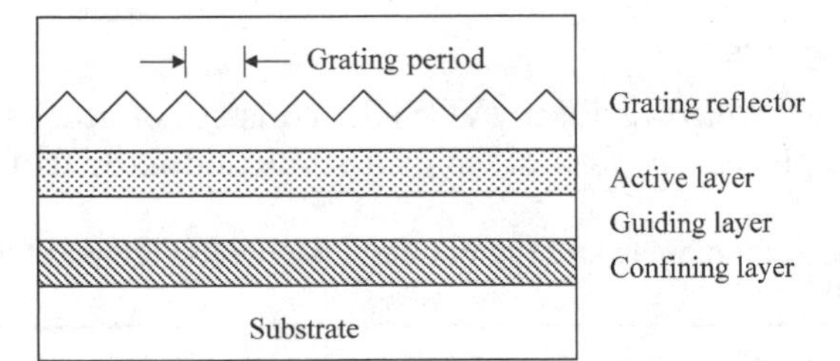

Figure 6.11. In a distributed-feedback (DFB) laser, a series of closely spaced reflectors provides light feedback for lasing.

TABLE 6.2. Typical Performance Parameter Values of DFB Lasers

Center wavelength, nm	Transmitter type	Threshold current, mA	Linewidth	Wavelength drift, pm/°C	Peak optical output power, mW
1310	Uncooled	15	1 nm	100	0.5–1
1550	Cooled	12–25	2 MHz (10^{-5} pm)	1	2–5

6.3.3. Tunable lasers

In many networks there is a need to tune the wavelength of a laser transmitter. This is especially important in a multiwavelength network that uses many laser transmitters operating on a grid of closely spaced wavelengths on the same fiber. Since each laser needs to emit at a precise wavelength (within a fraction of a nanometer), the ability to tune the laser precisely is essential. A number of different technologies have been considered for making tunable lasers, each having certain advantages and limitations with respect to tuning range, tuning speed, power output, and control complexity. Table 6.3 lists some example technologies and typical performance parameter values. Tuning is achieved by changing the temperature of the device (since the wavelength changes by approximately 0.1 nm/°C), by altering the injection current into the gain region (yielding a wavelength change of 0.01 to 0.04 nm/mA), or by using a voltage change to vary the orientation of a MEMS (microelectromechanical system) mirror to change the length of a lasing cavity.

The DFB laser has been around for a long time and is very reliable, but its tuning time is slow. The distributed Bragg reflector (DBR) laser has a simple structure, a high output power, and a modest tuning range. The GCSR (grating-assisted codirectional coupler with sampled reflector) laser and the SG+DBR (sampled grating+DBR) laser have fairly complex structures, but both have a large tuning range of 40 nm. The VMPS (VCSEL + MEMS + pump laser + solid-state optical amplifier) offers simple tuning over a large range, but the hybrid packaging structure is complex. For further details on these lasers and other structures, see the referenced article by R.-C. Yu.

6.3.4. Vertical cavity surface emitting laser

A *vertical cavity surface emitting laser* (VCSEL) consists of stacks of up to 30 thin mirroring layers placed on both sides of a semiconductor wafer to form a

TABLE 6.3. Tunable Laser Technologies and Typical Parameters (T = Temperature; C = Current; V = Voltage)

Technology	Tuning	Tuning range, nm	Tuning time, ms	Output power, mW	Advantages	Limits
DFB	T	3–9	10,000	2	Proven reliability	Slow; small range
DBR	C	8–10	10	5	Simple device; high power out	Intermediate tuning range
GCSR	C	40	10	2	Wide tuning range	Complex control
SG+DBR	C	40	10	2	Wide tuning range	Complex control
VMPS	V	32	10	10–20	Wide tuning; high output	Complex hybrid packaging

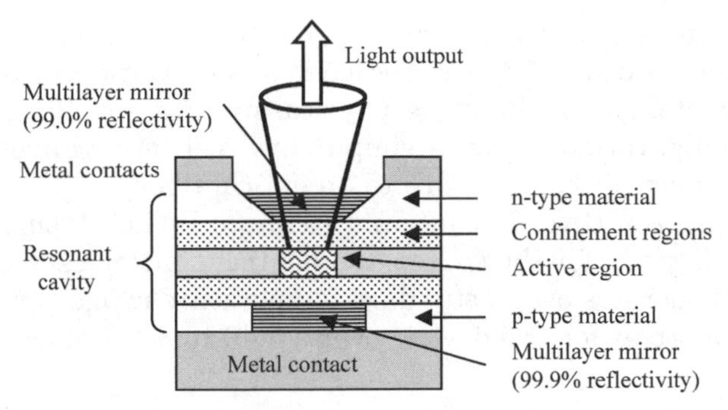

Figure 6.12. Basic architecture of a vertical cavity surface emitting laser (VCSEL).

lasing cavity. Figure 6.12 shows a general VCSEL structure. The layers of mirrors selectively reflect a narrow range of wavelengths, with the bottom stack having a 99.9 percent reflectivity. The top has a 99.0 percent reflectivity which allows the light to exit from the top. A circular metal contact on the bottom creates circular, low-divergent output light beams for easy and highly efficient coupling into optical fibers. Because of manufacturing difficulties, VCSELs were developed first at 850 nm and used for applications such as Gigabit and 10-Gigabit Ethernet in a LAN environment. However, now VCSELs are available through the L-band region.

6.3.5. Pump lasers

In addition to the light sources used for data transmission, laser diodes are needed for supplying external energy to optical amplifiers. The optical output power of these pump lasers ranges from 200 to 500 mW depending on the specific application. The lasers emit at wavelengths ranging from 1350 to 1520 nm for optical amplification in the S-, C-, and L-bands. Chapter 11 addresses optical amplifiers and their associated pump lasers in greater detail.

6.4. Optical Transmitters

As mentioned earlier, generally a light source is part of a transmitter package. This package provides the following: a mounting block for the light source, a holder for attaching a light-coupling fiber, a means for maintaining the temperature at a fixed value, and various control electronics. In some cases the transmitter package also contains an external modulator for very high-speed applications.

6.4.1. LED transmitters

The low power output (typically −16 dBm coupled into a 62.5-μm fiber) and slow response time characteristics of an LED compared to a laser diode limit its use to short-distance, low-speed (up to 200 Mbps over a few kilometers) applications using multimode fibers. An example would be OC-3/STM-1 links between nearby buildings, which operate at 155 Mbps. For local-area network use, an LED transmitter often is contained in the same physical package as the corresponding receiver. This configuration allows a simple, low-cost, and convenient way to interface a printed-circuit board (PCB) to an optical fiber.

A typical LED transmitter will contain a 1310-nm InGaAsP surface emitting LED and driver circuitry for the source. Such a transmitter uses a single 3.3-V power supply and operates over a standard temperature range of 0 to +70°C, or optionally (for a more expensive device) over a −40 to 85°C range.

6.4.2. Laser transmitters

Since the wavelength of a laser drifts by approximately 0.1 nm/°C and because the output efficiency changes with temperature, a standard method of stabilizing these parameters is to use a miniature *thermoelectric cooler*. This device uses a temperature sensor and an electronic controller to maintain the laser at a constant temperature. In addition there normally is a photodiode in the package that monitors the optical power level emitted by the laser, since this level can change with factors such as age or bias voltage. The module also may contain a second photodiode for precisely monitoring the peak output wavelength. Figure 6.13 shows a typical structure for the laser, thermoelectric cooler, and monitoring photodiode.

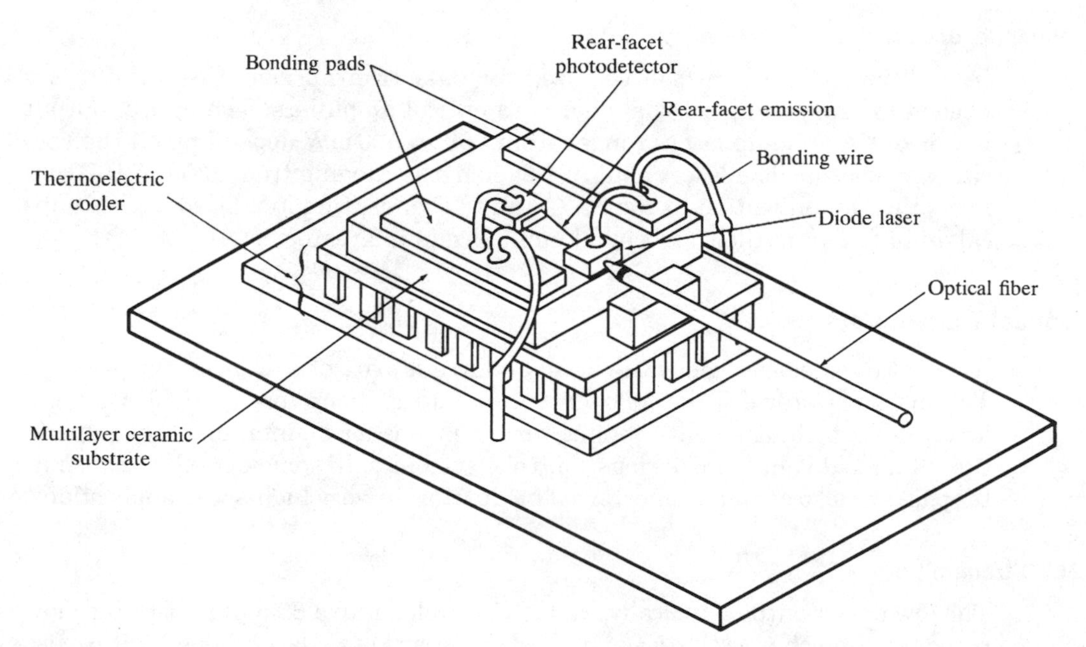

Figure 6.13. Construction of a laser transmitter that has a thermoelectric cooler and a monitoring photodiode.

6.4.3. Transmitter controllers

Figure 6.14 illustrates various functions that a commercially available optical transmitter controller might perform with respect to maintaining laser stability and operating an external modulator (See, for example, www.pinephotonics.com). The external modulator can be in a separate package, or it can be integrated within the laser package (see Sec. 6.5). The transmitter control functions include the following:

- The temperature control function uses a temperature sensor and a thermoelectric cooler to set the laser diode temperature at a particular value (say, 25°C) and to maintain it to within a 0.02°C stability.
- The wavelength controller can stabilize the laser output wavelength to a few picometers (say, 5 pm = 0.005 nm).
- The laser bias control works in conjunction with an optical power monitoring photodiode to continuously adjust the drive current for the laser.
- The transmitter output power controller works in conjunction with a variable optical attenuator (VOA) to maintain a constant optical output level from the external modulator.
- The alarm processor provides alerts for abnormal operation conditions.
- The transmission control signal processor enables both low-speed and high-speed interfaces for data exchange with a motherboard microprocessor.

A typical size for such a device is 50 × 50 × 8 mm (2.0 × 2.0 × 0.3 in), and it normally resides on a PCB along with the laser package, a modulator driver, the external modulator if it is a separate package, and some peripheral electronics.

6.4.4. Transmitter packages

Two popular transmitter package configurations that can be mounted on a PCB are the *butterfly package* with fiber pigtails and the industry-standardized *small form*

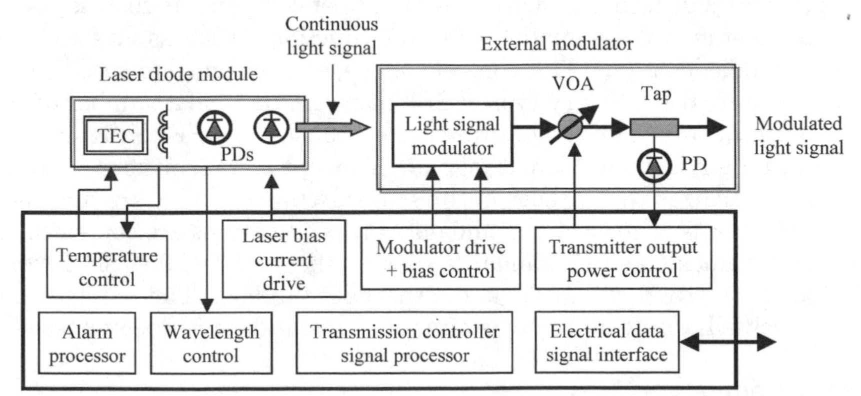

Figure 6.14. Functions that an optical transmitter controller might perform with respect to maintaining laser stability and operating an external modulator.

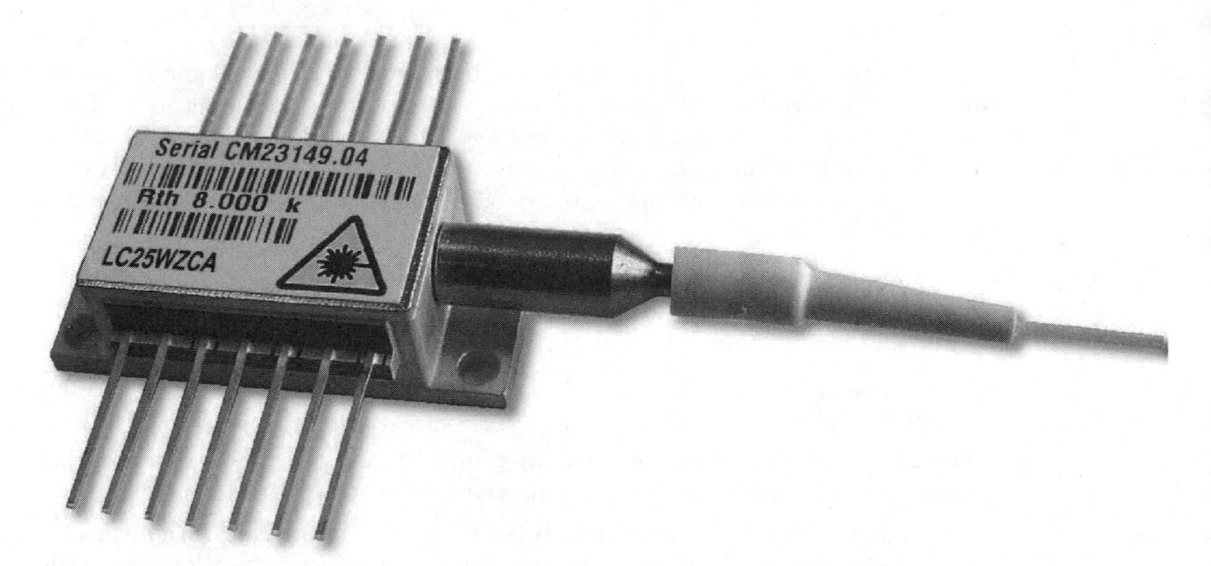

Figure 6.15. A typical industry standard 14-pin hermetically sealed butterfly package for housing a laser diode module. (*Photo courtesy of Bookham Technology; www.bookham.com.*)

factor (SFF) package that has optical fiber connectors integrated into the package. The term *pigtail* or *flylead* refers to a short length of optical fiber that is attached to a device in the factory. This allows easy attachment of a connector for coupling of the device to a transmission fiber. Chapter 8 has more details on fiber-to-fiber coupling.

Figure 6.15 shows a typical industry standard 14-pin hermetically sealed butterfly package for housing a laser diode module, such as that shown in Fig. 6.13. The package contains the laser diode, an optical isolator, a thermoelectric cooler (TEC), a power-monitor photodiode, and a wavelength-monitor photodiode. The *optical isolator* prevents unwanted optical power reflections from entering the laser diode and causing output instabilities. Although the 14-pin package has become a de facto standard for optical fiber components such as laser diodes, there is no standard designation as to the functions (such as bias voltage, monitoring photodiode output, TEC control voltage, etc.) assigned to specific pin numbers.

Figure 6.16 shows a typical SFF package. In 1998 a number of vendors signed a *multisource agreement* (MSA) to establish the SFF configuration as an internationally compatible package for fiber optic transmission modules. The MSA standardized the package outline, the circuit board layout, and the pin function definitions. The electrical and optical specifications are compatible with parameter values listed in standards such as the ITU-T G.957 SDH recommendation and the IEEE 802.3z Gigabit Ethernet standard. The optical connector is not specified, but it normally is an industry standard connector (see Chap. 8).

6.5. External Modulators

When direct modulation is used in a laser transmitter, the process of turning the drive current on and off produces a widening of the laser linewidth.

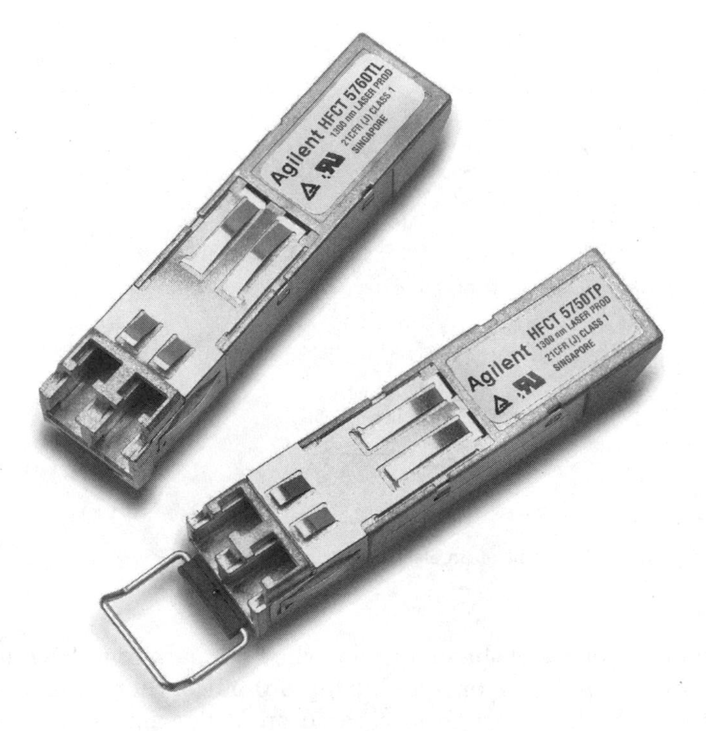

Figure 6.16. A typical industry-standardized small-form-factor package. (*Photo courtesy of Agilent Technologies; www.agilent.com.*)

This phenomenon is referred to as *chirp* and makes directly modulated lasers undesirable for operation at data rates greater than 2.5 Gbps (see Sec. 15.5). As shown in Fig. 6.17, for these applications it is preferable to use an external modulator either that is integrated physically in the same package with the light source or that can be a separate device. The two main device types are the electrooptical modulator and the electroabsorption modulator.

The *electrooptical* (EO) modulator typically is made of lithium niobate ($LbNiO_3$). In an EO modulator the light beam is split in half and then sent through two separate paths, as shown in Fig. 6.18. A high-speed electric signal then changes the phase of the light signal in one of the paths. This is done in such a manner that when the two halves of the signal meet again at the device output, they will recombine either constructively or destructively. The constructive recombination produces a bright signal and corresponds to a 1 pulse. On the other hand, destructive recombination results in the two signal halves canceling each other, so there is no light at the output. This of course corresponds to a 0 pulse. $LbNiO_3$ modulators are separately packaged devices and can be up to 12 cm (about 5 in) long.

The *electroabsorption modulator* (EAM) typically is constructed from indium phosphide (InP). It operates by having an electric signal change the

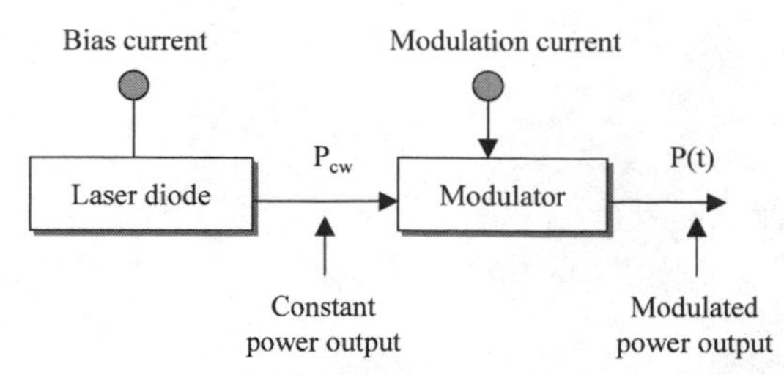

Figure 6.17. Operational concept of an external modulator.

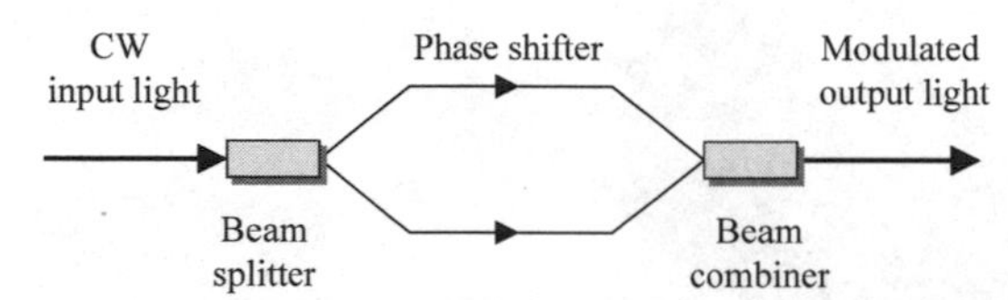

Figure 6.18. Operational concept of an electrooptical lithium niobate modulator.

transmission properties of the material in the light path to make it either transparent during a 1 pulse or opaque during a 0 pulse. Since InP is used as the material for an EAM, it can be integrated onto the same substrate as a DFB laser diode chip. The complete laser plus modulator module then can be put in a standard butterfly package, thereby reducing drive voltage, power, and space requirements compared to a separate laser and $LbNiO_3$ modulator.

6.6. Summary

Light sources and their associated transmitters for optical fiber communications come in many different sizes and power ratings, ranging from low-speed, inexpensive LEDs to various types of high-power, high-speed lasers with sophisticated control circuitry. The key characteristics of light sources include the operating wavelengths of various device materials, the source output spectral width (the range of wavelengths that are contained in an optical output pulse), and the device modulation capability (how fast a device can be turned on and off).

Compared to the other light sources used in optical communications, LEDs are much less expensive and easier to use in transmitter designs. However, because of their relatively low power output, broad emission pattern, and slow turn-on time, their use is limited to low-speed (less than 200-Mbps), short-distance (up to a few kilometers) applications using multimode fibers.

The four main laser types are the Fabry-Perot (FP) laser, the distributed-feedback (DFB) laser, tunable lasers, and the vertical cavity surface emitting laser (VCSEL). Key properties include high optical output powers (greater than 1 mW), narrow linewidths (a fraction of a nanometer, except for the FP laser), and highly directional output beams for efficient coupling of light into fiber cores. The FP laser is popular for short- and intermediate-distance (up to 15-km) applications running at data rates of up to 622 Mbps. Single-mode DFB lasers are used extensively in high-speed transmission systems. A number of different technologies are being considered for making tunable lasers, each having certain advantages and limitations with respect to tuning range, tuning speed, power output, and control complexity. VCSELs were developed first at 850 nm for applications such as Gigabit and 10-Gigabit Ethernet in a LAN environment, but now also are available at longer wavelengths.

Transmitters for optical communications come in a wide range of complexity. At the simple and inexpensive end, the small-form-factor package has been standardized for moderate-speed (up to 622-Mbps), intermediate-distance applications. Dense WDM (DWDM) applications require a highly sophisticated laser transmitter that includes a temperature controller to maintain a 0.02°C stability, a wavelength-locking controller to stabilize the wavelength to a few picometers, optical power level sensing and control functions, an alarm processor that provides alerts for abnormal operation conditions, and optionally a built-in light modulator.

Further Reading

1. G. Keiser, *Optical Fiber Communications*, 3d ed., McGraw-Hill, Burr Ridge, Ill., 2000, Chap. 4.
2. E. Garmire, "Sources, modulators, and detectors for fiber optic communication systems," Chap. 4 in *Fiber Optics Handbook*, McGraw-Hill, New York, 2002.
3. R.-C. Yu, "Tunable lasers," *Fiberoptic Product News*, vol. 16, pp. 28–33, February 2002 (www.fpnmag.com).
4. H. Shakouri, "Wavelength lockers make fixed and tunable lasers precise," *WDM Solutions*, January 2002 (www.wdm-solutions.com).
5. H. Volterra, "Indium phosphide addresses 10-Gb/s metro demand," *Laser Focus World*, vol. 38, pp. 61–66, April 2002 (www.laserfocusworld.com).
6. T. Rahban, "PWM (pulse width modulation) temp controller for thermo-electric modules," *Lightwave*, vol. 19, pp. 81–84, August 2002 (www.lightwaveonline.com).

Chapter

7

Photodiodes and Receivers

Since somebody has to listen to what is being said, at the end of an optical transmission link there must be a receiver. The function of this device is to interpret the information contained in the optical signal. An optical receiver consists of a photodetector and various amounts of electronics. Depending on how sophisticated the receiving and associated signal processing functions are in the receiver, these electronics can range from some simple amplification functions for relatively strong, clean signals to hordes of complex circuitry if the receiver needs to interpret weak, distorted signals at high data rates.

In this chapter we first examine the performance characteristics of photodetectors without going deeply into the physics of their operation. We then look at how these devices are used in receivers by considering factors such as random noises associated with the photodetection process, the concept of signal-to-noise ratio, and the probability of errors occurring in a data stream.

7.1. The *pin* Photodiode

The first element of the receiver is a photodetector. The *photodetector* senses the light signal falling on it and converts the variation of the optical power to a correspondingly varying electric current. Since the optical signal generally is weakened and distorted when it emerges from the end of the fiber, the photodetector must meet strict performance requirements. Among the most important of these are

- A high sensitivity to the emission wavelength range of the received light signal
- A minimum addition of noise to the signal
- A fast response speed to handle the desired data rate

In addition, the photodetector needs to

- Be insensitive to temperature variations
- Be compatible with the physical dimensions of the fiber
- Have a reasonable cost compared to that of other system components
- Have a long operating lifetime

Semiconductor-based photodiodes are the main devices that satisfy this set of requirements. The two types of devices used are called a *pin* photodiode and an *avalanche photodiode* (APD). This section addresses characteristics of *pin* photodiodes, and Sec. 7.2 discusses the APD.

7.1.1. Operation of a *pin* photodiode

The most common photodetector is the semiconductor *pin* photodiode, shown schematically in Fig. 7.1. The device structure consists of *p* and *n* semiconductor regions separated by a very lightly *n*-doped intrinsic (*i*) region. In normal operation a reverse-bias voltage is applied across the device so that no free electrons or holes exist in the intrinsic region.

Recall from Sec. 6.1 that electrons in semiconductor materials are allowed to reside in only two specific energy bands, as shown in Fig. 7.2. The two allowed bands are separated by a forbidden region called an *energy gap*. The energy difference between the top and bottom bands is referred to as the *bandgap energy*.

Now suppose an incident photon comes along that has an energy greater than or equal to the bandgap energy of the semiconductor material. This photon can give up its energy and excite an electron from the valence band to the conduction band. This process, which occurs in the intrinsic region, generates free (mobile) electron-hole pairs. These charge carriers are known as *photocarriers*, since they are generated by a photon. The electric field across the device causes the photocarriers to be swept out of the intrinsic region, thereby giving rise to a current flow in an external circuit. This current flow is known as the *photocurrent*.

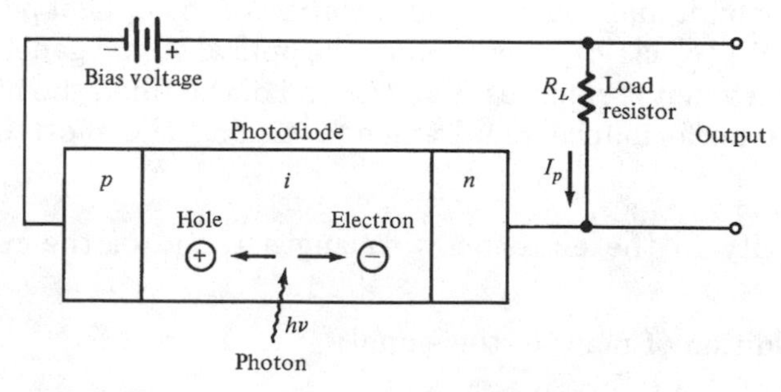

Figure 7.1. Schematic of a *pin* photodiode circuit with an applied reverse bias.

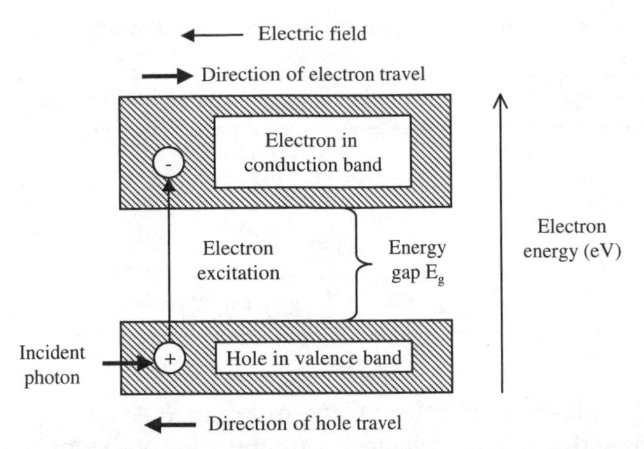

Figure 7.2. A photon gives up its energy to excite an electron from the valence band to the conduction band.

An incident photon is able to boost an electron to the conduction band only if it has an energy that is greater than or equal to the bandgap energy. This means that beyond a certain wavelength, the light will not be absorbed by the material since the wavelength of a photon is inversely proportional to its energy. The longest wavelength at which this occurs is called the *cutoff wavelength*, which is designated by λ_c. If E_g is the bandgap energy expressed in electron volts (eV), then λ_c is given in micrometers by

$$\lambda_c = \frac{hc}{E_g} = \frac{1.240}{E_g} \tag{7.1}$$

Here $h = 6.6256 \times 10^{-34}\,\text{J}\cdot\text{s}$ is Planck's constant, and c is the speed of light. There also is a lower bound on the wavelength at which the photodetection response cuts off. This is a result of high-energy photons being absorbed very close to the photodetector surface. Since the lifetimes of the electron-hole pairs that are generated close to the surface are very short, they recombine quickly before they can be collected by the photodetector circuitry. Consequently a photodetector has a certain wavelength range over which it may be used.

7.1.2. Photodetector materials

The choice of a photodetector material is important since its bandgap properties determine the wavelength range over which the device will operate. Early optical fiber systems used photodetectors made of silicon (Si), germanium (Ge), or gallium arsenide (GaAs), since these materials were available and respond well to photons in the 800- to 900-nm region. Of these, Si and GaAs are used most widely, since Ge has higher noise levels than other materials. However, Si and GaAs are not sensitive for wavelengths beyond 1100 nm where long-distance communication links operate. Therefore the ternary and

TABLE 7.1. Operating Wavelength Ranges for Several Different Photodetector Materials

Material	Energy gap, eV	λ_{cutoff}, nm	Wavelength range, nm
Silicon	1.17	1060	400–1060
Germanium	0.775	1600	600–1600
GaAs	1.424	870	650–870
InGaAs	0.73	1700	900–1700
InGaAsP	0.75–1.35	1650–920	800–1650

quaternary materials indium gallium arsenide (InGaAs) and InGaAsP were developed. Of these, InGaAs is used most commonly for both long-wavelength *pin* and avalanche photodiodes. Table 7.1 summarizes the wavelength ranges over which these materials are sensitive and the corresponding cutoff wavelength.

7.1.3. Quantum efficiency

An important characteristic of a photodetector is its *quantum efficiency*, which is designated by the symbol η (eta). The quantum efficiency is the number of electron-hole pairs that are generated per incident photon of energy $h\nu$ and is given by

$$\eta = \frac{\text{number of eletron-hole pairs generated}}{\text{number of incident photons}} = \frac{I_p/q}{P_0/h\nu} \tag{7.2}$$

Here q is the electron charge, I_p is the average photocurrent generated by a steady photon stream of average optical power P_0 incident on the photodetector, and $\nu = c/\lambda$ is the light frequency. In a practical photodiode, 100 photons will create between 30 and 95 electron-hole pairs. This gives a quantum efficiency ranging from 30 to 95 percent.

7.1.4. Responsivity

The performance of a photodiode may be characterized by its *responsivity* $\mathcal{R}$. This is related to the quantum efficiency by

$$\mathcal{R} = \frac{I_p}{P_0} = \frac{\eta q}{h\nu} \tag{7.3}$$

This parameter is quite useful since it specifies the photocurrent generated per unit optical power. Figure 7.3 shows typical *pin* photodiode responsivities as a function of wavelength. Representative values are 0.65 A/W for silicon at 900 nm and 0.45 A/W for germanium at 1300 nm. For InGaAs, typical values are 0.9 A/W at 1300 nm and 1.0 A/W at 1550 nm.

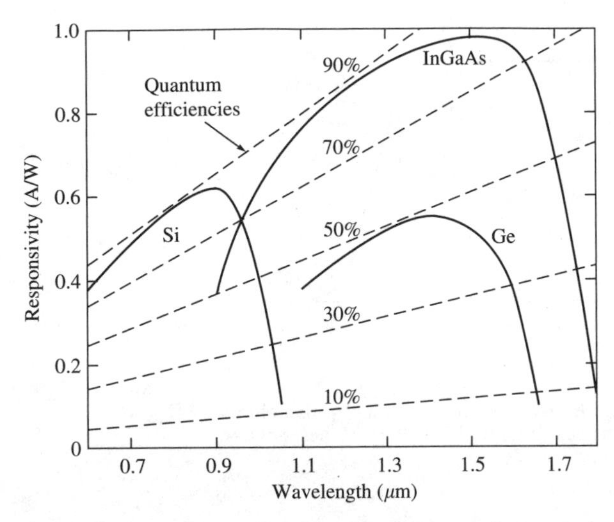

Figure 7.3. Comparison of the responsivity and quantum efficiency as a function of wavelength for *pin* photodiodes constructed of different materials.

Example Photons of energy $h\nu = 1.53 \times 10^{-19}$ J are incident on a photodiode which has a responsivity of 0.65 A/W. If the optical power level is 10 μW, then from Eq. (7.3) the photocurrent generated is

$$I_p = \mathcal{R}P_0 = (0.65\ \text{A/W})(10\ \mu\text{W}) = 6.5\ \mu\text{A}$$

In most photodiodes the quantum efficiency is independent of the power level falling on the detector at a given photon energy. Thus the responsivity is a linear function of the optical power. That is, the photocurrent I_p is directly proportional to the optical power P_0 incident on the photodetector. This means that the responsivity $\mathcal{R}$ is constant at a given wavelength. Note, however, that the quantum efficiency is not a constant at all wavelengths, since it varies according to the photon energy. Consequently, the responsivity is a function of the wavelength and of the photodiode material. For a given material, as the wavelength of the incident photon becomes longer, the photon energy becomes less than that required to excite an electron from the valence band to the conduction band. The responsivity thus falls off rapidly beyond the cutoff wavelength, as can be seen in Fig. 7.3.

7.1.5. Speed of response

Photodiodes need to have a fast response speed in order to properly interpret high data rate signals. If the detector output does not track closely the variations of the incoming optical pulse shape, then the shape of the output pulse will be distorted. This will affect the link performance since it may introduce errors in interpreting the optical signal. The detector response speed is measured in

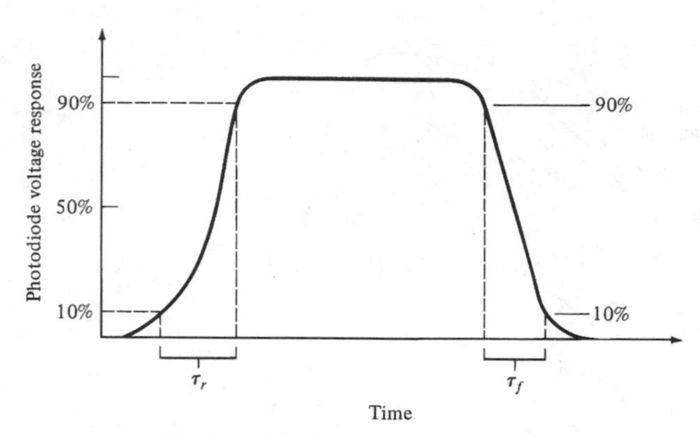

Figure 7.4. Photodetector response to an optical pulse showing the 10 to 90 percent rise time and the 90 to 10 percent fall time.

terms of the time it takes the output signal to rise from 10 percent to 90 percent of its peak value when the input to the photodiode is turned on instantaneously. This is shown in Fig. 7.4 and is known as the *10- to 90-percent rise time.* Similarly, the time it takes the output to drop from its 90 percent to its 10 percent value is known as the *fall time.*

The rise and fall times depend on factors such as how much of the light is absorbed at a specific wavelength, the width of the intrinsic region, various photodiode and electronics capacitance values, and various detector and electronic resistances. As a result, the rise and fall times are not necessarily equal in a receiver. For example, large capacitance values can cause a long decay tail to appear in the falling edge of the output pulse, thereby creating long fall times.

7.1.6. Bandwidth

The response speeds of the photodiode and the electronic components result in a gradual drop in the output level beyond a certain frequency. The point at which the output has dropped to 50 percent of its low-frequency value is called the *3-dB point.* At this point only one-half as much signal power is getting through the detector compared to lower frequencies. The 3-dB point defines the receiver *bandwidth* (sometimes referred to as the *3-dB bandwidth*), which is the range of frequencies that a receiver can reproduce in a signal. If the rise and fall times are equal, the 3-dB bandwidth (in megahertz) can be estimated from the rise time by the relationship

$$\text{Bandwidth, MHz} = \frac{350}{\text{rise time, ns}} \tag{7.4}$$

where the rise time is expressed in units of nanoseconds.

7.2. Avalanche Photodiodes

An avalanche photodiode (APD) internally multiplies the primary signal photocurrent before it enters the input circuitry of the following amplifier. The multiplication effect is achieved by applying a very high electric field across the photodiode. When a photon-generated electron encounters this high electric field, it can acquire sufficient energy to kick more electrons from the valence to the conduction band, thereby creating secondary electron-hole pairs. These secondary pairs also get accelerated to higher energies and therefore can generate even more electron-hole pairs. This increases receiver sensitivity since the photocurrent is multiplied prior to encountering the electrical noise associated with the receiver circuitry. The process is called *avalanche multiplication*, and hence the device is called an avalanche photodiode.

Since the avalanche process is random, the mean number of electron-hole pairs created is a measure of the carrier multiplication. This is called the *gain* and is designated by M (or sometimes by G or G_m). The value of M can be made quite large, but larger gains increase the noise currents of the device because of larger variations in the photocurrent. Thus an APD has a *noise figure* $F(M)$ that is associated with the random nature of the avalanche process (see Sec. 7.4).

Analogous to the *pin* photodiode, the performance of an APD is characterized by its responsivity $\mathcal{R}_{\text{ADP}}$. Thus in an APD the multiplied photocurrent I_M is given by

$$I_M = \mathcal{R}_{\text{ADP}} P_0 = M \mathcal{R} P_0 \tag{7.5}$$

where $\mathcal{R}$ is the unity-gain responsivity.

7.3. Comparisons of Photodetectors

To see the differences between various types of photodetectors, let us look at some generic operating characteristics of Si, Ge, and GaAs photodiodes. Tables 7.2 and 7.3 list the performance values for *pin* and avalanche photodiodes, respectively. The values were derived from various vendor data sheets and from performance numbers reported in the literature. They are given as guidelines for comparison purposes. Detailed values on specific devices can be obtained from photodetector and receiver module suppliers.

TABLE 7.2. Generic Operating Parameters of Si, Ge, and GaAs *pin* Photodiodes

Parameter	Symbol	Unit	Si	Ge	InGaAs
Wavelength range	λ	nm	400–1100	800–1650	1100–1700
Responsivity	$\mathcal{R}$	A/W	0.4–0.6	0.4–0.5	0.75–0.95
Dark current	I_D	nA	1–10	50–500	0.5–2.0
Rise time	τ_r	ns	0.5–1	0.1–0.5	0.05–0.5
Bandwidth	B	GHz	0.3–0.7	0.5–3	1–2
Bias voltage	V_B	V	5	5–10	5

TABLE 7.3. Generic Operating Parameters of Si, Ge, and GaAs Avalanche Photodiodes

Parameter	Symbol	Unit	Si	Ge	InGaAs
Wavelength range	λ	nm	400–1100	800–1650	1100–1700
Avalanche gain	M	—	20–400	10–200	10–40
Dark current	I_D	nA	0.1–1	50–500	10–50 @ M = 10
Rise time	τ_r	ns	0.1–2	0.5–0.8	0.1–0.5
Gain × bandwidth	$M \cdot B$	GHz	100–400	2–10	20–250
Bias voltage	V_B	V	150–400	20–40	20–30

For short-distance applications, Si devices operating around 850 nm provide relatively inexpensive solutions for most links. Longer links usually require operation in the higher spectral bands (O-band through L-band). In these bands one normally uses InGaAs-based photodiodes.

7.4. Optical Receiver

An *optical receiver* consists of a photodetector and electronics for amplifying and processing the signal. In the process of converting the optical signal power emerging from the end of an optical fiber to an electric signal, various noises and distortions will unavoidably be introduced due to imperfect component responses. This can lead to errors in the interpretation of the received signal.

The most meaningful criterion for measuring the performance of a digital communication system is the *average error probability*. In an analog system the fidelity criterion usually is specified in terms of a peak *signal-to-noise ratio*. The calculation of the error probability for a digital optical communication receiver differs from that of its electronic counterpart. This is a result of the discrete quantum nature of the optical signal and also because of the probabilistic character of the gain process when an avalanche photodiode is used.

This section first gives a definition of the signal-to-noise ratio and then looks at the origins of noises. Given this knowledge, one then can characterize the performance of an optical fiber communication system. Chapter 14 looks at this topic in greater detail.

7.4.1. Photodetector noise

In fiber optic communication systems, the photodiode must detect very weak optical signals. Detection of the weakest possible optical signals requires that the photodetector and its associated electronic amplification circuitry be optimized so that a specific signal-to-noise ratio is maintained. The term *noise* describes unwanted components of a signal that tend to disturb the transmission and processing of the signal in a physical system. Noise is present in every communication system and represents a basic limitation on the transmission and detection of signals. In electric circuits noise is caused by the spontaneous fluctuations of current or voltage. Intuitively one can see that if the received signal level is close to the noise level, it is difficult to get a good reproduction of the original signal that was sent. As an analogy, if people are talking to one another in a noisy room, it is hard to understand what others are saying if their voice levels are only a little bit higher than the background noise level.

The *signal-to-noise ratio* SNR (also designated by *S/N*) at the output of an optical receiver is defined by

$$\text{SNR} = \frac{\text{signal power from photocurrent}}{\text{photodetector noise power} + \text{amplifier noise power}} \tag{7.5}$$

The noises in the receiver arise from the statistical nature of the randomness of photon-to-electron conversion process and the electronic noise in the receiver amplification circuitry. Section 7.4.2 describes what these are.

To achieve a high SNR, the numerator in Eq. (7.5) should be maximized and the denominator should be minimized. Thus, the following conditions should be met:

1. The photodetector must have a high quantum efficiency to generate a large signal power.
2. The photodetector and amplifier noises should be kept as low as possible.

For most applications it is the noise currents in the receiver electronics that determine the minimum optical power that can be detected, since the photodiode quantum efficiency normally is close to its maximum possible value.

The sensitivity of a photodetector in an optical fiber communication system is describable in terms of the *minimum detectable optical power*. This is the optical power necessary to produce a photocurrent equal to the total noise current or, equivalently, to yield a SNR of 1.

7.4.2. Noise sources

The main noises associated with *pin* photodiode receivers are quantum or shot noise and dark current associated with photodetection, and thermal noise occurring in the electronics.

Shot noise arises from the statistical nature of the production and collection of photoelectrons. It has been found that these statistics follow a Poisson process. Since the fluctuation in the number of photocarriers created is a fundamental property of the photodetection process, it sets the lower limit on the receiver sensitivity when all other conditions are optimized.

The photodiode *dark current* arises from electrons and holes that are thermally generated at the *pn* junction of the photodiode. This current continues to flow through the bias circuit of the device when no light is incident on the photodetector. In an APD these liberated carriers also get accelerated by the electric field across the device and therefore are multiplied by the avalanche mechanism.

Thermal noise arises from the random motion of electrons that is always present at any finite temperature.

Noise Calculations

Thermal Noise Consider a resistor that has a value R at a temperature T. If I_{thermal} is the thermal noise current associated with the resistor, then in a bandwidth or frequency range B_e its variance $\sigma^2_{\text{thermal}}$ is

$$\langle I^2_{\text{thermal}} \rangle = \sigma^2_{\text{thermal}} = \frac{4k_B T}{R} B_e \tag{7.6}$$

Here $k_B = 1.38054 \times 10^{-23}$ J/K is Boltzmann's constant, and the symbol $\langle x \rangle$ designates the mean value of the quantity x. Typical values of the quantity $(\langle I^2_{\text{thermal}} \rangle / B_e)^{1/2}$ are 1 pA/$\sqrt{\text{Hz}}$. Note that in practice the receiver bandwidth B_e varies from $1/(2T_{\text{bit}})$ to $1/T_{\text{bit}}$, where T_{bit} is the bit period.

Shot Noise For a photodiode the variance of the shot noise current I_{shot} in a bandwidth B_e is

$$\langle I^2_{\text{shot}} \rangle = \sigma^2_{\text{shot}} = 2qI_p B_e M^2 F(M) \tag{7.7}$$

where I_p is the average photocurrent as defined in Eq. (7.3), M is the gain of an APD, and $F(M)$ is the APD noise figure. For a *pin* photodiode, M and $F(M)$ are equal to 1.

Dark Current If I_D is the primary (not multiplied) dark current, then its variance is given by

$$\langle I^2_D \rangle = \sigma^2_D = 2qI_D M^2 F(M) B_e \tag{7.8}$$

Again, for a *pin* photodiode, M and $F(M)$ are equal to 1.

Signal-to-Noise Ratio Since the various noise currents are not correlated, the total mean-square photodetector noise current $\langle I^2_N \rangle$ can be written as

$$\langle I^2_N \rangle = \langle I^2_{\text{shot}} \rangle + \langle I^2_D \rangle + \langle I^2_{\text{thermal}} \rangle \tag{7.9}$$

From Eq. (7.3) we know $(I_P)^2 = (\mathcal{R}P_0)^2 M^2$ is the mean-square signal current. Then by using Eq. (7.5) the SNR is equal to

$$\text{SNR} = \frac{I^2_p M^2}{\langle I^2_N \rangle} = \frac{I^2_p M^2}{2q(I_p + I_D) M^2 F(M) B_e + 4k_B T B_e / R} \tag{7.10}$$

7.5. Summary

The function of an optical receiver is to interpret the information contained in the optical signal. A receiver consists of a photodetector and various amounts of electronics. The electronics can range from some simple amplification functions for relatively strong, clean signals to complex circuitry if the receiver needs to interpret weak, distorted signals at high data rates.

Semiconductor-based photodiodes are the main devices that satisfy optical receiver requirements. The two basic types of devices used are called a *pin* photodiode and an avalanche photodiode (APD).

The choice of a photodetector material determines the wavelength range over which the device will operate. Material choices include Si, Ge, or GaAs, InGaAs, and InGaAsP. Of these the most widely used are Si for operation in the 850-nm region and InGaAs for both long-wavelength *pin* and avalanche photodiodes. Table 7.1 summarizes the wavelength ranges over which various materials are sensitive.

Other important characteristics of a photodetector include

- Its quantum efficiency, which is the number of electron-hole pairs that are generated per incident photon of energy $h\nu$
- Its responsivity, which specifies the photocurrent generated per unit optical power
- Its response speed, which is particularly important to properly interpret high data rate signals
- The gain and its associated noise figure for avalanche photodiodes

Tables 7.2 and 7.3 list representative performance values for *pin* and avalanche photodiodes, respectively. The values were derived from various vendor data sheets and from performance numbers reported in the literature. They are given as guidelines for comparison purposes.

Further Reading

1. D. Neamen, *Semiconductor Physics and Devices*, 3d ed., McGraw-Hill, New York, 2003.
2. S. O. Kasap, *Optoelectronics and Photonics*, Prentice Hall, Upper Saddle River, N.J., 2001.
3. G. Keiser, *Optical Fiber Communications*, 3d ed., McGraw-Hill, Burr Ridge, Ill., 2000, Chap. 6.
4. S. R. Forrest, "Optical devices for lightwave communications," in S. E. Miller and I. P. Kaminow, eds., *Optical Fiber Telecommunications—II*, Academic, New York, 1988.
5. M. C. Brain and T. P. Lee, "Optical receivers for lightwave communication systems," *J. Lightwave Technology*, vol. 3, pp. 1281–1300, December 1985.

Chapter

8

Connectors and Splices

Now that we have looked at the major pieces of an optic fiber link, the next concern is how to hook them together mechanically and optically. From an optical point of view the challenges are how to launch optical power into a particular type of fiber and how to couple optical power from one fiber into another. Launching light from a source into a fiber entails considerations such as the numerical aperture, core size, and core-cladding refractive index differences of the fiber, plus the size, radiance, and angular power distribution of the optical source. Mechanical factors include highly precise alignment of fibers, low loss and repeatability of connections, and ruggedness of fiber-to-fiber joints.

This chapter first considers the issues involved in coupling light from a source into an optical fiber. The next topic covers the conditions that need to be taken into account in making a fiber-to-fiber joint. Here the difficulty lies in how to precisely align fibers that are roughly the diameter of a human hair (which is 50 to 100 μm in diameter) so that only a minute fraction of light is lost across a fiber joint. A further topic notes that no matter what alignment and coupling scheme is used, the end faces of the fiber must be prepared properly. The final two sections deal with connectors and splices. As will be seen in that discussion, a splice is a permanent joint between two fibers, whereas connectors are mounted on the ends of fiber cables so they can be plugged and unplugged easily and often.

8.1. Source-to-Fiber Power Coupling

In practice, many source suppliers offer devices with a short length of optical fiber (1 m or less) already attached to the source in an optimum power coupling configuration. This section of fiber commonly is referred to as a *flylead* or a *pigtail*. The power-launching problem from these pigtailed sources thus reduces to a simpler one of coupling optical power from one fiber into another. So let us look at how to achieve optimum light coupling from a source into a fiber flylead.

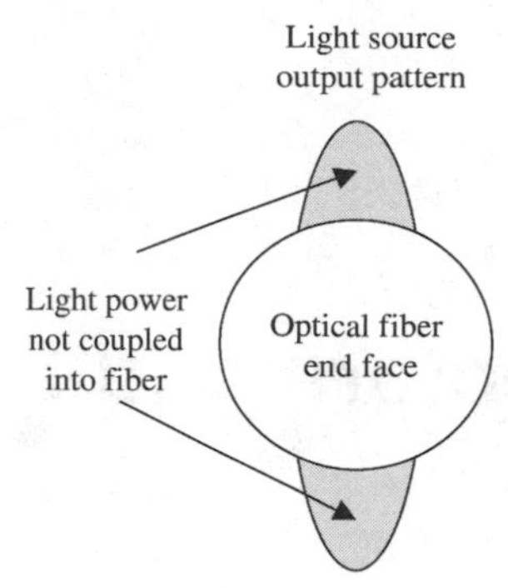

Figure 8.1. Overlap areas of a light source pattern and a fiber end face. Light in the visible shaded area is not coupled into the fiber.

8.1.1. Source output pattern

To determine the light-accepting capability of a fiber, the spatial radiation pattern of the source must be known first. For surface emitting sources the output is a symmetric cone, but the angle of the cone can vary from one device type to another. For edge emitting sources the output tends to be an asymmetric cone; that is, one will see an elliptical pattern when looking down into the cone. As shown in Fig. 8.1, since the core is circular, this will result in a certain amount of power being lost in trying to launch the light into a fiber.

Radiance A convenient and useful measure of the optical output from a light source is its radiance (or *brightness*) B at a given diode drive current. *Radiance* is the optical power radiated into a unit solid angle per unit of emitting surface area. It is specified in terms of watts per square centimeter per steradian [$\mathrm{W/(cm^2 \cdot sr)}$]. Since the optical power that can be coupled into a fiber depends on the radiance (i.e., on the spatial distribution of the optical power), the radiance of a source rather than the total output power is the important parameter when one is considering source-to-fiber coupling efficiencies.

For surface emitting sources the output power is emitted into a circularly symmetric cone. In this case the power delivered at an angle θ, measured relative to a line perpendicular to the emitting surface, varies as $\cos^n \theta$. The brightness of the emission pattern for a circular source thus follows the relationship $B = B_0 \cos^n \theta$. Here B_0 is the radiance along the line perpendicular to the emitting surface. When $n = 1$, the source emits in a *lambertian* pattern, which means it is equally bright when viewed from any direction.

Figure 8.2 compares a lambertian pattern from an LED with that emitted by a laser diode that has $n = 180$. The much narrower output beam from the laser allows significantly more light to be coupled into an optical fiber.

Edge emitting sources have a more complex light emission pattern. The emission cone is asymmetric so that in radial coordinates the radiance pattern may be approximated by

$$\frac{1}{B(\theta,\phi)} = \frac{\sin^2\phi}{B_0 \cos^T \theta} + \frac{\cos^2\phi}{B_0 \cos^L \theta} \tag{8.1}$$

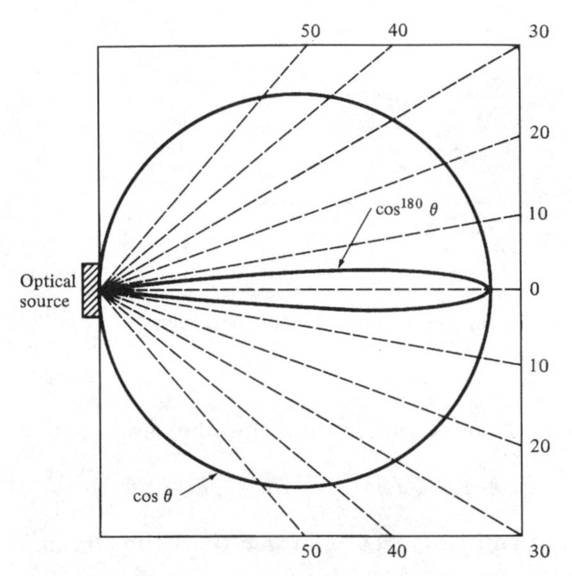

Figure 8.2. Radiance patterns for a lambertian source and a highly directional laser diode. Both sources have their peak output normalized to unity.

Relative to the *pn* junction of the source, the integers T and L are the transverse and lateral power distribution coefficients, respectively. For edge emitting lasers, both T and L can take on values over 100, which describes a very narrow output beam.

8.1.2. Power coupling calculation

The size of the source relative to the core diameter and the numerical aperture (NA) of the fiber are the two main factors to consider when one is calculating the amount of power that a specific source can couple into a fiber flylead. Clearly, if the source emitting area is larger than the end-face area of the fiber, only that fraction of power which falls directly on the fiber end face has a chance of being coupled into the fiber.

We say "has a chance" since, in order to be coupled, the emitted light has to fall within the acceptance cone of the fiber, as Fig. 8.3 illustrates. This shows a light source that is smaller than the fiber end-face area, but which emits into an angular region that goes beyond the fiber acceptance angle. Thus light outside the acceptance angle is lost. As described in Chap. 4, the numerical aperture is a standard way of representing the light acceptance angle of a multimode step-index fiber.

As a simple example, let us consider light from a surface emitting LED that has a lambertian output being coupled into a step-index fiber. Assume the source is circular with a radius r_{LED} and that the fiber has a radius a. Let P_S be the power emitted by the source into a hemisphere, and let P_F be the light coupled into the

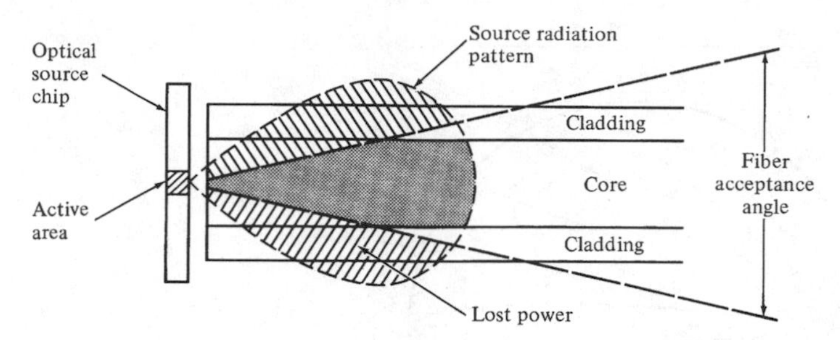

Figure 8.3. Schematic of an optical source coupled to a fiber. Light outside the acceptance angle is lost.

fiber. Then if the source area is smaller than the fiber end face, we have

$$P_F = P_S\,(\mathrm{NA})^2 \qquad \text{for } r_{\mathrm{LED}} \leq a \tag{8.2}$$

When the radius of the emitting area is larger than the radius a of the fiber core, we have

$$P_F = (a/r_{\mathrm{LED}})^2 P_S\,(\mathrm{NA})^2 \qquad \text{for } r_{\mathrm{LED}} > a \tag{8.3}$$

Thus if the source is larger than the fiber end face, the power coupled decreases by the ratio of the fiber-to-source areas. The ratio P_F/P_S is defined as the *coupling efficiency*.

For detailed power coupling calculations, the reader is referred to the book *Optical Fiber Communications*, by Keiser.

8.1.3. Lensed fibers

If a source emitting area is larger than the fiber core area, then the resulting optical power coupled into the fiber is the maximum that can be achieved. This is a result of fundamental energy and radiance conversion principles (also known as the *law of brightness*). However, if the source emitting area is smaller than the fiber core area, a miniature lens may be placed between the source and the fiber to improve the power coupling efficiency. The function of this lens is to magnify the source emitting area to match exactly the core area of the fiber end face. If the emitting area is increased by a magnification factor M, the solid angle within which optical power is coupled to the fiber is increased by the same factor.

Figure 8.4 shows several possible lensing schemes. These include a rounded fiber end, a small glass sphere that is in contact with both the fiber and the source, a larger spherical lens used to image the source on the core area of the fiber end face, a cylindrical lens that might be formed from a short section of fiber, a combination of a spherical-surfaced source and a spherical-ended fiber, and a taper-ended fiber. A popular method is to fabricate a miniature lens on the end of a fiber to improve the light coupling efficiency.

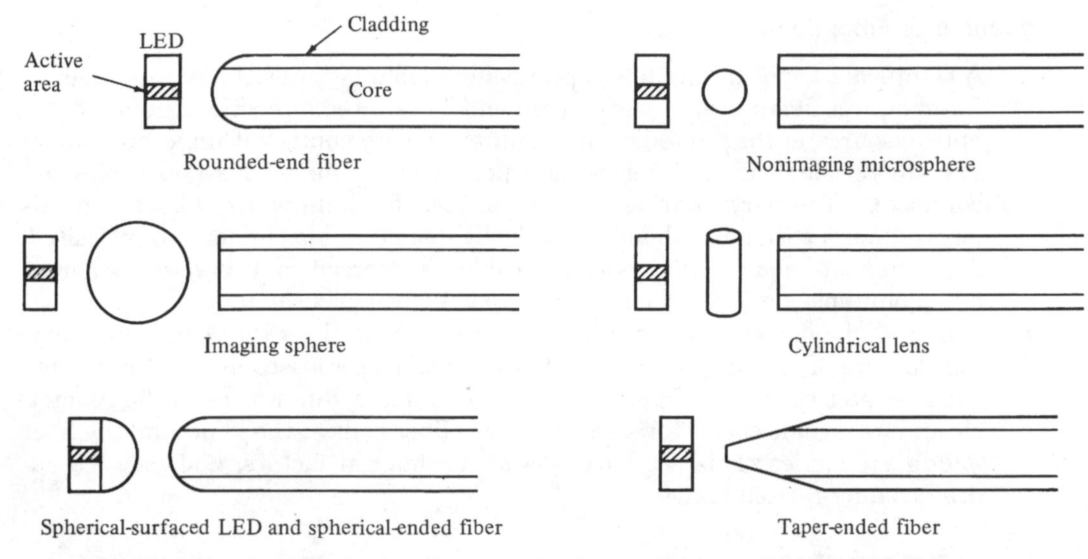

Figure 8.4. Examples of possible lensing schemes used to improve source-to-fiber coupling efficiency.

8.1.4. End-face reflections

When a flat fiber end is brought close to a light source, a fraction of the light is reflected off the fiber surface and goes back into the source. If the medium between the source and the fiber has a refractive index n and the fiber core has an index n_1, then the amount of power reflected off the fiber end face is given by

$$R = \left(\frac{n_1 - n}{n_1 + n} \right)^2 \tag{8.4}$$

where R is the *Fresnel reflection* or the *reflectivity* of the fiber core end face. The *reflection coefficient*, which is given by the ratio $r = (n_1 - n)/(n_1 + n)$, relates the amplitude of the reflected optical wave to the amplitude of the incident optical wave.

As an example, if the outside medium is air with $n = 1.00$ and $n_1 = 1.45$ for glass, then 3.4 percent of the light is reflected in the reverse direction. Since this can create instabilities in laser diodes, fibers generally have their end faces coated with an *antireflection material* to prevent possibly disruptive optical power reflection in the reverse direction.

An alternative to eliminating light reflections back into the source is to polish the end of the fiber flylead at an small angle. This angle is typically 8°. With these angle-polished fiber ends, instead of being reflected straight back into the core, light that bounces off the end faces leaves at an angle and is not directed back to where it came from.

8.2. Mechanics of Fiber Joints

A significant factor in any fiber optic system installation is the requirement to interconnect fibers in a low-loss manner. These interconnections occur at the optical source, at the photodetector, at intermediate points within a cable where two fibers join, and at intermediate points in a link where two cables are connected. The particular technique selected for joining the fibers depends on whether a permanent bond or an easily demountable connection is desired. A permanent bond (usually within a cable) is referred to as a *splice*, whereas a demountable joint at the end of a cable is known as a *connector*.

Every joining technique is subject to certain conditions that can cause varying degrees of optical power loss at the joint. These losses depend on factors such as the mechanical alignments of the two fibers, differences in the geometric and waveguide characteristics of the two fiber ends at the joint, and the fiber end-face qualities. This section looks at mechanical factors, and Sec. 8.3 addresses fiber-related losses.

8.2.1. Mechanical misalignments

The core of a standard multimode fiber nominally is 50 to 100 μm in diameter, which is equivalent to the thickness of a human hair (without body-enhancing gel). Single-mode fibers have core diameters on the order of 9 μm. This is about the size of the soft underbelly down hair of Himalayan mountain goats, which is used to make fashionable pashmina fabrics. Owing to this microscopic size, *mechanical misalignment* is a major challenge in joining two fibers. Power losses result from misalignments because the radiation cone of the emitting fiber does not match the acceptance cone of the receiving fiber. The magnitude of the power loss depends on the degree of misalignment.

Figure 8.5 illustrates the three fundamental types of misalignment between two fibers. *Axial displacement* (also called *lateral displacement*) results when the axes of the two fibers are offset by a distance d. *Longitudinal separation* occurs when the fibers have the same axis but have a gap s between their end faces. *Angular misalignment* results when the two axes form an angle so that the fiber end faces are no longer parallel.

8.2.2. Misalignment effects

The most common misalignment occurring in practice, which also causes the greatest power loss, is *axial displacement*. This axial offset reduces the overlap

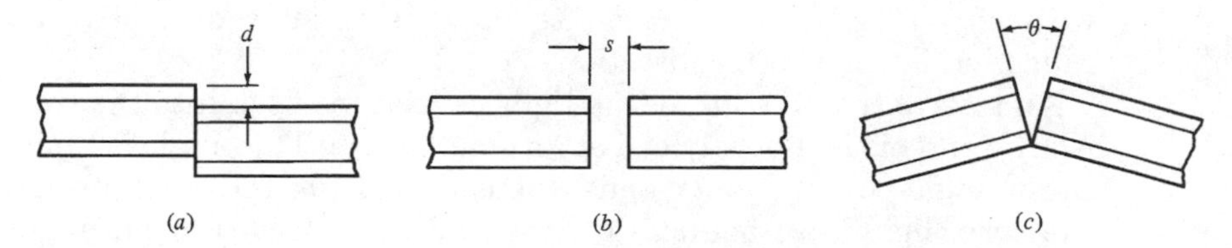

Figure 8.5. Axial, longitudinal, and angular misalignments between two fibers.

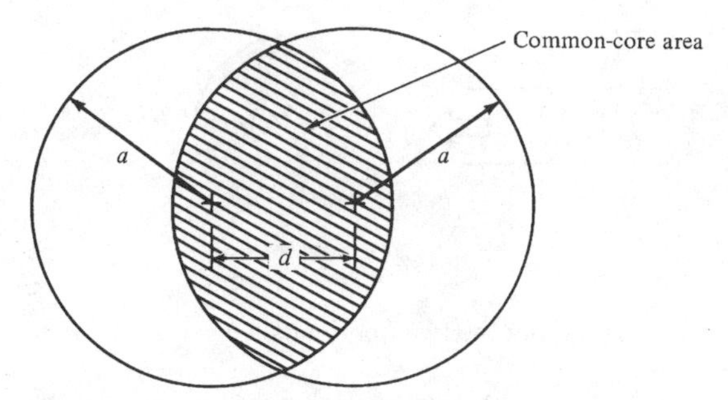

Figure 8.6. Axial offset reduces the common-core area of the two fiber end faces.

area of the two fiber-core end faces, as illustrated in Fig. 8.6, and consequently reduces the amount of optical power that can be coupled from one fiber into another. In practice, axial offsets of less than 1 µm are achievable, which result in losses of less than 0.1 dB for multimode fibers and 0.3 dB for single-mode fibers.

Axial Offset Loss The calculation of an expression for the axial offset loss for both step-index and graded-index fibers is straightforward but somewhat mathematically involved. When the axial misalignment d is small compared to the core radius a, an approximate expression for the total power P_T accepted by the receiving fiber that is accurate to within 1 percent for $d/a < 0.4$ is

$$P_T \approx P\left(1-\frac{8d}{3\pi a}\right) \tag{8.5}$$

where P is the power emerging from the emitting fiber. The coupling loss for fiber offsets then is given by

$$L_{\text{offset}} = -10 \log \frac{P_T}{P} \tag{8.6}$$

Figure 8.7 shows the effect of separating the two fiber ends longitudinally by a gap s. Not all the optical power emitted in the ring of width x will be intercepted by the receiving fiber. In most connectors the fiber ends are separated intentionally by a small gap. This prevents the ends from rubbing against each other and becoming damaged during connector engagement. Typical gaps range from 0.025 to 0.10 mm, which results in losses of less than 0.1 dB for a 50-µm fiber.

Separation Loss If the emitting and receiving fibers are identical step-index fibers, the loss resulting from a gap s between them is

$$L_{\text{gap}} = -10 \log\left(\frac{a}{a+s\tan\theta_c}\right)^2 = -10 \log\left[\frac{a}{a+s\arcsin(\text{NA}/n)}\right]^2 \tag{8.7}$$

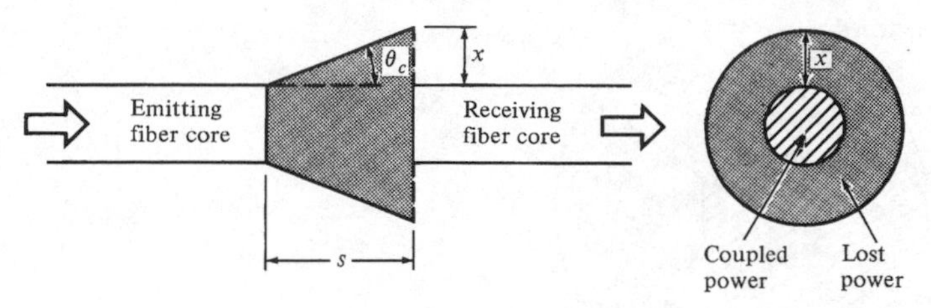

Figure 8.7. The loss effect when separating the two fiber ends longitudinally by a gap s.

where a is the fiber radius, θ_c is the critical angle, NA is the numerical aperture of the fiber, and n is the refractive index of the material between the fiber ends (usually either air or index-matching gel).

When the axes of two joined fibers have an angular misalignment at the joint, the optical power that leaves the emitting fiber outside the solid acceptance angle of the receiving fiber will be lost. Obviously, the larger the angle, the greater the loss will be. Typical angular misalignments in a standard mated connector are less than 1°, for which the associated loss is less than 0.5 dB.

8.3. Fiber-Related Losses

In addition to mechanical misalignments, differences in the geometric and waveguide characteristics of any two mated fibers can have a profound effect on the joint loss. The differences include variations in core diameter, core-area ellipticity, numerical aperture, and core-cladding concentricity of each fiber. Since these are manufacturer-related variations, the user has little control over them, except to specify certain tolerances in these parameters when purchasing the fiber. For a given percentage mismatch between fiber parameters, differences in core sizes and numerical apertures have a significantly larger effect on joint losses than mismatches in the refractive-index profile or core ellipticity.

Core area mismatches. For simplicity let the subscripts E and R refer to the emitting and receiving fibers, respectively. If the axial numerical apertures and the core index profiles are equal [that is, $\mathrm{NA}_E(0) = \mathrm{NA}_R(0)$ and $\alpha_E = \alpha_R$], but the fiber diameters d_E and d_R are not equal, then the coupling loss is

$$L_F(d) = \begin{cases} -10 \log \left(\dfrac{d_R}{d_E} \right)^2 & \text{for } d_R < d_E \\ 0 & \text{for } d_R \geq d_E \end{cases} \tag{8.8}$$

Core area mismatches can occur when one is trying to connect a 62.5-μm fiber to one with a 50-μm core, for example. In this case, going from the larger to the smaller fiber results in a 1.9-dB loss, or 36 percent of the power. A much more serious loss occurs when one inadvertently tries to couple light from a multimode to a single-mode fiber. For example, if one connects a 62.5-μm multimode fiber to a 9-μm single-mode fiber, the loss from the area mismatch will be 17 dB, or almost 98 percent of the light.

Numerical aperture mismatches. In the case of multimode fibers, if the diameters and the index profiles of two coupled fibers are identical but their axial numerical apertures are different, then the joint loss from this effect is

$$L_F(\mathrm{NA}) = \begin{cases} -10 \log \left(\dfrac{\mathrm{NA}_R}{\mathrm{NA}_E} \right)^2 & \text{for } \mathrm{NA}_R < \mathrm{NA}_E \\ 0 & \text{for } \mathrm{NA}_R \geq \mathrm{NA}_E \end{cases} \tag{8.9}$$

Core index mismatches. If the fiber diameters and the axial numerical apertures are the same but the core refractive-index profiles differ in the joined fibers, then the joint loss is

$$L_F(\alpha) = \begin{cases} -10 \log \dfrac{\alpha_R(\alpha_E + 2)}{\alpha_E(\alpha_R + 2)} & \text{for } \alpha_R < \alpha_E \\ 0 & \text{for } \alpha_R \geq \alpha_E \end{cases} \tag{8.10}$$

This relationship comes about because for $\alpha_R < \alpha_E$ the number of modes that the receiving fiber can support is less than the number of modes in the emitting fiber. If $\alpha_E < \alpha_R$, then the receiving fiber captures all modes from the emitting fiber.

8.4. End-Face Preparation

One of the first steps that must be followed before fibers are connected or spliced is to prepare the fiber end faces properly. In order not to have light deflected or scattered at the joint, the fiber ends must be flat and smooth and must have the proper angle relative to the axis (either perpendicular or at an 8° angle). Common end preparation techniques include a grinding and polishing method and a controlled-fracture procedure.

8.4.1. Grinding and polishing

Conventional *grinding and polishing* techniques can produce a very smooth surface. Normally this is done in a controlled environment such as a laboratory or a factory, but polishing machines also are available for use in the field to attach connectors and to perform emergency repairs. The procedure employed is to use successively finer abrasives to polish the fiber end face. The end face is polished with each successive abrasive until the scratches created by the previous abrasive are replaced by the finer scratches of the present abrasive.

The number of abrasives used in this step-down approach depends on the degree of smoothness that is desired. Fiber inspection and cleanliness are important during each step of fiber polishing. This inspection is done visually by the use of a standard microscope at 200 to 400 times magnification. Figure 8.8 gives an example of a polishing machine that can prepare several fibers simultaneously.

8.4.2. Controlled fracture

The *controlled-fracture* techniques are based on score-and-break methods for cleaving fibers. In this operation, the fiber to be cleaved is first scratched to create a stress concentration at the surface. The fiber is then bent over a curved surface while tension is applied simultaneously, as shown in Fig. 8.9. This action produces a stress distribution across the fiber. The maximum stress occurs at the scratch point so that a crack starts to propagate through the fiber, resulting in a highly smooth and perpendicular end face.

A number of different tools based on the controlled-fracture technique are available commercially for both factory and field uses.

8.4.3. End-face quality

The controlled-fracture method requires careful control of the curvature of the fiber and of the amount of tension applied. If the stress distribution across the

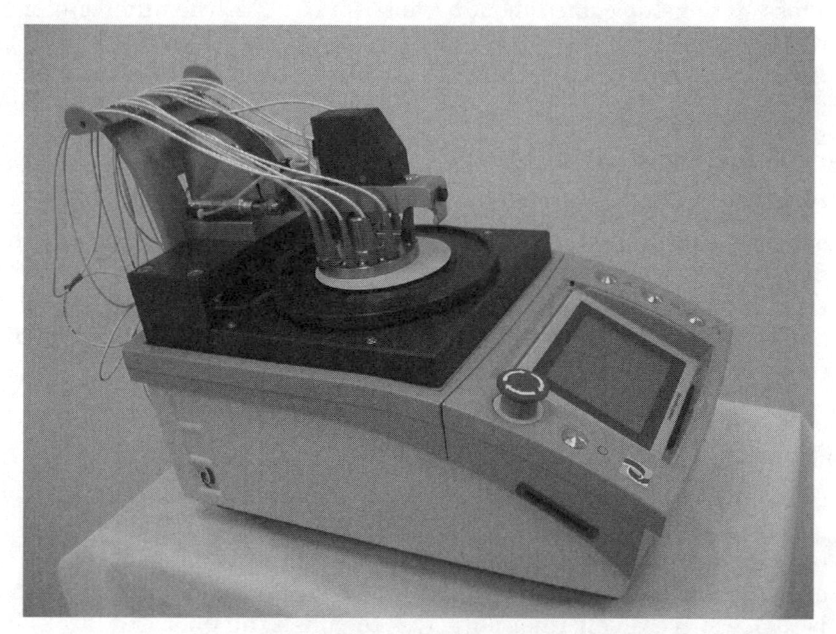

Figure 8.8. Example of a polishing machine that can prepare several fibers simultaneously. (*Photo courtesy of Domaille Engineering; www.Domaille Engineering. com.*)

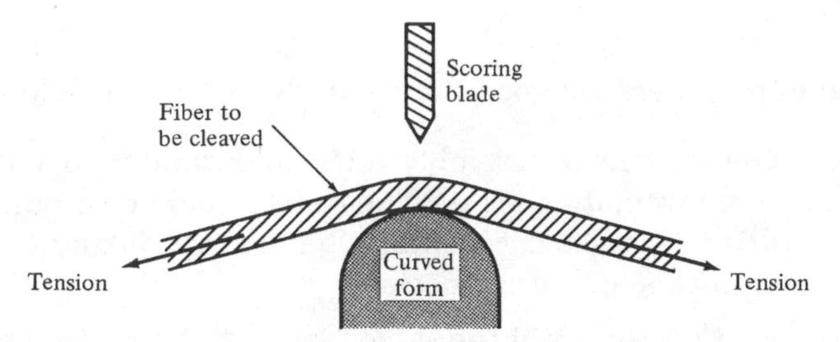

Figure 8.9. Controlled-fracture technique for cleaving fibers.

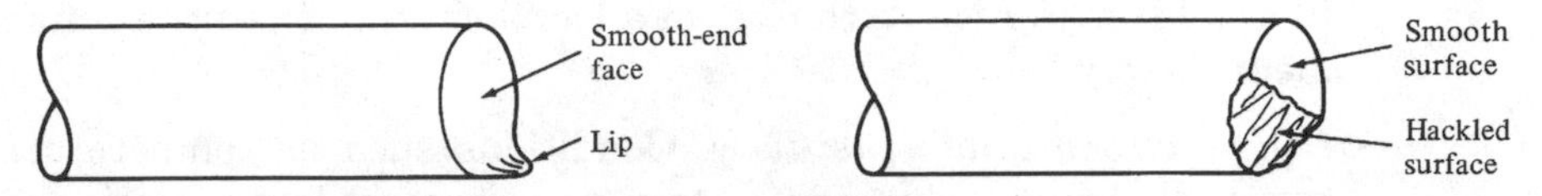

Figure 8.10. Two examples of improperly cleaved fiber ends.

crack is not controlled properly, the fracture propagating across the fiber can fork into several cracks. This forking produces defects such as a lip or a hackled portion on the fiber end, as shown in Fig. 8.10. The TIA/EIA Fiber Optic Test Procedures (FOTPs) 57B and 179 define these and the following other common end-face defects:

- *Lip*. This is a sharp protrusion from the edge of a cleaved fiber that prevents the cores from coming in close contact. Excessive lip height can cause fiber damage.
- *Roll-off*. This rounding off of the edge of a fiber is the opposite condition to lip formation. It also is known as *breakover* and can cause high insertion or splice loss.
- *Chip*. A chip is a localized fracture or break at the end of a cleaved fiber.
- *Hackle*. Figure 8.10 shows this as severe irregularities across a fiber end face.
- *Mist*. This is similar to hackle but much less severe.
- *Spiral or step*. These are abrupt changes in the end-face surface topology.
- *Shattering*. This is the result of an uncontrolled fracture and has no definable cleavage or surface characteristics.

8.5. Optical Connector Features

A wide variety of optical fiber connectors are available for numerous different applications. Their uses range from simple single-channel fiber-to-fiber connectors in a benign location to rugged multichannel connectors used under the ocean or for harsh military field environments. Here and in Sec. 8.6 we will concentrate on connectors used within a telecommunication or data communication facility.

8.5.1. Design requirements

Some principal requirements of good connector design are as follows:

- *Coupling loss*. The connector assembly must maintain stringent alignment tolerances to ensure low mating losses. The losses should be around 2 to 5 percent (0.1 to 0.2 dB) and must not change significantly during operation and after numerous connects and disconnects.
- *Interchangeability*. Connectors of the same type must be compatible from one manufacturer to another.
- *Ease of assembly*. A service technician should be able to install the connector in a field environment, that is, in a location other than the connector attachment factory.
- *Low environmental sensitivity*. Conditions such as temperature, dust, and moisture should have a small effect on connector loss variations.
- *Low cost and reliable construction*. The connector must have a precision suitable to the application, but it must be reliable and its cost must not be a major factor in the system.
- *Ease of connection*. Except for certain unique applications, one should be able to mate and disconnect the connector simply and by hand.

8.5.2. Connector components

Connectors are available in designs that screw on, twist on, or snap in place. The twist-on and snap-on designs are the ones used most commonly. The designs include both single-channel and multichannel assemblies for cable-to-cable and cable-to-circuit-card connections.

The majority of connectors use a butt-joint coupling mechanism, as illustrated in Fig. 8.11. The elements shown in this figure are common to most connectors. The key components are a long, thin stainless steel, glass, ceramic, or plastic cylinder, known as a *ferrule*, and a precision sleeve into which the ferrule fits. This sleeve is known variably as an *alignment sleeve*, an *adapter*, or a *coupling*

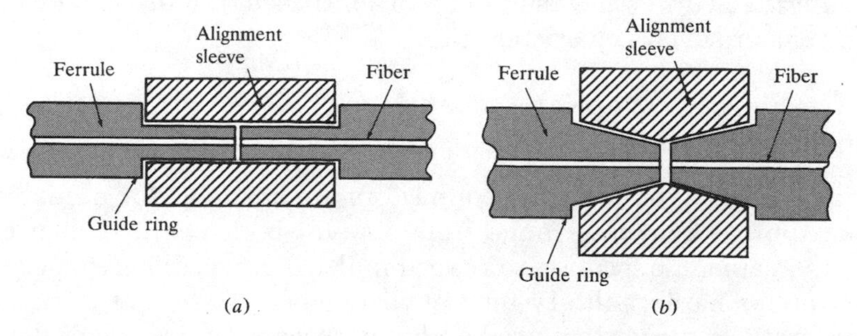

Figure 8.11. Example ferrules inserted into alignment sleeves for precision alignment of two fibers: (*a*) straight sleeve and (*b*) tapered sleeve.

receptacle. The center of the ferrule has a hole that precisely matches the size of the fiber cladding diameter. Typically, the hole size is $125.0 \pm 1.0\,\mu m$. The fiber is secured in the hole with epoxy, and the end of the ferrule is polished flat to a smooth finish. The mechanical challenges in fabricating a ferrule include maintaining both the dimensions of the hole diameter and the position of the hole relative to the ferrule outer surface.

Around 95 percent of the ferrules used in optical connectors are made of ceramic material due to some of the desirable properties they possess. These include low insertion loss required for optical transmission, remarkable strength, small elasticity coefficient, easy control of product characteristics, and strong resistance to changes in environmental conditions such as temperature. This is also why ferrules commonly are referred to as *ceramic ferrules*.

As shown in Fig. 8.11, since both ends of a cable have the same ferrule structure, an adapter is used to align the two fiber ends. Two popular adapter designs are the straight-sleeve and the biconical (or tapered-sleeve) mechanisms. In the *straight-sleeve connector*, the length of the sleeve and a guide ring on the ferrule determine the end separation of the two fibers. The *biconical connector* uses a tapered sleeve to accept and guide tapered ferrules. Again, the sleeve length and the guide rings on the ferrule maintain a given fiber end separation. Of these two coupling mechanisms, the straight-sleeve connector is in much wider use.

Note that an adapter also can be used to mate one type of connector to another. The only precaution in any case is to make sure that either the two fibers being joined have similar characteristics or that one is not trying to couple light from a multimode fiber into a single-mode fiber. However, it is not a problem to have light from a single-mode fiber couple into a multimode fiber. For example, this often is done at the receiving end of a link where a multimode flylead on a photodetector receives light from a single-mode transmission fiber.

Normally a connector also has some type of strain relief mechanism called a *boot*, which shields the junction of the connector body and the cable from bends and pulls. The connector figures in Sec. 8.6 show examples of boots.

8.6. Optical Connector Types

Since it is taking some time to establish standards in the optical connector industry, several large companies developed their own particular design. This has resulted in numerous connector styles and configurations. Typically, the different connector types are designated by combinations of two or three letters. The main ones are ST, SC, FC, LC, MU, MT-RJ, MPO, and variations on MPO. Therefore the vendors refer to these connectors as, for example, SC-type connectors or simply SC connectors. Each of these connector types is described in greater detail below. However, to understand the purpose of the seemingly arbitrary letter designations, let us first take a brief look at their origins:

- ST is derived from the words *straight tip*, which refers to the ferrule configuration.

- The letters SC were coined by NTT to mean *subscriber connector* or *square connector*, although now the connectors are not known by those names.
- A connector designed specifically for Fibre Channel applications was designated by the letters FC.
- Since Lucent developed a specific connector type, they obviously nicknamed it the LC connector; that is, LC can be considered an acronym for *Lucent connector*, although it is not called a Lucent connector.
- The letters MU were selected by NTT to indicate a *miniature unit.*
- The designation MT-RJ is an acronym for *media termination—recommended jack.*
- The letters MPO were selected to indicate a *multiple-fiber, push-on/pull-off* connecting function.

Table 8.1 summarizes several popular connector types and lists their main features and applications. More details and illustrations are given below. The optical SMA single-fiber connector is based on the electrical SMA connector design and was one of the first to appear. To improve on the screw-on mounting features of the SMA design, twist-on and snap-on single-fiber connectors such as the ST, SC, and FC were devised. Further emphasis on reducing connector sizes for higher packaging densities resulted in many concepts for small-form-factor (SFF) single-fiber connector types, such as the LC, MU, and MT-RJ designs. The biggest difference among the SFF connectors is whether they use ceramic or plastic ferrules.

As their name implies, *small-form-factor* connectors have a small size. This is important because as more fiber is being used in private networks, more electronics are being squeezed into smaller and smaller spaces. The SFF connectors are designed for fast termination in the field, but most require special tool sets for installation. SFF connectors also make smaller fiber network interface cards (NICs) practical for computer workstations and servers. However, since there are no standards for SFF connectors, network designers have the option to choose the connector best suited for their needs.

Another recent development was the introduction of an inexpensive, high-performance, compact *multiple-fiber connector*. These connectors save space by providing up to at least 12 potential connections within a single ferrule that has the same size as single-fiber SFF connectors. This means that one such multiple-fiber connector can replace up to 12 single-fiber connectors. These components are known by various acronyms such as MPO, MTP (trademark of US Conec), and MPX.

ST connector. The ST connector, which is shown in Fig. 8.12, is very popular for both data communication and telecommunication applications. It utilizes a precision zirconia (zirconium dioxide) ceramic ferrule. For multimode fibers this yields a typical insertion loss of 0.4 dB when using a manual polishing

TABLE 8.1. Popular Fiber Optic Connectors with Their Features and Applications

Connector type	Features	Applications
ST	Uses a ceramic ferrule and a rugged metal housing. It is latched in place by twisting	Designed for distribution applications using either multimode or single-mode fibers
SC	Designed by NTT for snap-in connection in tight spaces. Uses a ceramic ferrule in simplex or duplex plastic housings for either multimode or single-mode fibers	Widely used in Gigabit Ethernet, ATM, LAN, MAN, WAN, data communication, Fibre Channel, and telecommunication networks
FC	The preassembled, one-piece body design and prepolished ferrules enable fast, low-loss, and inexpensive terminations	Designed for terminating jumper and other cables in Fibre Channel and other telecommunication applications
SMA	Based on the threaded metallic housing of the proven SMA electrical connector	Designed for multimode data communication applications; also used for instrumentation connections
LC	SFF connector that uses a standard RJ-45 telephone plug housing and ceramic ferrules	Available in simplex and duplex configurations for CATV, LAN, MAN, and WAN applications
MU	SFF connector based on a 1.25-mm ceramic ferrule and a single free-floating ferrule	Suitable for board-mounted applications and for distribution-cable assemblies
MT-RJ	SFF connector with two fibers in one molded plastic ferrule and an improved RJ-45 latch mechanism	Applications are for MANs and LANs, such as horizontal optical cabling to the desktop
ESCON	Has a shockproof plastic housing with a retractable shroud to protect the ceramic ferrules from damage	Designed by IBM for connections between mainframe computers and peripheral networks and equipment
MPO	Can house up to 12 multimode or single-mode optical fibers in a single compact ferrule	Allows high-density connections between network equipment in telecommunication rooms

method or 0.2 dB when using an automated fiber polisher. Single-mode connectors typically achieve a 0.3-dB insertion loss and a 40-dB return loss using a simple manual polishing method.

The ST connector employs a rugged metal bayonet coupling ring with radial ramps to allow easy engagement to the studs of the mating adapter. Pairs of ST connectors are available in two colors for jacketed fiber, one with a beige boot and the other with a black boot (shown as the flexible black sleeve in Fig. 8.12). These two colors allow easy identification of the fibers in forming a duplex jumper cable. The connector is mated by pushing it into place and then twisting it to engage a spring-loaded bayonet socket.

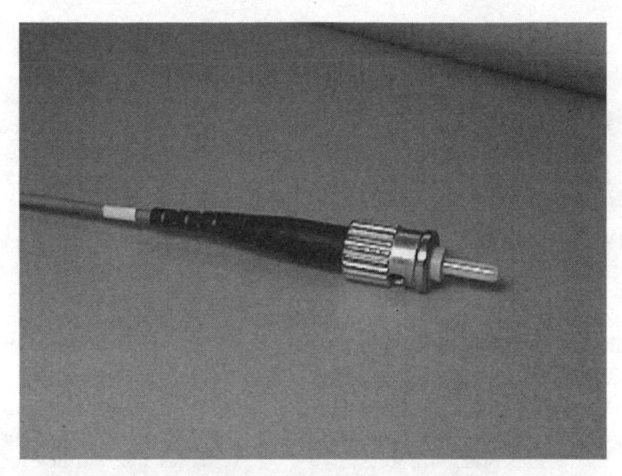

Figure 8.12. Example of an ST connector. (*Photo courtesy of Fitel Interconnectivity Corporation; www.fitelconn.com.*)

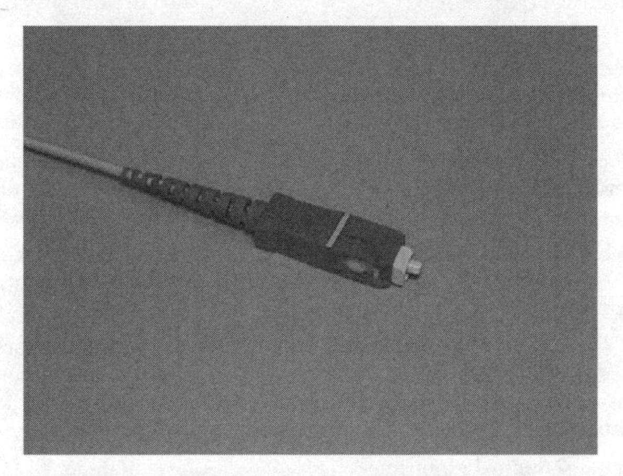

Figure 8.13. Example of an SC connector with a strain-relief boot. (*Photo courtesy of Fitel Interconnectivity Corporation; www.fitelconn.com.*)

SC connector. The SC connector was developed by NTT (Nippon Telegraph and Telephone) of Japan. The connector is mated by a simple snap-in method and can be disconnected by pushing in on a tab and then pulling the connector off. SC connectors are available in either simplex or duplex configurations, as shown in Fig. 8.13. Analogous to the ST connector, the SC connector uses a ceramic ferrule and has similar loss characteristics. The mating and loss features allow the connector to be used in tight spaces where turning an outer connector housing, such as on an ST connector, is not practical. An example of such spaces is a patch panel where there is a high packing density of connections.

SC simplex connectors have an outer plastic housing that is color-coded in accordance with ANSI/TIA/EIA-568-B.3 and ISO/IEC 11801-1.2 requirements.

This specifies beige for multimode fibers and blue for single-mode fibers. Duplex SC connectors are used widely on two-fiber patch cords and have been designed with a keying mechanism to maintain fiber optic cabling polarity. The word *polarity* means that the ends of a cable are matched properly for transmit and receive functions. Basically the duplex configuration combines two standard SC connectors in a common duplex plastic housing. This housing is keyed to maintain fiber polarity and provides smooth insertion and removal of the connector pairs.

FC connector. FC optical fiber connectors were designed specifically for telecommunication applications. This connector uses a threaded coupling mechanism, as illustrated in Fig. 8.14. Connector mating is done by means of bulkhead feedthrough adapters. These adapters combine a metal housing and either a precision ceramic or a rugged metal alignment sleeve. The connectors come in a preassembled, one-piece body design and have prepolished ferrules. This enables quick and economical terminations in both factory and field settings. This also ensures a uniform end-face geometry, thereby yielding consistent optical performance. Nominal insertion losses are less than 0.15 dB with single-mode fibers and less than 0.34 dB for multimode fibers.

An FC connector can be mated with an SC connector by one of two methods. In the first method, to directly connect them, a hybrid FC/SC connector adapter will accept an FC connector on one side and an SC connector on the other side. Both connectors should be flat-polished (not angle-polished) for this method to work. The second method uses a hybrid jumper cable and two connector adapters. A *hybrid jumper* is a cable with two different connectors on each end. These connectors should match the two connectors that are to be mated. This will allow regular connector adapters to be used in the patch panels to connect

Figure 8.14. Example of an FC connector. (*Photo courtesy of Fitel Interconnectivity Corporation; www.fitelconn.com.*)

the system. This method does add one extra connector pair loss to the system budget. This method also allows the use of angle-polished connectors.

SMA optical connector. The SMA 905 or 906 series of lightwave connectors were among the original optical fiber connectors designed for multimode data communication applications. Their applications include local-area networks, data processing networks, active device termination, premise installations, and connections for instrumentation. They use the proven and durable compact metal housing from the standard SMA electrical connector that was developed in the 1960s. The SMA connectors use a threaded interface and have typical insertion losses of less than 0.25 dB for multimode fibers. SMA 905 connectors have a straight ferrule while the SMA 906 connectors have a stepped design which uses a plastic sleeve for alignment.

LC connector. Lucent developed the LC connector to meet the growing demand for small, high-density fiber optic connectivity on equipment bays, in distribution panels, and on wall plates. Two common and proven technologies were combined in the LC connector. These are the industry-standard RJ-45 telephone plug interface and ceramic ferrule technology. The advantage of the RJ-45 housing is that it provides a reliable and robust latching mechanism for the LC connector. The LC connector has a six-position tuning feature to achieve very low insertion loss performance by optimizing the alignment of the fiber cores.

The LC connectors are available in both simplex and duplex configurations. They are available in industry-standard beige and blue and will accommodate 900-μm buffered fiber and 1.60-mm (0.063-in) and 2.40-mm (0.094-in) jacketed cable. Duplex and simplex LC adapters with either ceramic or metal sleeves are available from a variety of manufacturers.

MU connector. The MU connector is an SFF connector that was developed by NTT. Basically it can be considered as a smaller version of the SC-type connector. It is based on a 1.25-mm ceramic ferrule and utilizes a single free-floating ferrule independent of the backbone. It has a plastic housing and uses a push-pull latching mechanism. It is available in simplex, duplex, and higher-count-channel styles. The MU connector is suitable both for board-mounted applications and for distribution-cable assemblies to allow connections in simplex networks.

MT-RJ connector. The MT-RJ is an SFF connector with two fibers in one precision-molded plastic ferrule, as shown in Fig. 8.15. It was designed to meet the desire for an interface technology that is significantly lower in cost and size than the duplex SC connector. The MT-RJ uses an improved version of an industry-standard RJ-45 type of latch. The MT-RJ connector performs equally well for multimode and single-mode fiber applications. Its principal application is for horizontal cabling needs to the desktop. Products in the MT-RJ connector

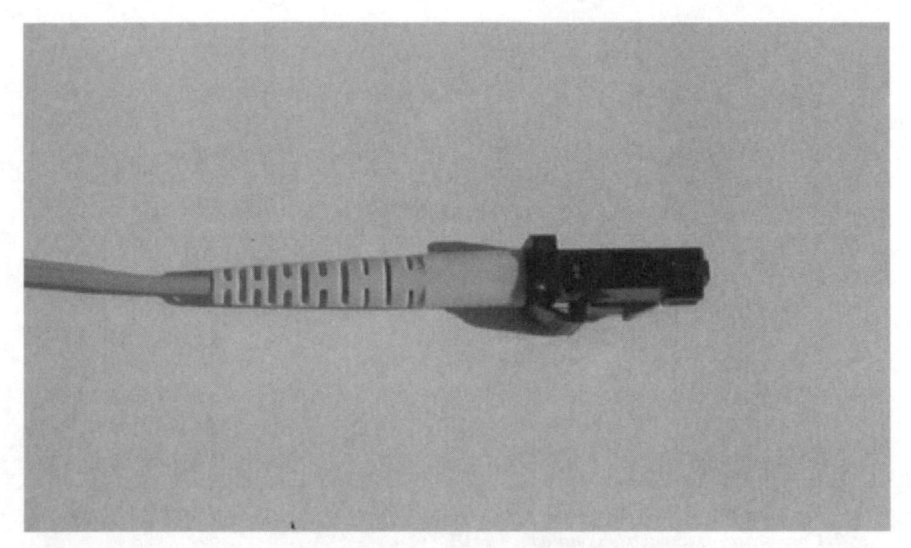

Figure 8.15. Example of an MT-RJ connector clearly showing the strain-relief boot. (*Photo courtesy of Photonics Comm Solutions; www.photonicscomm.com.*)

solution family include field-mountable connectors, duplex jumper cables, hybrid jumper cables, and flylead terminations.

ESCON connector. IBM developed the *Enterprise Systems Connection* (ESCON is a registered trademark of the IBM Corporation) duplex connector as a high-speed communication interconnect between mainframe computers and storage devices, peripheral control units, cluster controllers, and other networks. It has a shockproof plastic housing with a retractable shroud to protect the ceramic ferrules from damage.

Multiple-fiber connectors. The MPO connector, shown in Fig. 8.16, is one of several variations of compact multiple-fiber connectors. They all use a simple push-pull latching mechanism for easy and intuitive insertion and removal. The end of the MPO connector may be polished flat or at an 8° angle. The MTO connector is the same size as the SC; but since it can accommodate a maximum of 12 fibers, it provides up to 12 times the density, thereby offering savings in circuit card and rack space.

8.7. Optical Splices

A fiber *splice* is a permanent or temporary low-loss bond between two fibers. Such a bond can be made by using either fusion splicing or mechanical splicing. Most splices are permanent and typically are used to create long optical links or in situations where frequent connection and disconnection is not needed. Temporary splices may be necessary or convenient when one is making

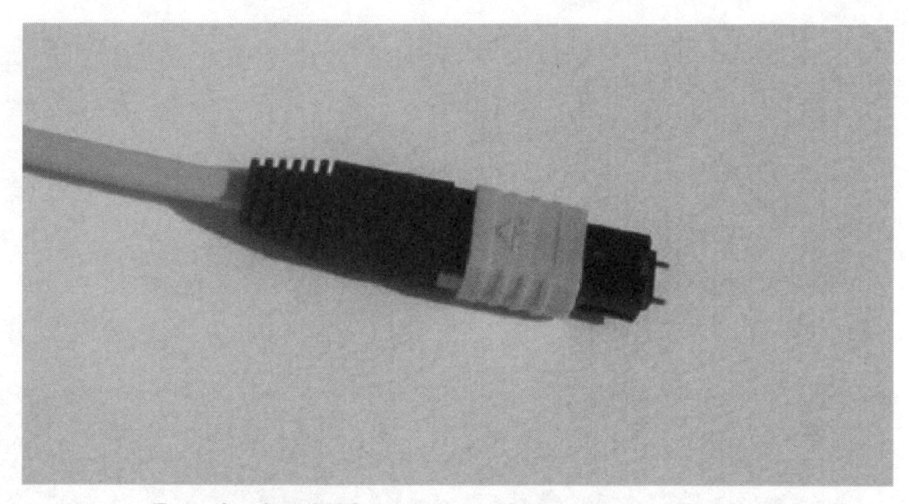

Figure 8.16. Example of an MPO connector. (*Photo courtesy of Fitel Interconnectivity Corporation; www.fitelconn.com.*)

emergency cable repairs or doing testing during installation or troubleshooting. Here we will first look at general splicing issues and then examine fusion and mechanical splicing methods.

8.7.1. Splicing issues

In making and evaluating optical fiber splices, one must take into account the physical differences in the two fibers, fiber misalignments at the joint, and the mechanical strength of the splice.

The physical differences in fibers that lead to splice losses are the same as those discussed above for connectors and result in what is called *intrinsic loss*. These fiber-related differences include variations in core diameter, core-area ellipticity, numerical aperture, and core-cladding concentricity of each fiber. *Extrinsic losses* depend on how well the fibers are prepared and the care taken to make the splice. The factors here include fiber misalignments at the joint, the smoothness and cleanliness of the fiber end faces, and the skill of the splice equipment operator. When the fiber bonding is done properly using high-quality equipment, the total splice loss typically is 0.05 to 0.10 dB for fusion splicing and around 0.5 dB for mechanical splices.

Those loss numbers naturally are for splicing similar types of fibers. For example, suppose a technician makes the mistake of assuming that two arbitrary, say, blue-jacketed fibers are identical, when in reality one is a multimode fiber and the other is a single-mode fiber. After the splicing of these two fibers, the attenuation measured when going from the single-mode to the multimode fiber may be 0.1 dB. However, it will be a nasty surprise to find that the attenuation in the other direction (multimode to single-mode path) is almost 20 dB! Even in a LAN environment where there may be a mixture of 50- and 62.6-μm fibers, inadvertently splicing two different fiber types can lead to unexpectedly high losses.

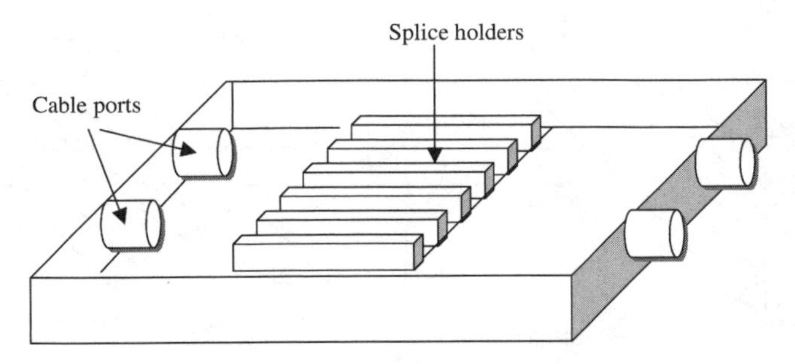

Figure 8.17. Illustration of splice holders stored in a splice tray.

When one is preparing a fiber for splicing, it is necessary to expose the cladding of the fiber by stripping away about 1 cm of the buffer coating. This stripping is done with tools that are designed especially not to nick or damage the fiber. Once a fiber is stripped, it is cleaved to yield a uniform, perpendicular surface that will allow maximum light transmission to the other fiber. At this point the fibers are ready for splicing.

Once the splice is made, it is encapsulated in a shielding mechanism that adds mechanical strength and protection from contaminants. Normally the spliced fiber then is stored in a *splice tray* or *splice closure*. As shown by a simple example in Fig. 8.17, this is a special housing that helps organize fibers when a multiple-fiber cable is spliced and that protects the splices from strains and environmental contaminants. Most splice trays have a splice holder in the center of the tray into which the operator can snap the encapsulated splice. A series of such splice holders may be stacked inside the enclosure. A cover fits over the unit so that it is sealed from the environment. Two basic configurations of splice closures are available commercially depending on whether the unit is to be mounted inside on a wall or used in an above-grade, underground, or outside aerial application.

8.7.2. Splicing methods

Fusion splices are made by thermally bonding prepared fiber ends, as illustrated in Fig. 8.18. In this method the fiber ends are first aligned and then butted together. This is done either in a grooved fiber holder or under a microscope with a micromanipulator. The butt joint then is heated with an electric arc or a laser pulse so that the fiber ends are melted momentarily and hence bonded. This technique can produce very low splice losses (typically averaging less than 0.1 dB). Care must be taken in this technique, since defect growth and residual stress induced near the bonded joint can produce a weak splice. However, skilled operators using modern automated splicers usually alleviate these concerns.

In *mechanical splicing* the assembly process involves stripping and cleaving the fibers, inserting them into a splice mechanism until they touch, and then securing them in place. The securing process is done by either clamping the

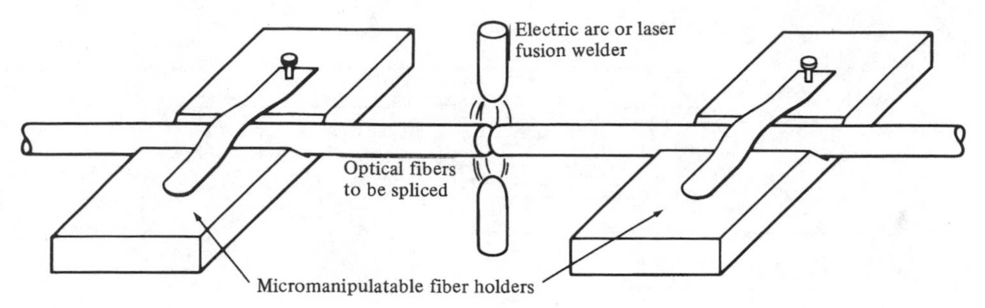

Figure 8.18. Fusion splicing of optical fibers.

fibers or gluing them to the splicing structure. To prevent light reflections within the splice, a special gel that has a refractive index close to that of glass can be injected into the space between the spliced fibers. This material is referred to as *index-matching gel*.

Most mechanical splice structures consist of a thermoplastic housing that is versatile and accepts fibers ranging in size from 250 to 900 μm. In addition they have the ability to tune or reposition the fibers to achieve optimal light coupling performance during installation. The goals of mechanical splice manufacturers are to have splicing structures that allow easy and quick assemblies of less than a minute or two after the fibers have been cleaved. Ideally the assembly should require no heat shrink, curing, crimping, or gluing in order to reduce installation times and cost.

8.8. Summary

In practice, most active optical devices are available with a short length of optical fiber, called a flylead or a pigtail, already attached in an optimum power coupling configuration. The power coupling problem from these pigtailed devices thus reduces to a simpler one of coupling optical power from one fiber into another.

Techniques for joining optical fibers are subject to various power loss conditions at the joint. These losses depend on factors such as the mechanical alignments of the two fibers, differences in the geometric and waveguide characteristics of the two fiber ends at the joint, and the fiber end-face qualities. Careful splicing can produce joint losses of less than 0.1 dB, whereas high-quality multimode and single-mode connectors have losses of less than 0.3 dB.

One of the first steps that must be followed before fibers are connected or spliced is to prepare the fiber end faces properly. In order not to have light deflected or scattered at the joint, the fiber ends must be flat and smooth and must have the proper angle relative to the axis (either perpendicular or at a nominal 8° angle). Common end preparation techniques include a grinding and polishing method and a controlled-fracture procedure.

A wide variety of optical fiber connectors are available for numerous different applications. Their uses range from simple single-channel fiber-to-fiber connectors in a benign location to rugged multichannel connectors used underwater or for harsh military field environments. Connectors are available in designs that screw on, twist on, or snap in place. The twist-on and push-on designs are the ones used most commonly. The designs include both single-channel and multichannel assemblies for cable-to-cable and cable-to-circuit-card connections.

Whereas a connector is a joint that can be mated and disconnected many times, a fiber splice is a permanent or temporary low-loss bond between two fibers. Such a bond can be made by either fusion splicing or mechanical splicing. Various splicing instruments are available for this function. Most splices are permanent and typically are used to create long optical links or in situations where frequent connection and disconnection is not needed. Temporary splices may be necessary or convenient when one is making emergency cable repairs or doing testing during installation or troubleshooting.

Further Reading

1. G. Keiser, *Optical Fiber Communications*, 3d ed., McGraw-Hill, Burr Ridge, Ill., 2000, Chap. 5.
2. TIA/EIA-455-57B (FOTP-57B), *Preparation and Examination of Optical Fiber Endface for Testing Purposes*, September 2000.
3. TIA/EIA-455-179 (FOTP-179), *Inspection of Cleaved Fiber Endfaces by Interferometry*, May 1988.
4. Bob Chomycz, *Fiber Optic Installer's Field Manual*, McGraw-Hill, New York, 2000, Chap. 11, "Splicing and termination."
5. M. Kihara, S. Nagasawa, and T. Tanifuji, "Return loss characteristics of optical fiber connectors," *J. Lightwave Tech.*, vol. 14, pp. 1986–1991, September 1996.

Chapter

9

Passive Optical Components

In addition to fibers, light sources, and photodetectors, many other components are used in a complex optical communication network to split, route, process, or otherwise manipulate light signals. The devices can be categorized as either passive or active components. *Passive optical components* do not hum or wink or blink, since they require no external source of energy to perform an operation or transformation on an optical signal. Just as a filter in a coffee pot or a sprayer head in a shower just sit there while performing very important functions, passive components carry out their unique processes without any physical or electrical action. For example, a passive optical filter will allow only a certain wavelength to pass through it while absorbing or reflecting all others, and an optical splitter divides the light entering it into two or more, smaller optical power streams. *Active components* require some type of external energy either to perform their functions or to be used over a wider operating range than a passive device, thereby offering greater flexibility.

Although optical fibers and connectors are passive elements, one usually considers them separately from other passive optical components. Some basic passive functions and the devices which enable them are as follows:

- Transfer energy: optical fibers
- Attenuate light signals: optical attenuators, isolators
- Influence the spatial distribution of a light wave: directional coupler, star coupler, beam expander
- Modify the state of polarization: polarizer, half-wave plates, Faraday rotator
- Redirect light: circulator, mirror, grating
- Reflect light: fiber Bragg gratings, mirror
- Select a narrow spectrum of light: optical filter, grating
- Convert light wave modes: fiber gratings, Mach-Zehnder interferometer

- Combine or separate independent signals at different wavelengths: WDM device

The passive components described in this chapter include optical couplers, isolators, circulators, filters, gratings, and wavelength multiplexers. Some of these passive devices also can be configured as active devices, which Chap. 10 addresses. Other passive devices used for wavelength division multiplexing (WDM) are described in Chap. 12. These include arrayed waveguide gratings (AWGs), bulk diffraction gratings, and interleavers.

9.1. Optical Couplers

The concept of a *coupler* encompasses a variety of functions, including splitting a light signal into two or more streams, combining two or more light streams, tapping off a small portion of optical power for monitoring purposes, or transferring a selective range of optical power from one fiber to another. When one is discussing couplers, it is customary to refer to them in terms of the number of input ports and output ports on the device. For example, a device with two inputs and two outputs is called a *2 × 2 coupler.* In general, an $N \times M$ coupler has $N \geq 2$ input ports and $M \geq 2$ output ports. The coupling devices can be fabricated either from optical fibers or by means of planar optical waveguides using material such as lithium niobate ($LiNbO_3$) or InP.

The couplers described in this section include 2 × 2 couplers, tap couplers, star couplers, and the Mach-Zehnder interferometer.

9.1.1. Basic 2 × 2 coupler

The 2 × 2 coupler is a simple fundamental device that we will use here to demonstrate the operational principles of optical couplers. These are known as *directional couplers*. A common construction is the fused-fiber coupler illustrated in Fig. 9.1. This is fabricated by twisting together, melting, and pulling two single-mode fibers so they get fused together over a uniform section of length W. Each input and output fiber has a long tapered section of length L, since the transverse dimensions are reduced gradually down to that of the coupling region when the fibers are pulled during the fusion process. This device is known as a *fused biconical tapered coupler*. As shown in Fig. 9.1, P_0 is the input power on the top fiber (which we will take as the primary fiber in a link), P_1 is the throughput power, and P_2 is the power coupled into the second fiber. Parameters P_3 and P_4 are extremely low optical signal levels (-50 to -70 dB or, equivalently, factors of 10^{-5} to 10^{-7} below the input power level). These result from backward reflections and scattering due to packaging effects and bending in the device.

To understand how power is coupled from one fiber to the other, recall Fig. 4.3. This figure shows that the power distributions of any given mode are not confined

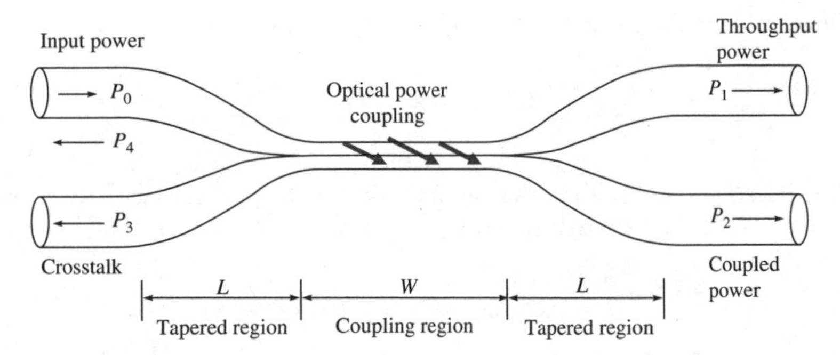

Figure 9.1. Cross-sectional view of a fused-fiber coupler having a coupling region W and two tapered regions of length L. The total span $2L + W$ is the coupler draw length.

completely to the fiber core, but instead extend partially into the cladding. Therefore if two fiber cores are brought close together, the tail of the power distribution in one fiber will extend into the adjacent fiber core. Consequently, some of the optical power will transfer to the adjacent fiber through evanescent coupling. The amount of optical power coupled from one fiber to another can be varied by changing the coupling length W or the distance between the two fiber cores.

9.1.2. Coupler performance

In specifying the performance of an optical coupler, one usually indicates the percentage division of optical power between the output ports by means of the *splitting ratio* or *coupling ratio*. Referring to Fig. 9.1, with P_0 being the input power and P_1 and P_2 the output powers, we have

$$\text{Coupling ratio} = \frac{P_2}{P_1 + P_2} \times 100\% \tag{9.1}$$

By adjusting the parameters so that power is divided evenly, with one-half of the input power going to each output, one creates a *3-dB coupler*. A coupler could also be made in which almost all the optical power at 1550 nm goes to one port and almost all the energy around 1310 nm goes to the other port. Such couplers also are used to combine 980- or 1480-nm laser pump powers along with a C-band signal in erbium-doped fiber amplifiers, which Chap. 11 describes in greater detail.

Loss Categories In the above analysis, we have assumed for simplicity that the device is lossless. However, in any practical coupler there is always some light that is lost when a signal goes through it. The two basic losses are excess loss and insertion loss. The *excess loss* is defined as the ratio of the input power to the total output power.

Thus, in decibels, the excess loss for a 2 × 2 coupler is

$$\text{Excess loss} = 10 \log\left(\frac{P_0}{P_1 + P_2}\right) \tag{9.2}$$

The *insertion loss* refers to the loss for a particular port-to-port path. For example, for the path from input port i to output port j, we have, in decibels,

$$\text{Insertion loss} = 10 \log\left(\frac{P_i}{P_j}\right) \tag{9.3}$$

Another performance parameter is *crosstalk*, which measures the degree of isolation between the input at one port and the optical power scattered or reflected back into the other input port. That is, it is a measure of the optical power level P_3 shown in Fig. 9.1:

$$\text{Crosstalk} = 10 \log\left(\frac{P_3}{P_0}\right) \tag{9.4}$$

Example A 2 × 2 biconical tapered fiber coupler has an input optical power level of $P_0 = 200\,\mu\text{W}$. The output powers at the other three ports are $P_1 = 90\,\mu\text{W}$, $P_2 = 85\,\mu\text{W}$, and $P_3 = 6.3\,\text{nW}$. From Eq. (9.1), the coupling ratio is

$$\text{Coupling ratio} = \frac{85}{90 + 85} \times 100\% = 48.6\%$$

From Eq. (9.2), the excess loss is

$$\text{Excess loss} = 10 \log\left(\frac{200}{90 + 85}\right) = 0.58\text{ dB}$$

By using Eq. (9.3), the insertion losses are

$$\text{Insertion loss (port 0 to port 1)} = 10 \log \frac{200}{90} = 3.47\text{ dB}$$

$$\text{Insertion loss (port 0 to port 2)} = 10 \log \frac{200}{85} = 3.72\text{ dB}$$

The crosstalk is given by Eq. (9.4) as

$$\text{Crosstalk} = 10 \log\left(\frac{6.3 \times 10^{-3}}{200}\right) = -45\text{ dB}$$

Table 9.1 lists some typical power-splitting ratios for a fiber optic coupler operating at either 1310 or 1550 nm. The first number in the designation PP/SP is for the throughput channel (primary power), and the second number is for the coupled channel (secondary power). Typical excess losses are less than 0.1 dB.

TABLE 9.1. Typical Coupling Ratios and Insertion Losses in Optical Fiber Couplers

Coupling ratio (primary/secondary)	Insertion loss (PP/SP), dB
50/50	3.4/3.4
60/40	2.5/4.4
70/30	1.8/5.6
80/20	1.2/7.6
90/10	0.6/10.8
95/05	0.4/14.6

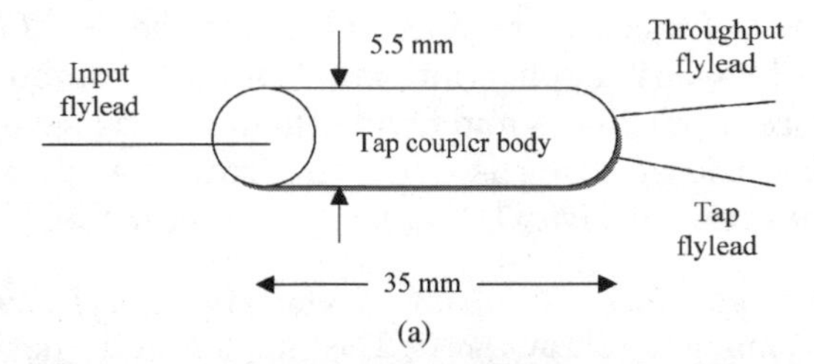

(a)

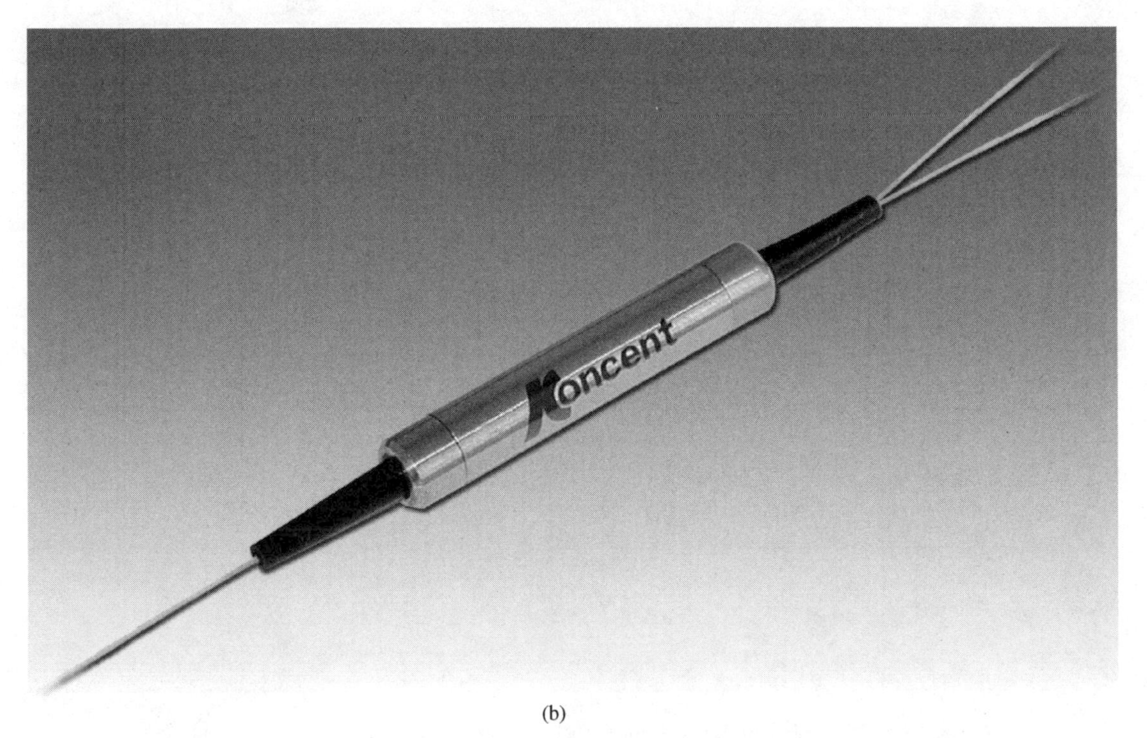

(b)

Figure 9.2. A typical package for a tap coupler. (*Photo courtesy of Koncent Communication; www.koncent.com.*)

Since the couplers are available in different grade qualities, vendor data sheets will have exact values for a specific coupler configuration.

9.1.3. Tap coupler

To monitor the light signal level or quality in a link, one can use a 2 × 2 device that has a coupling fraction of around 1 to 5 percent, which is selected and fixed during fabrication. This is known as a *tap coupler*. Nominally the tap coupler is packaged as a three-port device with one arm of the 2 × 2 coupler being terminated inside the package. This termination might use the specialty attenuating fiber described in Sec. 4.7. Figure 9.2 shows a typical package for such a tap coupler, and Table 9.2 lists some representative specifications.

9.1.4. Star coupler

An $N \times N$ *star coupler* is a more general form of the 2 × 2 coupler, as shown in Fig. 9.3. In the broadest application, star couplers combine the light streams from two or more input fibers and divide them among several output fibers. In the general case, the splitting is done uniformly for all wavelengths, so that each of the N outputs receives $1/N$ of the power entering the device.

Star Splitting Loss In an ideal star coupler, the optical power from any input is evenly divided among the output ports. The total loss of the device consists of its split-

TABLE 9.2. Representative Specifications for a 2 × 2 Tap Coupler

Parameter	Unit	Specification
Tap ratio	%	1 to 5
Insertion loss	dB	0.5
Return loss	dB	55
Power handling	mW	1000
Flylead length	m	1
Size (diameter × length)	mm	5.5 × 35

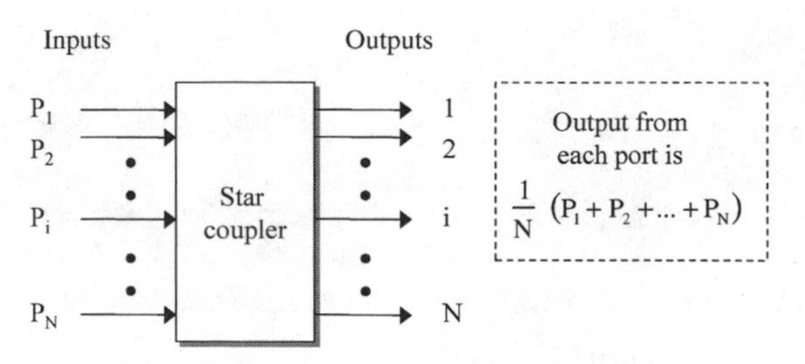

Figure 9.3. Basic star coupler concept for combining or splitting optical powers.

ting loss plus the excess loss in each path through the star. The *splitting loss* is given in decibels by

$$\text{Splitting loss} = -10 \log \frac{1}{N} = 10 \log N \tag{9.5}$$

Similar to Eq. (9.2), for a single input power P_{in} and N output powers, the excess loss in decibels is given by

$$\text{Fiber star excess loss} = 10 \log \frac{P_{\text{in}}}{\sum_{i=1}^{N} P_{\text{out},i}} \tag{9.6}$$

The insertion loss and crosstalk can be found from Eqs. (9.3) and (9.4), respectively.

9.1.5. Mach-Zehnder interferometer

Wavelength-dependent multiplexers can be made by using Mach-Zehnder interferometry techniques. These devices can be either active or passive. Here we first look at passive multiplexers. Figure 9.4 illustrates the constituents of an individual *Mach-Zehnder interferometer* (MZI). This 2×2 MZI consists of three stages: an initial 3-dB directional coupler which splits the input signals equally into two separate paths, a central section where one of the waveguides is longer by ΔL to give a wavelength-dependent phase shift between the two arms, and another 3-dB coupler which recombines the signals at the output. The function of this arrangement is that, by splitting the input beams and introducing a specific phase shift in one of the paths, the recombined signals will interfere constructively at one output and destructively at the other. Thus, if the input to port 1 is at λ_1 and the input to port 2 is at λ_2, then both signals finally emerge from only one output port.

Reciprocally, if independent signals at wavelengths λ_1 and λ_2 enter the bottom port on the right-hand side of Fig. 9.4 (this time going from right to left through the device), then the signal at λ_1 will exit the device from the top port on the left-hand side, and the signal at λ_2 will exit the MZI coupler from the bottom port.

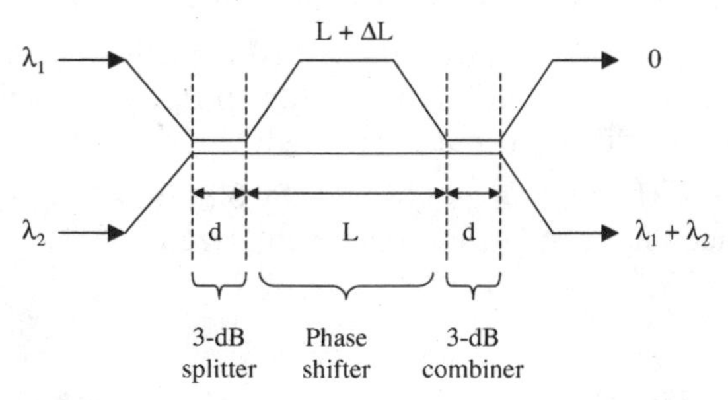

Figure 9.4. Layout of a basic 2×2 Mach-Zehnder interferometer.

MZI Equation To have input signals at wavelengths λ_1 and λ_2 that enter ports 1 and 2, respectively, emerge from only one output port, the length difference in the interferometer arms should be

$$\Delta L = \left[2n_{\text{eff}}\left(\frac{1}{\lambda_1} - \frac{1}{\lambda_2}\right)\right]^{-1} = \frac{c}{2n_{\text{eff}}\,\Delta\nu} \tag{9.7}$$

where n_{eff} is the effective refractive index of the MZI material and $\Delta\nu$ is the frequency separation of the two wavelengths.

Example Assume that the input wavelengths of a 2×2 silicon MZI are separated by 100 GHz (or equivalently, 0.75 nm at 1550 nm). With $n_{\text{eff}} = 1.5$ in a silicon waveguide, we know from Eq. (9.7) that the waveguide length difference must be

$$\Delta L = \frac{3 \times 10^8 \text{ m/s}}{2(1.5) \times 10^{11}/\text{s}} = 1 \text{ mm}$$

By using basic 2×2 MZI modules, any size $N \times N$ multiplexers (with $N = 2^n$) can be constructed.

9.2. Isolators and Circulators

In a number of applications it is desirable to have a passive optical device that is nonreciprocal; that is, it works differently when its inputs and outputs are reversed. Two examples of such a device are isolators and circulators. To understand the operation of these devices, we need to recall some facts about polarization and polarization-sensitive components from Chap. 3.

- Light can be represented as a combination of a parallel vibration and a perpendicular vibration, which are called the two *orthogonal plane polarization states* of a light wave.
- A *polarizer* is a material or device that transmits only one polarization component and blocks the other.
- A *Faraday rotator* is a device that rotates the state of polarization (SOP) of light passing through it by a specific angular amount.
- A device made from birefringent materials (called a *walk-off polarizer*) splits the light signal entering it into two orthogonally (perpendicularly) polarized beams, which then follow different paths through the material.
- A *half-wave plate* rotates the SOP of a lightwave by 90°; for example, it converts right circularly polarized light to left circularly polarized light.

9.2.1. Optical isolators

Optical isolators are devices that allow light to pass through them in only one direction. This is important in a number of instances to prevent scattered or reflected light from traveling in the reverse direction. One common application of an optical isolator is to keep such light from entering a laser diode and possibly causing instabilities in the optical output.

Many design configurations of varying complexity exist for optical isolators. The simple ones depend on the state of polarization of the input light. However, such a design results in a 3-dB loss (one-half the power) when unpolarized light is passed through it, since it blocks one-half of the input signal. In practice, the optical isolator should be independent of the SOP since light in an optical link normally is not polarized.

Figure 9.5 shows a design for a polarization-independent isolator that is made of three miniature optical components. The core of the device consists of a 45° Faraday rotator that is placed between two wedge-shaped birefringent plates or walk-off polarizers. These plates could consist of a material such as YVO_4 or TiO_2, as described in Chap. 3. Light traveling in the forward direction (left to right in Fig. 9.5) is separated into ordinary and extraordinary rays by the first birefringent plate. The Faraday rotator then rotates the polarization plane of each ray by 45°. After exiting the Faraday rotator the two rays pass through the second birefringent plate. The axis of this polarizer plate is oriented in such a way that the relationship between the two types of rays is maintained. Thus when they exit the polarizer, they both are refracted in an identical parallel direction. Going in the reverse direction (right to left), the relationship of the ordinary and extraordinary rays is reversed when exiting the Faraday rotator due to the nonreciprocity of the Faraday rotation. Consequently, the rays diverge when they exit the left-hand birefringent plate and are not coupled to the fiber anymore.

Table 9.3 lists some operational characteristics of commercially available isolators. The packages have similar configurations to the tap coupler shown in Fig. 9.2.

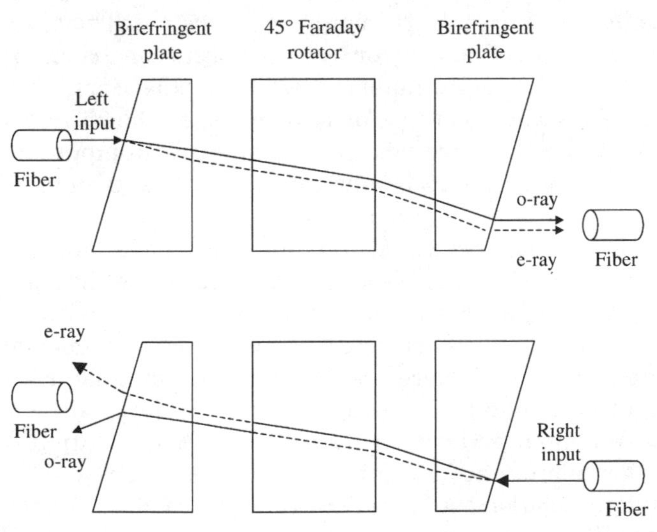

Figure 9.5. Design and operation of a polarization-independent isolator made of three miniature optical components.

TABLE 9.3. Typical Parameter Values of Commercially Available Isolators

Parameter	Unit	Value
Central wavelength λ_c	nm	1310, 1550
Peak isolation	dB	40
Isolation at $\lambda_c \pm 20\,\text{nm}$	dB	30
Insertion loss	dB	<0.5
Polarization-dependent loss	dB	<0.1
Polarization mode dispersion	ps	<0.25
Size (diameter × length)	mm	6 × 35

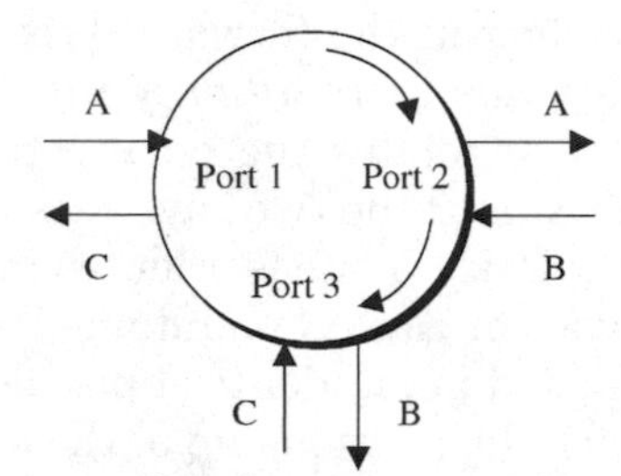

Figure 9.6. Operational concept of a three-port circulator.

9.2.2. Optical circulators

An *optical circulator* is a nonreciprocal multiport passive device that directs light sequentially from port to port in only one direction. The operation of a circulator is similar to that of an isolator except that its construction is more complex. Typically it consists of a number of walk-off polarizers, half-wave plates, and Faraday rotators and has three or four ports, as shown in Fig. 9.6. To see how it works, consider the three-port circulator. Here an input on port 1 is sent out on port 2, an input on port 2 is sent out on port 3, and an input on port 3 is sent out on port 1.

Similarly, in a four-port device, ideally one could have four inputs and four outputs. We say *ideally* since that description assumes the circulator is perfectly symmetric. In practice, in many applications it is not necessary to have four inputs and four outputs. Furthermore, such a perfectly symmetric circulator is rather complex to fabricate. Therefore, in a four-port circulator it is common to have three input ports and three output ports by making port 1 an input-only port, ports 2 and 3 input and output ports, and port 4 an output-only port.

A variety of circulators are available commercially. These devices have low insertion loss, high isolation over a wide wavelength range, minimal polarization-dependent loss (PDL), and low polarization mode dispersion (PMD). Table 9.4 lists some operational characteristics of commercially available circulators.

TABLE 9.4. Typical Parameter Values of Commercially Available Circulators

Parameter	Unit	Value
Wavelength band	nm	C-band: 1525–1565 L-band: 1570–1610
Insertion loss	dB	<0.6
Channel isolation	dB	>40
Optical return loss	dB	>50
Operating power	mW	<500
Polarization-dependent loss	dB	<0.1
Polarization mode dispersion	ps	<0.1
Size (diameter × length)	mm	5.5 × 50

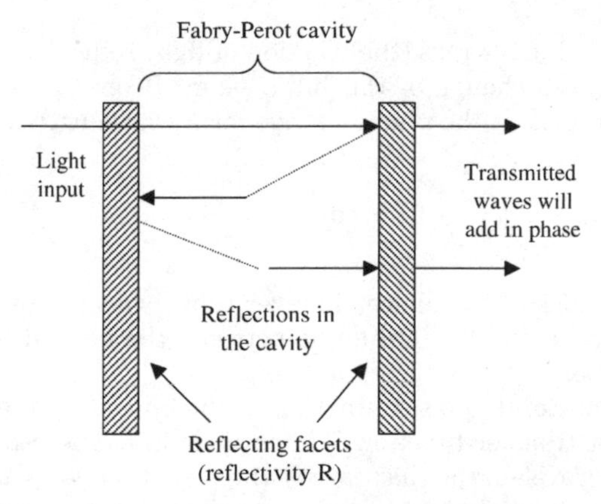

Figure 9.7. Structure of a Fabry-Perot interferometer or an etalon.

Optical circulators are used in optical amplifiers, add/drop multiplexers, and dispersion compensation modules. For example, Sec. 12.2 describes their use as part of an add/drop multiplexer in a wavelength division multiplexing (WDM) system.

9.3. Dielectric Thin-Film Filters

A dielectric *thin-film filter* (TFF) is used as an *optical bandpass filter*. This means that it allows a particular, very narrow wavelength band to pass straight through it and reflects all others. The basis of these devices is a classical Fabry-Perot filter structure, which is a cavity formed by two parallel, highly reflective mirror surfaces, as shown in Fig. 9.7. This structure is called a *Fabry-Perot interferometer* or an *etalon*. It also is known as a *thin-film resonant cavity filter*.

To see how it works, consider a light signal that is incident on the left surface of the etalon. After the light passes through the cavity and hits the inside surface on the right, some of the light leaves the cavity and some light is reflected. The amount of light that is reflected depends on the reflectivity R of the surface. If the round-trip distance between the two mirrors is an integral multiple of a wavelength λ (that is, λ, 2λ, 3λ, etc.), then all light at those wavelengths which passes through the right facet *adds in phase*. This means that these wavelengths *interfere constructively* in the device output beam, so they add in intensity. These wavelengths are called the *resonant wavelengths* of the cavity. The etalon reflects all other wavelengths.

Etalon Theory The transmission T of an ideal etalon in which there is no light absorption by the mirrors is an Airy function given by

$$T = \left[1 + \frac{4R}{(1-R)^2} \sin^2 \frac{\phi}{2}\right]^{-1} \tag{9.8}$$

where R is the *reflectivity* of the mirrors (the fraction of light reflected by the mirror) and ϕ is the round-trip phase change of the light beam. If one ignores any phase change at the mirror surface, then the phase change for a wavelength λ is

$$\phi = \frac{2\pi}{\lambda} 2nD \cos\theta \tag{9.9}$$

where n is the refractive index of the dielectric layer that forms the mirror, D is the distance between the mirrors, and θ is the angle between the normal to the surface and the incoming light beam.

Figure 9.8 gives a plot of Eq. (9.8) as a function of the optical frequency $f = 2\pi/\lambda$. This shows that the power transfer function T is periodic in f. The peaks, called the *passbands*, occur at those wavelengths that satisfy the condition $N\lambda = 2nD$, where N is an integer. Thus in order for a single wavelength to be selected by the filter from a particular spectral range, all the wavelengths must lie between two successive passbands of the filter transfer function. If some wavelengths lie outside this range, then the filter would transmit several wavelengths. The distance between adjacent peaks is

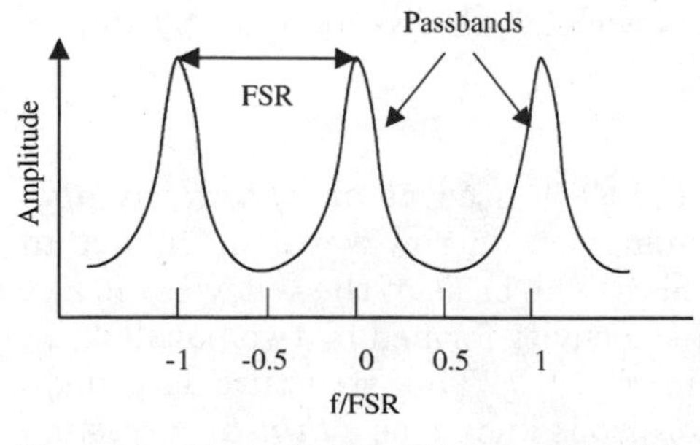

Figure 9.8. Example of an Airy function. The distance between adjacent peaks is called the free spectral range, or FSR.

called the *free spectral range* (FSR). This is given by

$$\mathrm{FSR} = \frac{\lambda^2}{2nD} \tag{9.10}$$

Another important parameter is the measure of the full width of the passband at its half-maximum value, which is designated by FWHM (*full-width half-maximum*). This is of interest in WDM systems for determining how many wavelengths can lie within the FSR of the filter. The ratio FSR/FWHM gives an approximation of the number of wavelengths that a filter can accommodate. This ratio is known as the *finesse* F of the filter and is given by

$$F = \frac{\pi\sqrt{R}}{1-R} \tag{9.11}$$

In practice one uses a multilayer filter consisting of a stack of several dielectric thin films separated by cavities, as shown in Fig. 9.9. The layers typically

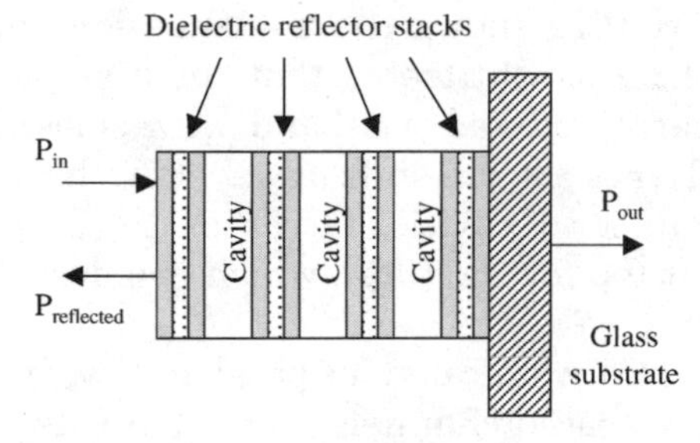

Figure 9.9. A multilayer filter consists of a stack of several dielectric thin films separated by cavities.

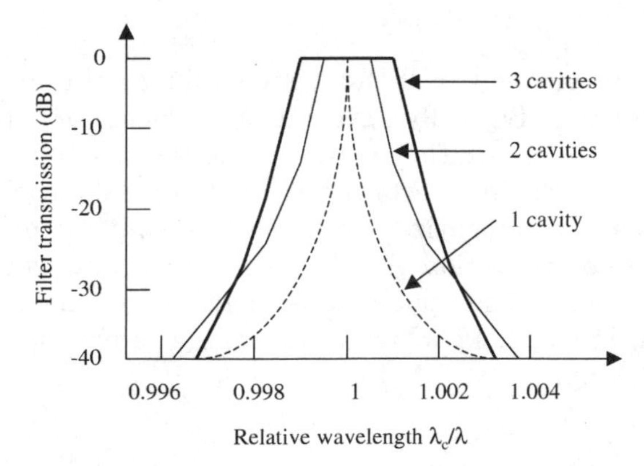

Figure 9.10. Transmission characteristics of three different thin-film filter structures relative to the peak wavelength.

TABLE 9.5. Typical Parameter Values of Commercially Available 50-GHz Thin-Film Filters

Parameter	Unit	Value
Channel passband	GHz	> ±10 at 0.5 dB
Insertion loss at f_c ±10 GHz	dB	<3.5
Polarization-dependent loss	dB	<0.20
Isolation, adjacent channels	dB	>25
Isolation, nonadjacent channels	dB	>40
Optical return loss	dB	>45
Polarization mode dispersion	ps	<0.2
Chromatic dispersion	ps/nm	<50

are deposited on a glass substrate. Each dielectric layer acts as a nonabsorbing reflecting surface, so that the structure is that of a series of cavities each of which is surrounded by mirrors. Figure 9.10 shows the transmission characteristic relative to the peak wavelength. This illustrates that for a single cavity the transmission function has a sharply peaked passband with sides that roll off smoothly. This is not very useful since a small shift in wavelength results in rapidly changing filtering. As the number of cavities increases, the passband of the filter sharpens up to create a flat top for the filter, which is a desirable characteristic for a practical filter.

Thin-film filters are available in a wide range of passbands varying from 50 to 800 GHz and higher for widely spaced channels. Table 9.5 lists some operational characteristics of commercially available 50-GHz multilayer dielectric thin-film filters for use in fiber optic communication systems.

9.4. Gratings

A grating is an important element in WDM systems for combining and separating individual wavelengths. Basically a *grating* is a periodic structure or perturbation in a material. This variation in the material has the property of reflecting or transmitting light in a certain direction depending on the wavelength. Thus gratings can be categorized as either transmitting or reflecting. Here we will concentrate on reflection gratings, since these are widely used in optical fiber communications. The applications of these gratings will be discussed in greater detail in Chap. 12 when we examine the principles of wavelength division multiplexing.

9.4.1. Grating principle

Figure 9.11 defines key parameters for a reflection grating. Here θ_i is the incident angle of the light, θ_d is the diffracted angle, and Λ (lambda) is the *period*

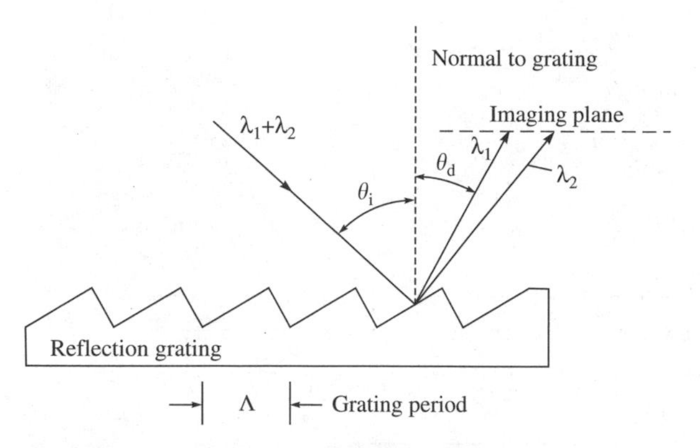

Figure 9.11. Basic parameters in a reflection grating.

of the grating (the periodicity of the structural variation in the material). In a transmission grating consisting of a series of equally spaced slits, the spacing between two adjacent slits is called the *pitch* of the grating. Constructive interference at a wavelength λ occurs in the imaging plane when the rays diffracted at the angle θ_d satisfy the *grating equation*, given by

$$\Lambda(\sin\theta_i - \sin\theta_d) = m\lambda \tag{9.12}$$

Here m is called the *order* of the grating. In general, only the first-order diffraction condition $m = 1$ is considered. (Note that in some texts the incidence and refraction angles are defined as being measured from the same side of the normal to the grating. In this case, the sign in front of the term $\sin\theta_d$ changes.) A grating can separate individual wavelengths since the grating equation is satisfied at different points in the imaging plane for different wavelengths.

9.4.2. Fiber Bragg gratings

Devices called *Bragg gratings* are used extensively for functions such as dispersion compensation, stabilizing laser diodes, and add/drop multiplexing in optical fiber systems. One embodiment is to create a *fiber Bragg grating* (FBG) in an optical fiber. This can be done by using two ultraviolet light beams to set up a periodic interference pattern in a section of the core of a germania-doped silica fiber. Since this material is sensitive to ultraviolet light, the interference pattern induces a permanent periodic variation in the core refractive index along the direction of light propagation. This index variation is illustrated in Fig. 9.12, where n_1 is the refractive index of the core of the fiber, n_2 is the index of the cladding, and Λ is the period of the grating. If an incident optical wave at λ_0 encounters a periodic variation in refractive index along the direction of propagation, λ_0 will be reflected if the following condition is met: $\lambda_0 = 2n_{\text{eff}}\Lambda$, where n effective (n_{eff}) is the average weighting of the two indices of refraction n_1 and n_2. When a specific wavelength λ_0 meets this condition, that wavelength will get reflected and all others will pass through.

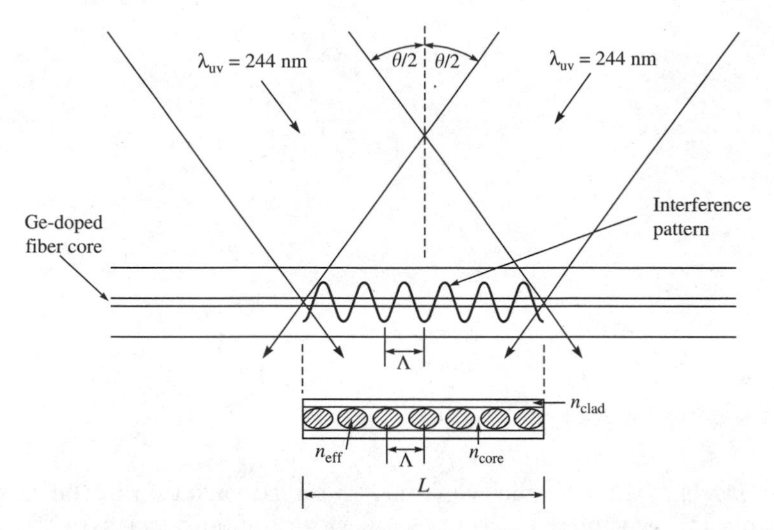

Figure 9.12. Formation of a Bragg grating in a fiber core by means of two intersecting ultraviolet light beams.

TABLE 9.6. Typical Parameter Values of Commercially Available Fiber Bragg Gratings

Parameter	Typical values		
Channel spacing	25 GHz	50 GHz	100 GHz
Reflection bandwidth	>0.08 nm @ −0.5 dB	>0.15 nm @ −0.5 dB	>0.3 nm @ −0.5 dB
	<0.2 nm @ −3 dB	<0.4 nm @ −3 dB	<0.75 nm @ −3 dB
	<0.25 nm @ −25 dB	<0.5 nm @ −25 dB	<1 nm @ −25 dB
Transmission bandwidth	>0.05 nm @ −25 dB	>0.1 nm @ −25 dB	>0.2 nm @ −25 dB
Adjacent channel isolation	>30 dB		
Insertion loss	<0.25 dB		
Central λ tolerance	<±0.05 nm @ 25°C		
Thermal λ drift	<1 pm/°C (for an athermal design)		
Package size	5 mm (diameter) × 80 mm (length)		

Fiber Bragg gratings are available in a wide range of reflection bandwidths from 25 GHz and higher. Table 9.6 lists some operational characteristics of commercially available 25-, 50-, and 100-GHz fiber Bragg gratings for use in optical communication systems.

Figure 9.13 illustrates the meanings of the *adjacent channel isolation* and the *reflection bandwidth* parameters listed in Table 9.6. Applications of fiber Bragg gratings can be found in WDM systems (see Chap. 12) and in dispersion compensation techniques (see Chap. 15).

In the FBG illustrated in Fig. 9.12, the grating spacing is uniform along its length. It is also possible to have the spacing vary along the length of the fiber,

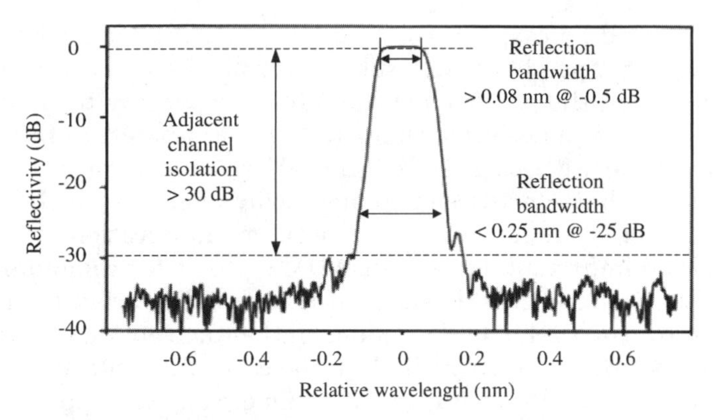

Figure 9.13. Illustration of the meanings of several filter parameters given in Table 9.6.

which means that a range of different wavelengths will be reflected by the FBG. This is the basis of what is known as a *chirped grating*. Section 15.2 describes the construction and application of such a device in greater detail.

9.5. Summary

Passive components include optical couplers, isolators, circulators, filters, gratings, and wavelength multiplexers. Optical couplers perform functions such as splitting a light signal into two or more streams, combining two or more light streams, tapping off a small portion of optical power for monitoring purposes, or transferring a selective range of optical power from one fiber to another. In general, an $N \times M$ coupler has $N \geq 2$ input ports and $M \geq 2$ output ports. For example, a device with two inputs and two outputs is called a 2×2 coupler.

Optical isolators are devices that allow light to pass through them in only one direction. This is important in a number of instances to prevent scattered or reflected light from traveling in the reverse direction. One common application of an optical isolator is to keep such light from entering a laser diode and possibly causing instabilities in the optical output. Table 9.3 lists some typical parameter values of commercially available isolators.

An *optical circulator* is a nonreciprocal multiport passive device that directs light sequentially from port to port in only one direction. The operation of a circulator is similar to that of an isolator except that its construction is more complex. A variety of circulators are available commercially. These devices have low insertion loss, high isolation over a wide wavelength range, minimal polarization-dependent loss (PDL), and low polarization mode dispersion (PMD). Table 9.4 lists some operational characteristics of commercially available circulators.

A dielectric thin-film filter (TFF) is used as an optical bandpass filter. This means that it allows a particular very narrow wavelength band to pass straight through it and reflects all others. The basis of these devices is a classical Fabry-Perot filter structure. In practice a TFF consists of a stack of several dielectric thin films separated by cavities. For a single cavity the transmission function has a sharply peaked passband with sides that roll off smoothly. As the number of cavities increases, the passband of the filter sharpens up to create a flat top for the filter, which is a desirable characteristic for a practical filter. Thin-film filters are available with passbands from 50 GHz and higher. Table 9.5 lists some operational characteristics of commercially available 50-GHz multilayer dielectric thin-film filters for use in fiber optic communication systems.

A grating is an important element in WDM systems for combining and separating individual wavelengths. Basically it is a periodic variation in a material, which reflects transmitted light depending on the wavelength. One embodiment is to create a fiber Bragg grating (FBG) in the core of an optical fiber. Table 9.6 lists some operational characteristics of commercially available 25-, 50-, and 100-GHz fiber Bragg gratings for use in optical communication systems.

Further Reading

1. J. J. Pan, F. Q. Zhou, and M. Zhou, "Thin films improve 50-GHz DWDM devices," *Laser Focus World*, vol. 38, pp. 111–116, May 2002 (www.laserfocusworld.com).
2. F. Chatain, "Fiber Bragg grating technology passes light to new passive components," *Lightwave*, vol. 18, pp. 186, 190–191, March 2001 (www.light-wave.com).
3. C. Zhou, P. Chan, J. Yian, and P. Kung, "Fiber Bragg gratings stretch metro applications," *WDM Solutions*, vol. 4, pp. 41–43, May 2002 (www.wdm-solutions.com).
4. S. DeMange, "Thin-film filters give flexibility to OADMs," *WDM Solutions*, vol. 4, pp. 31–34, June 2002 (www.wdm-solutions.com).

Chapter

10

Active Optical Components

Active components require some type of external energy either to perform their functions or to be used over a wider operating range than a passive device, thereby offering greater application flexibility. In that sense, optical sources, external modulators, and optical amplifiers can be considered as falling into the broad area of active devices. However, these are examined in separate chapters since they constitute major elements in an optical link. In addition, to get a full appreciation of the functions and applications of a particular active device, one needs a detailed discussion of how it is used in its intended system place. Therefore, in a number of cases we will only briefly look at the characteristics of a certain component and will defer a detailed description of its functions to a later chapter. Section 10.1 specifies which devices fall into this category.

The active devices described in this chapter include variable optical attenuators, tunable optical filters, dynamic gain equalizers, optical add/drop multiplexers, polarization controllers, and dispersion compensators. Many types of active optical components are based on using microelectromechanical systems (MEMS) technology. This is the topic of Sec. 10.2. The remainder of the chapter describes various active devices. Sometimes one can purchase a module that contains many different components but which has one particular function (e.g., dispersion compensators, add/drop multiplexers). These will be treated as active components for the purposes of this chapter.

10.1. Overview of Major Components

Before diving into device details, we first take an introductory look at various types and categories of active components to get an overview of the different functions they perform. In some cases the device characteristics are examined in either previous or forthcoming chapters. In other cases the devices will only be highlighted briefly in this chapter and will be described in greater detail in later chapters when examining their application.

- *Tunable optical sources*, which are described in Chap. 6, allow their emission wavelength to be tuned precisely to a particular optical frequency (or, equivalently, to a particular wavelength) by some external control mechanism.
- *Wavelength lockers* are important devices in WDM systems to maintain the output from a laser diode at a predefined ITU-T frequency with a precision of ± 1 GHz (or 8 pm). More details on this device are given in Chap. 12 on WDM.
- *External modulators* are described in Chap. 6. Such a device can be in the form of a separate external package, or it can be integrated into the laser diode package. These components allow the optical output to be modulated external to the light source at rates greater than 2.5 Gbps without significant distortion.
- *Photodetectors* are described in Chap. 7. A photodetector acts upon an optical signal by sensing the light signal falling on it and converting the variation of the optical power to a correspondingly varying electric current.
- *Optical amplifiers* are described in Chap. 11. These devices operate completely in the optical domain to boost the power level of optical signals. They work over a broad spectral range and boost the amplitudes of independent signals at all wavelengths in this band simultaneously. The fundamental optical amplification mechanisms are based on semiconductor devices, erbium-doped optical fibers, and the Raman effect in standard transmission fibers.
- *Variable optical attenuators* (VOAs) are used in multiple-wavelength links to adjust the power levels of individual wavelengths so that they closely have the same value. This chapter describes the construction and operation of VOAs.
- *Tunable optical filters* are key elements in a WDM system where one needs the flexibility to be able to select a specific wavelength for data receipt or performance monitoring. This chapter describes the construction and operation of tunable optical filters. Chapter 12 on WDM presents further details concerning the applications of these tunable filters.
- *Dynamic gain equalizers*, also called *dynamic channel equalizers* or *dynamic spectral equalizers*, provide dynamic gain equalization or blocking of individual channels across a given spectral band within a link in a WDM system. This chapter describes the construction and operation of some representative devices.
- *Optical add/drop multiplexers* (OADM) can be passive or active devices. Their function is to add or drop one or more selected wavelengths at a designated point in an optical network. This chapter describes an active OADM, and Chap. 17 describes switching applications in a network.
- *Polarization controllers* offer high-speed real-time polarization control in a closed-loop system that includes a polarization sensor and control logic. These devices dynamically adjust any incoming state of polarization to an arbitrary output state of polarization. This chapter describes their construction and operation.

- *Chromatic dispersion compensators* optically restore signals that have become degraded by chromatic dispersion, thereby significantly reducing bit error rates at the receiving end of a fiber span. This chapter describes the construction and operation of one representative device type. Chapter 15 illustrates further applications.
- *Optical performance monitors* track optical power, wavelength, and optical signal-to-noise ratio to check operational performance trends of a large number of optical channels and to identify impending failures. Chapter 18 looks at these devices in greater detail within the discipline of system maintenance and control.
- *Optical switches* that work completely in the optical domain have a variety of applications in optical networks, including optical add/drop multiplexing, optical cross-connects, dynamic traffic capacity provisioning, and test equipment. Chapter 17 looks at switching applications in greater detail.
- *Wavelength converters* are used in WDM networks to transform data from one incoming wavelength to a different outgoing wavelength without any intermediate optical-to-electrical conversion. Chapter 11 on optical amplifiers describes devices and techniques for doing this.

10.2. MEMS Technology

MEMS is the popular acronym for *microelectromechanical systems*. These are miniature devices that combine mechanical, electrical, and optical components to provide sensing and actuation functions. MEMS devices are fabricated using integrated-circuit compatible batch-processing techniques and range in size from micrometers to millimeters. The control or actuation of a MEMS device is done through electrical, thermal, or magnetic means such as microgears or movable levers, shutters, or mirrors. The devices are used widely for automobile air bag deployment systems, in ink-jet printer heads, for monitoring mechanical shock and vibration during transportation of sensitive goods, for monitoring the condition of moving machinery for preventive maintenance, and in biomedical applications, for patient activity monitoring and pacemakers.

Figure 10.1 shows a simple example of a MEMS *actuation method*. At the top of the device there is a thin suspended polysilicon beam that has typical length, width, and thickness dimensions of 80, 10, and 0.5 μm, respectively. At the bottom there is a silicon ground plane which is covered by an insulator material. There is a gap of nominally 0.6 μm between the beam and the insulator. When a voltage is applied between the silicon ground plane and the polysilicon beam, the electric force pulls the beam down so that it makes contact with the lower structure.

MEMS technologies also are finding applications in lightwave systems for variable optical attenuators, tunable optical filters, tunable lasers, optical add/drop multiplexers, optical performance monitors, dynamic gain equalizers, optical switches, and other optical components and modules. For example, Fig. 10.2 illustrates the use of MEMS technology to make a tunable VCSEL. This is

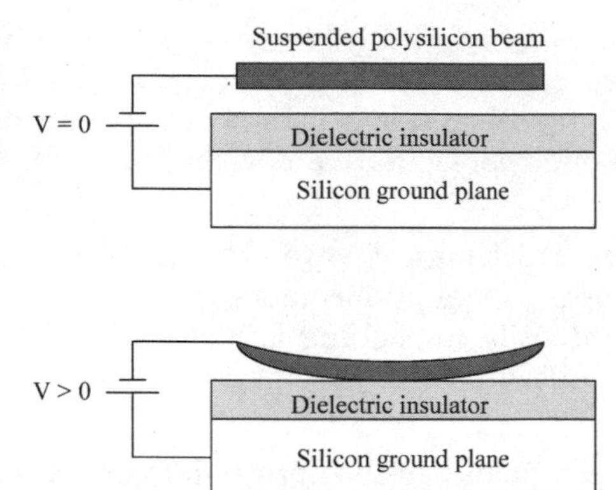

Figure 10.1. A simple example of a MEMS actuation method. The top shows an off position, and the bottom shows an on position.

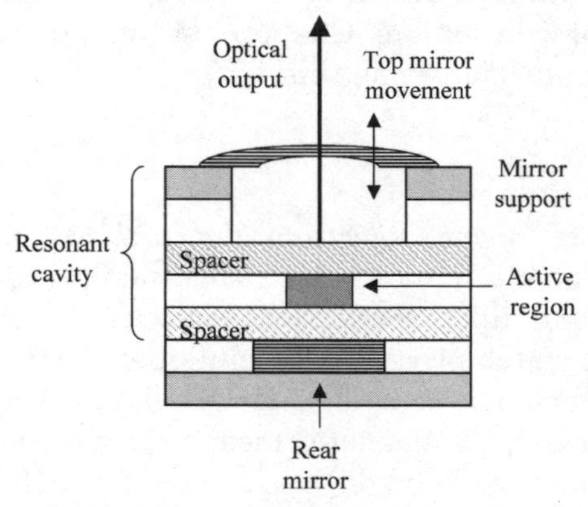

Figure 10.2. The use of MEMS technology to make a tunable VCSEL.

a slight variation on the VCSEL structure illustrated in Fig. 6.12. In the configuration shown in Fig. 10.2, the top curved external mirror membrane can be moved up or down. This action changes the length of the lasing cavity and hence changes the wavelength of the light emitted by the laser.

Initially MEMS devices were based on standard silicon technology, which is a very stiff material. Since some type of electric force typically is used to bend or deflect one of the MEMS layers to produce the desired mechanical motion, stiffer materials require higher voltages to achieve a given mechanical deflection. To reduce these required forces, recently MEMS devices started being made with highly compliant polymeric materials that are as much as 6 orders of magnitude less stiff than silicon. This class of components is referred to as *compliant MEMS*, or CMEMS. This technology employs a soft, rubberlike material called an *elastomer* (from the words *elastic* and *polymer*). Elastomeric material can be stretched as much as 300 percent, as opposed to less than 1 percent for silicon. As a result, CMEMS devices require much lower voltages to achieve a given mechanical deflection, and for equivalent voltages their mechanical range of motion is much larger than that with silicon MEMS.

Depending on the technology used, MEMS devices may have some reliability issues if the manufacturer does not follow strict fabrication procedures. One major failure problem is *stiction*, which is the tendency of two silicon surfaces to stick to each other permanently. A second concern is how to keep contaminants such as dust and saw-generated particles away from MEMS structures during *singulation*, which is the process of cutting up a large fabricated wafer into individual MEMS devices. A third issue involves packaging the devices so that no liquid, vapor, particles, or other contaminants are present to cause malfunctions in the moving parts of the MEMS device.

10.3. Variable Optical Attenuators

Precise active signal-level control is essential for proper operation of DWDM networks. For example, all wavelength channels exiting an optical amplifier need to have the same gain level, certain channels may need to be blanked out to perform network monitoring, span balancing may be needed to ensure that all signal strengths at a user location are the same, and signal attenuation may be needed at the receiver to prevent photodetector saturation. A *variable optical attenuator* (VOA) offers such dynamic signal-level control. This device attenuates optical power by various means to control signal levels precisely without disturbing other properties of a light signal. That means a VOA should be polarization-independent, attenuate light independently of the wavelength, and have a low insertion loss. In addition, it should have a dynamic range of 15 to 30 dB (a control factor ranging from 30 to 1000).

The control methods include mechanical, thermooptic, MEMS, or electrooptic techniques. The mechanical control methods are reliable but have a low dynamic range and a slow response time. Thermooptic methods have a high dynamic range, but are slow and require the use of a thermoelectric cooler (TEC), which may not be desirable. The two most popular control methods are MEMS-based and electrooptic-based techniques. For MEMS techniques an electrostatic actuation method is the most common and well developed, since integrated-circuit processes offer a wide selection of conductive and insulating materials. In this method a voltage change across a pair of electrodes provides an electrostatic actuation force. This requires lower power levels than other methods and is the fastest.

When wavelengths are added, dropped, or routed in a WDM system, a VOA can manage the optical power fluctuations of this and other simultaneously propagating wavelength signals. Figure 10.3 shows a typical VOA package. Large power handling capabilities are possible with bigger package sizes. Table 10.1 shows some representative operational parameter values for a VOA.

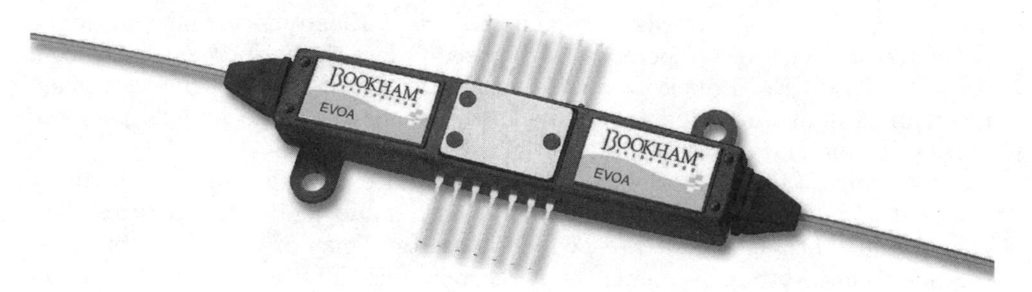

Figure 10.3. Typical VOA package for moderate optical power handling capabilities. (*Photo courtesy of Bookham Technology; www.bookham.com.*)

TABLE 10.1. Some Representative Operational Parameter Values for a Typical VOA

Parameter	Specification
Insertion loss	<1.8 dB
Attenuation range	>25 dB (up to 60 dB possible)
PDL @ 25-dB attenuation	<0.3 dB
Maximum optical power per channel	>150 mW (up to 500 mW possible)
Optical return loss	>42 dB
Insertion loss	<0.7 dB

10.4. Tunable Optical Filters

Tunable optical filters are key components for dense WDM optical networks. Several different technologies can be used to make a tunable filter. Two key ones are MEMS-based and Bragg-grating-based devices. MEMS actuated filters have the advantageous characteristics of a wide tuning range and design flexibility.

Fiber Bragg gratings are wavelength-selective reflective filters with steep spectral profiles and flat tops. This is shown in Fig. 10.4 for a 100-GHz filter that has a reflection bandwidth of less than 0.8 nm at 25 dB below the peak. Standard uniformly spaced fiber gratings have large sidelobes, which lie only about 9 dB below the central peak. However, these sidelobes can be reduced to be at least 30 dB below the peak by using a special apodization mask to precisely control the UV beam shape in the Bragg grating fabrication process. *Apodization* is a mathematical technique used to reduce ringing in an interference pattern, thus giving a large central waveform peak with low sidelobes.

As Sec. 9.4 describes, one way of creating a fiber grating is by using ultraviolet light to set up a periodic interference pattern in a section of the core of a germania-doped silica fiber. This pattern induces a permanent periodic variation in the refractive index along the core. Tunable optical filters based on fiber Bragg gratings involve a stretching and relaxation process of the fiber grating. Since glass is a slightly stretchable medium, as an optical fiber is stretched with the

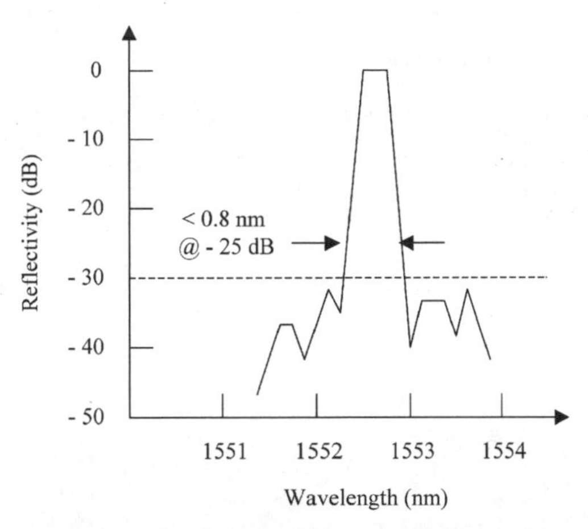

Figure 10.4. Fiber Bragg gratings are wavelength-selective reflective filters with steep spectral profiles.

grating inside of it, the spacing of the index perturbations and the refractive index will change. This process induces a change in the Bragg wavelength, thereby changing the center wavelength of the filter. Such optical filters can be made for the S-, C-, and L-bands and for operation in the 1310-nm region.

Tunable Grating Technology Before it is stretched, the center wavelength λ_c of a fiber Bragg grating filter is given by $\lambda_c = 2n_{\text{eff}}\Lambda$, where n_{eff} is the effective index of the fiber containing the grating and Λ (lambda) is the period of the index variation of the grating. When the fiber grating is elongated by a distance $\Delta\Lambda$, the corresponding change in the center wavelength is $\Delta\lambda_c = 2n_{\text{eff}}\,\Delta\Lambda$.

The stretching can be done by thermomechanical, piezoelectric, or stepper-motor means, as shown in Fig. 10.5. The thermomechanical methods might use a bimetal differential-expansion element which changes its shape as its temperature varies. In the figure the high-expansion bar changes its length more with temperature than the low-expansion frame, thereby leading to temperature-induced length variations in the fiber grating. This method is inexpensive, but it is slow, takes time to stabilize, and has a limited tuning range. The piezoelectric technique uses a material that changes its length when a voltage is applied (the *piezoelectric effect*). This method provides precise wavelength resolution but is more expensive, complex to implement, and has a limited tuning range. The stepper-motor method changes the length of the fiber grating by pulling or relaxing one end of the structure. It has a moderate cost, is reliable, and has a reasonable tuning speed.

One tunable filter version is a tunable variation on the classical structure that has been used widely for interferometer applications. The device consists of two sets of epitaxial layers that form a single Fabry-Perot cavity. Its operation is based on allowing one of the two mirrors to be moved precisely by an actuator. This enables a variation of the distance between the two cavity mirrors, thereby resulting in the selection of different wavelengths to be filtered. This

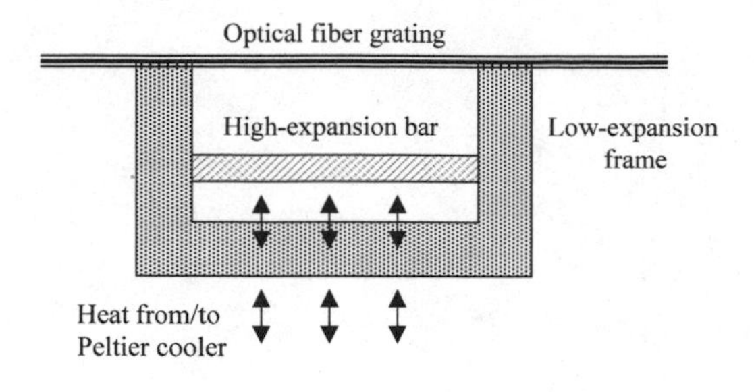

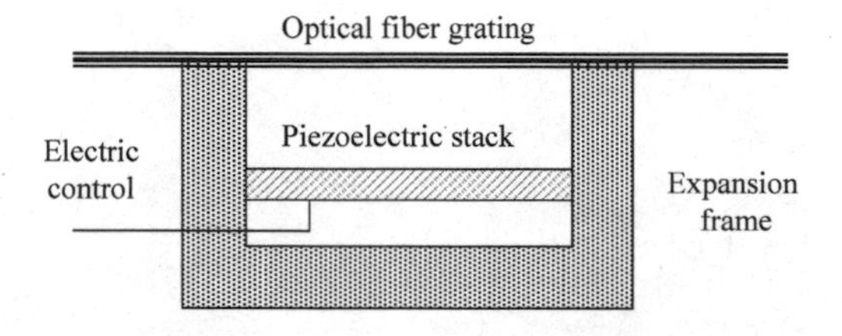

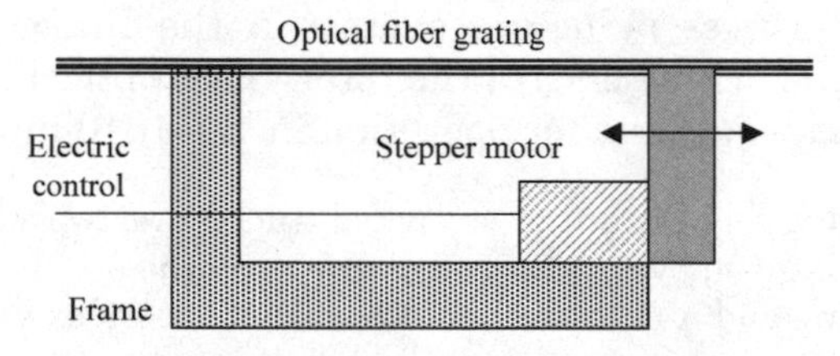

Figure 10.5. Three methods for adjusting the wavelength of a tunable Bragg grating.

distance variation can be achieved by any of the three methods described above for tunable gratings.

Table 10.2 lists some representative performance parameters of a tunable optical filter (see Sec. 9.4 for definitions of some of the filter terms). Applications of these devices include gain-tilt monitoring in optical fiber amplifiers, optical performance monitoring in central offices, channel selection at the receive side of a WDM link, and suppression of amplified spontaneous emission (ASE) noise in optical amplifiers (see Chap. 11).

10.5. Dynamic Gain Equalizers

A *dynamic gain equalizer* (DGE) is used to reduce the attenuation of the individual wavelengths within a spectral band. These devices also are called *dynamic channel equalizers* (DCEs) or *dynamic spectral equalizers*. The function of a DGE is equivalent to filtering out individual wavelengths and equalizing them on a channel-by-channel basis. Their applications include flattening the nonlinear

TABLE 10.2. Some Typical Performance Parameters of a Tunable Optical Filter

Parameter	Specification
Tuning range	100 nm typical
Free spectral range (FSR)	150 nm typical
Channel selectivity	100, 50, and 25 GHz
Bandwidth	<0.2 nm
Insertion loss	<3 dB across tuning range
Polarization-dependent loss (PDL)	<0.2 dB across tuning range
Tuning speed	10 nm/μs in both C- and L-bands
Tuning voltage	40 V

gain profile of an optical amplifier (such as an EDFA or the Raman amplifier described in Chap. 11), compensation for variation in transmission losses on individual channels across a given spectral band within a link, and attenuating, adding, or dropping selective wavelengths. For example, the gain profile across a spectral band containing many wavelengths usually changes and needs to be equalized when one of the wavelengths is suddenly added or dropped on a WDM link. Note that certain vendors distinguish between a DGE for flattening the output of an optical amplifier and a DCE which is used for channel equalization or add/drop functions. Depending on the application, certain operational parameters such as the channel attenuation range may be different.

These devices operate by having individually tunable attenuators, such as a series of VOAs, control the gain of a small spectral segment across a wide spectral band, such as the C- or L-band. For example, within a 4-THz spectral range (around 32 nm in the C-band) a DGE can individually attenuate the optical power of 40 channels spaced at 100 GHz or 80 channels spaced at 50 GHz. The operation of these devices can be controlled electronically and configured by software residing in a microprocessor. This control is based on feedback information received from a performance-monitoring card that provides the parameter values needed to adjust and adapt to required link specifications. This allows a high degree of agility in responding to optical power fluctuations that may result from changing network conditions.

Figure 10.6 shows an example of how a DGE equalizes the gain profile of an erbium-doped fiber amplifier (EDFA).

10.6. Optical Add/Drop Multiplexers

The function of an *optical add/drop multiplexer* (OADM) is to insert (add) or extract (drop) one or more selected wavelengths at a designated point in an optical network. Figure 10.7 shows a simple OADM configuration that has four input and four output ports. Here the add and drop functions are controlled by MEMS-based miniature mirrors that are activated selectively to connect the

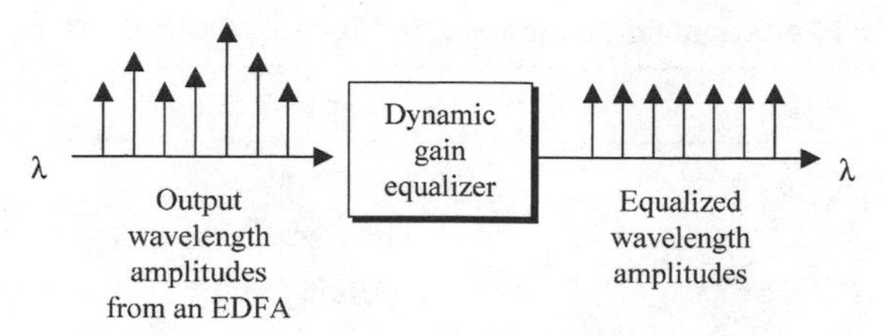

Figure 10.6. Example of how a DGE equalizes the gain profile of an erbium-doped fiber amplifier.

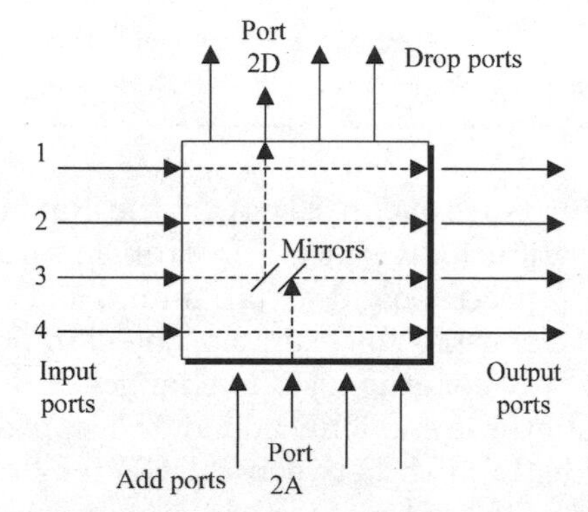

Figure 10.7. Example of adding and dropping wavelengths with a 4×4 OADM device that uses miniature switching mirrors.

desired fiber paths. When no mirrors are activated, each incoming channel passes through the switch to the output port. Incoming signals can be dropped from the traffic flow by activating the appropriate mirror pair. For example, to have the signal carried on wavelength λ_3 entering port 3 dropped to port 2D, the mirrors are activated as shown in Fig. 10.7. When an optical signal is dropped, another path is established simultaneously, allowing a new signal to be added from port 2A to the traffic flow.

There are many variations on optical add/drop device configurations depending on the switching technology used. However, in each case the operation is independent of wavelength, data rate, and signal format.

10.7. Polarization Controllers

Polarization controllers offer high-speed real-time polarization control in a closed-loop system that includes a polarization sensor and control logic. These devices dynamically adjust any incoming state of polarization to an arbitrary output state of polarization. For example, the output could be a fixed, linearly polarized state. Nominally this is done through electronic control voltages that are applied independently to adjustable polarization retardation plates.

Applications of polarization controllers include polarization mode dispersion (PMD) compensation, polarization scrambling, and polarization multiplexing.

10.8. Chromatic Dispersion Compensators

A critical factor in optical links operating above 2.5 Gbps is compensating for chromatic dispersion effects. This phenomenon causes pulse broadening which leads to increased bit error rates. An effective means of meeting the strict narrow dispersion tolerances for such high-speed networks is to start with a first-order dispersion management method, such as a dispersion-compensating fiber (see Chap. 15), that operates across a wide spectral range. Then fine-tuning can be carried out by means of a tunable dispersion compensator that works over a narrow spectral band to correct for any residual and variable dispersion.

The device for achieving this fine-tuning is referred to as a *dispersion-compensating module* (DCM). Similar to many other devices, this module can be tuned manually, remotely, or dynamically. *Manual tuning* is done by a network technician prior to or after installation of the module in telecommunications racks. By using network management software it can be adjusted *remotely* from a central management console by a network operator, if this feature is included in its design. *Dynamic tuning* is done by the module itself without any human intervention.

One method of achieving dynamic chromatic dispersion is through the use of a chirped fiber Bragg grating (FBG), as shown in Fig. 10.8. Here the grating spacing varies linearly over the length of the grating, which creates what is known as a *chirped grating*. This results in a range of wavelengths that satisfy the Bragg condition for reflection. In the configuration shown, the spacings decrease along the fiber which means that the Bragg wavelength decreases with distance along the grating length. Consequently, the shorter-wavelength components of a pulse travel farther into the fiber before being reflected. Thereby they experience greater delay in going through the grating than the longer-wavelength components. The relative delays induced by the grating on the different-frequency components of the pulse are the opposite of the delays caused by the fiber. This results in dispersion compensation, since it compresses the pulse. Section 15.2.2 has more details on these devices.

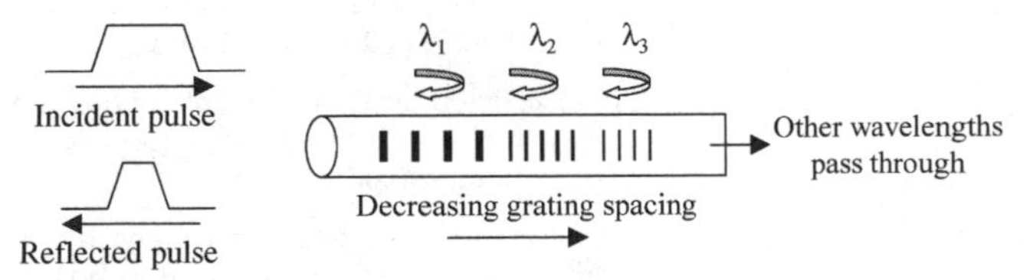

Figure 10.8. Dynamic chromatic dispersion compensation is accomplished through the use of a chirped fiber Bragg grating.

10.9. Summary

Dynamically tunable devices are essential for high-performance WDM networks. Their functions include tunable optical sources, variable optical attenuators, optical switches, optical filters, dynamic gain equalizers, add/drop multiplexers, dispersion-compensating modules, polarization controllers, and optical performance monitors. Table 10.3 gives a summary of some typical tunable devices and indicates where they are described in this book.

TABLE 10.3. Summary of Some Tunable Devices, Their Functions, and in Which Chapter to Find Application Details

Tunable device	Function	Chapter
Tunable laser diode	Emits optical power at a precisely tunable frequency (wavelength) over a certain spectral band	6
Wavelength locker	Precisely maintains the output from a laser diode at a predefined ITU-T frequency	12
External modulator	Allows the optical output from a laser diode to be modulated external to the source at rates greater than 2.5 Gbps	6
Optical amplifier	Operates completely in the optical domain to boost the power level of optical signals	11
Variable optical attenuator	Used in multiple-wavelength links to adjust the power levels of individual wavelengths	10
Optical filter	Selects a specific narrow wavelength range for data receipt or performance monitoring	9,10,12
Dynamic gain equalizer	Provides dynamic gain equalization or blocking of individual channels	10
Optical add/drop multiplexer	Adds and/or drops selected wavelengths at a designated point in an optical network	10,17
Polarization controller	Dynamically adjusts any incoming state of polarization to an arbitrary output state	10
Chromatic dispersion compensator	Optically restores signals that have become degraded by chromatic dispersion	10,15
Optical performance monitor	Tracks optical power, wavelength, and optical signal-to-noise ratio to check operational performance trends	18
Optical switch	Operates completely in the optical domain to switch an incoming signal from an input to an output port	17
Wavelength converter	Transforms one incoming wavelength to a different outgoing wavelength in WDM links	11

Further Reading

1. M. R. Tremblay and M. S. Ner, "MEMS-VOA: All-optical power management," *Fiberoptic Product News*, vol. 17, pp. 36–38, May 2002 (www.fpnmag.com).
2. P. DeDobbelaere, K. Falta, L. Fan, S. Gloeckner, and S. Patra, "Digital MEMS for optical switching," *IEEE Communications Magazine*, vol. 40, pp. 88–95, March 2002.
3. D. Krakauer, "Advanced MEMS usher in all-optical networks," *Communications System Design*, pp. 32–36, June 2002.
4. N. Cockroft, "Array-based VOAs offer compact signal control," *WDM Solutions*, pp. 82–86, June 2002 (www.wdm-solutions.com).
5. S. Blackstone, "Making MEMS reliable," *OE Magazine*, vol. 2, pp. 32–34, September 2002.
6. A. Neukermans and R. Ramaswami, "MEMS technology for optical networking applications," *IEEE Communications Magazine*, vol. 39, pp. 62–69, January 2001.
7. A. Willner, "Taming Dispersion in Tomorrow's High-Speed Networks," Phaethon Communications White Paper, March 2001 (http://www.phaethoncommunications.com).
8. R. Allan, "Highly accurate dynamic gain equalizer controls optical power precisely," *Electronic Design*, pp. 37–40, February 4, 2002.

Chapter 11

Optical Amplifiers

Scattering and absorption mechanisms in an optical fiber cause a progressive attenuation of light signals as they travel along a fiber. At some point the signals need to be amplified so that the receiver can interpret them properly. Traditionally this was done by means of a regenerator that converted the optical signal to an electrical format, amplified this electric signal, and then reconverted it to an optical format for further transmission along the link. The development of an optical amplifier circumvented the time-consuming function of a regenerator by boosting the level of a light signal completely in the optical domain. Thus optical amplifiers now have become indispensable components in high-performance optical communication links.

This chapter classifies the three fundamental optical amplifier types, defines their operational characteristics, and describes their basic applications. The three basic technologies are semiconductor optical amplifiers (SOAs), doped-fiber amplifiers (DFAs), and Raman amplifiers. SOAs are based on the same operating principles as laser diodes, whereas the other two types employ a fiber as the gain mechanism. Among the DFAs, erbium-doped fiber amplifiers (EDFAs) are used widely in the C- and L-bands for optical communication networks. In contrast to an EDFA which uses a specially constructed fiber for the amplification medium, a Raman amplifier makes use of the transmission fiber itself.

This chapter first looks at SOA technology since the same operational principles apply to all types of amplifiers. These discussions include external pumping principles, gain mechanisms, noise effects, and SOA characteristics. Next we look at the operational principles and applications of EDFA devices. Following this is a discussion of Raman amplification technology and use. Finally, Sec. 11.6 illustrates how an SOA may be used as a wavelength-converting device for deployment in optical networks.

11.1. Basic Applications

Optical amplifiers have found widespread use not only in long-distance, point-to-point optical fiber links, but also in multiaccess networks (where an optical signal is divided among many users) to compensate for signal-splitting losses. The features of optical amplifiers have led to many diverse applications, each having different design challenges. Figure 11.1 shows general applications of optical amplifiers, which are as follows:

1. *In-line optical amplifiers*. In a single-mode link, the effects of fiber dispersion may be small so that the main limitation to repeater spacing is fiber attenuation. Since such a link does not necessarily require a complete regeneration of the signal, simple amplification of the optical signal at periodic locations along the transmission path is sufficient. This function is known as *in-line amplification*. Thus an optical amplifier can be used to compensate for transmission loss and to increase the distance between regenerative repeaters, as the top part of Fig. 11.1 illustrates.
2. *Preamplifier*. The center schematic in Fig. 11.1 shows an optical amplifier being used as a front-end *preamplifier* for an optical receiver. Thereby a weak optical signal is amplified ahead of the photodetection process so that the signal-to-noise ratio degradation caused by thermal noise in the receiver electronics can be suppressed. Compared with other front-end devices such as avalanche photodiodes or optical heterodyne detectors, an optical preamplifier provides a larger gain factor and a broader bandwidth.

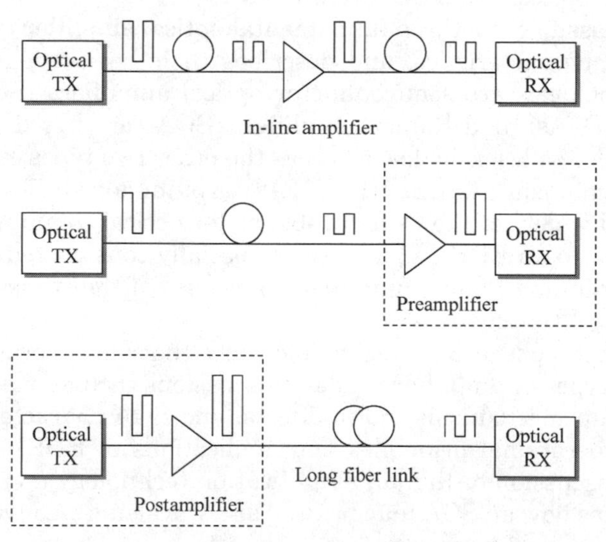

Figure 11.1. Three possible applications of optical amplifiers. (*a*) In-line amplifier to increase transmission distance; (*b*) preamplifier to improve receiver sensitivity; (*c*) postamplifier to boost transmitted power.

3. *Postamplifier*. Placing an amplification device immediately after the optical transmitter gives a boost to the light level right at the beginning of a fiber link, as the bottom schematic in Fig. 11.1 shows. This is known as a *postamplifier* (called *post* since it comes after the transmitter) and serves to increase the transmission distance by 10 to 100 km depending on the amplifier gain and fiber loss. As an example, using this boosting technique together with an optical preamplifier at the receiving end can enable continuous underwater transmission distances of 200 to 250 km.

11.2. Amplification Mechanism

All optical amplifiers increase the power level of incident light through a process of stimulated emission of radiation. Recall from Chap. 6 that *stimulated emission* occurs when some external stimulant, such as a signal photon, causes an excited electron sitting at a higher energy level to drop to the ground state. The photon emitted in this process has the same energy (i.e., the same wavelength) as the incident signal photon and is in phase with it. This means their amplitudes add to produce a brighter light. For stimulated emission to occur, there must be a *population inversion* of carriers, which means that there are more electrons in an excited state than in the ground state. Since this is not a normal condition, population inversion is achieved by supplying external energy to boost (pump) electrons to a higher energy level.

The "pumping" techniques can be optical or electrical. The basic operation is shown in Fig. 11.2. Here the device absorbs energy supplied from an external optical or electrical source called the *pump*. The pump supplies energy to electrons in an *active medium*, which raises them to higher energy levels to produce a population inversion. An incoming signal photon will trigger these excited electrons to drop to lower levels through a stimulated emission process, thereby producing an amplified signal.

One of the most important parameters of an optical amplifier is the *signal gain* or *amplifier gain* G, which is defined as

$$G = \frac{P_{\text{out}}}{P_{\text{in}}} \tag{11.1}$$

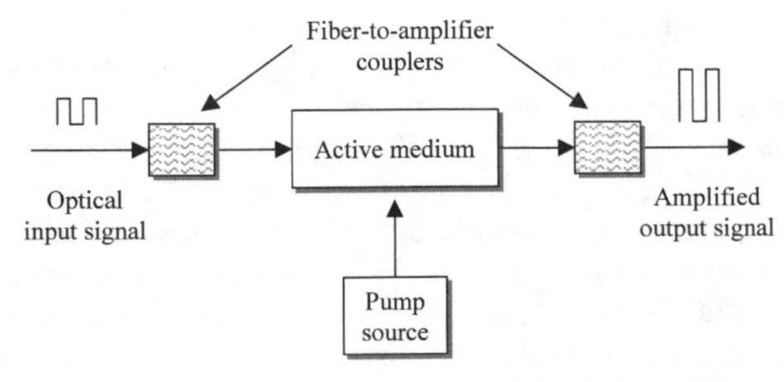

Figure 11.2. Basic operation of a generic optical amplifier.

where P_{in} and P_{out} are the input and output powers, respectively, of the optical signal being amplified. The gain generally is measured under small-signal conditions (with $P_{out} < 0$ dBm) and is expressed in decibel units as

$$G \text{ dB} = P_{out} \text{ dBm} - P_{in} \text{ dBm} \tag{11.2}$$

where the input and output powers are given in dBm units.

11.3. Semiconductor Optical Amplifiers

Semiconductor optical amplifiers (SOAs) are based on the same technology as laser diodes. In fact, an SOA is essentially an InGaAsP laser that is operating below its threshold point. The attractiveness of this is that SOAs can operate in every fiber wavelength band extending from 1280 nm in the O-band to 1650 nm in the U-band. Furthermore, since they are based on standard semiconductor technology, they can be integrated easily on the same substrate as other optical devices and circuits (e.g., couplers, optical isolators, and receiver circuits). Compared to a DFA, the SOA consumes less power, is constructed with fewer components, and can be housed compactly in a standard 14-pin butterfly package. SOAs have a more rapid gain response (on the order of 1 to 100 ps), which enables them to be used for switching and signal processing. Additional applications include optical signal conversion from one wavelength to another without having the signal enter the electrical domain.

Despite these advantages, the limitation of an SOA is that its rapid carrier response causes the gain at a particular wavelength to fluctuate with the signal rate for bit rates up to several gigabits per second. Since the gain at other wavelengths also fluctuates, this gives rise to crosstalk effects when a broad spectrum of wavelengths must be amplified. As a result, the SOA is not highly suitable for WDM applications.

11.3.1. SOA construction

Figure 11.3 shows a simple diagram of an SOA. Indium phosphide (InP) is used in an SOA for the substrate, and the active material consists of InGaAsP layers. Analogous to the construction of a laser diode, the gain wavelength can be selected between approximately 1100 and 1700 nm by varying the composition of the active InGaAsP material.

As mentioned earlier, an SOA is essentially a semiconductor laser, but without feedback from its input and output ports. Because of this feature it also is called a *traveling-wave* (TW) *amplifier*. This means that in contrast to a laser where the optical signal makes many passes through the lasing cavity, in the SOA the significant difference is that the optical signal travels through the device only once. During this passage the signal gains energy and emerges intensified at the other end of the device. Thus, since an SOA does not have the optical feedback mechanism that is necessary for lasing to take place, it can boost incoming signal levels but cannot generate a coherent optical output by itself.

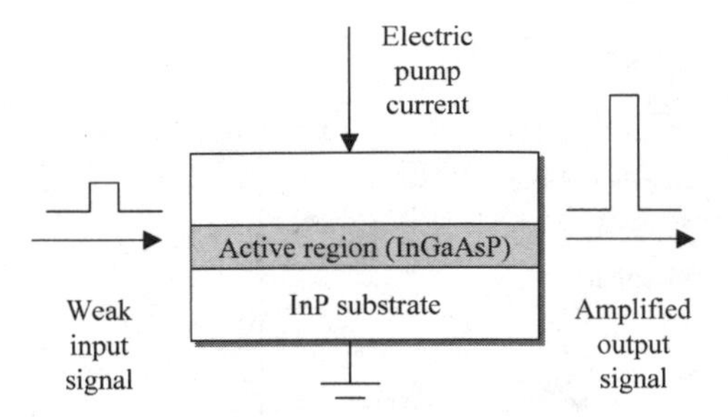

Figure 11.3. Simple diagram of a semiconductor optical amplifier (SOA).

As in the operation of laser diodes, *external current injection* is the pumping method used to create the population inversion needed for the operation of the gain mechanism in semiconductor optical amplifiers.

Whereas the end facets of a laser diode have a high reflectivity (more than 99 percent) so that the optical signal can oscillate in the lasing cavity, a facet reflectivity of less than 0.01 percent is necessary in an SOA to prevent oscillations. Typically this is achieved through a combination of antireflection coatings and angular end faces; that is, the optical waveguide in the SOA is tilted by a few degrees with respect to the facet. Any light reflected from the facet thus will propagate away from the waveguide.

11.3.2. Basic SOA parameters

The five basic parameters used to characterize SOAs are gain, gain bandwidth, saturation power, noise figure, and polarization sensitivity. The following describes the characteristics of these parameters.

- *Gain*. Figure 11.4 illustrates the dependence of the gain on the input power for three different bias conditions for a representative SOA. In the example here, at a bias current of 300 mA the zero-signal gain (or *small-signal gain*) is $G_0 = 26$ dB, which is a gain factor of about 400. The curves show that as the input signal power is increased, the gain first stays constant near the small-signal level. This flat region is called the *unsaturated region* of the SOA gain. For higher input powers above this region the gain starts to decrease. The higher power levels are still amplified, but the gain declines as the power is increased. This gain decline is caused by a reduction in carrier density due to high optical input power. After decreasing linearly over a certain range of input powers, the gain finally approaches an asymptotic value of 0 dB (a unity gain) for a very high power level.
- *Saturation power*. The region in which the gain value declines is called the *saturated gain region* of the SOA. The point at which the gain is reduced by 3 dB (a factor of 2) from the unsaturated value is called the *saturation power*. This is illustrated in Fig. 11.4 where for a 300-mA bias current the saturation

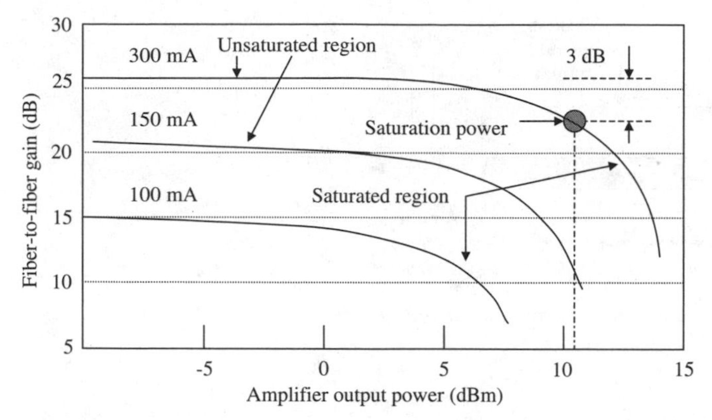

Figure 11.4. Dependence of gain on input power for three different bias conditions for a representative SOA. Definitions are illustrated for saturated and unsaturated regions and saturation power.

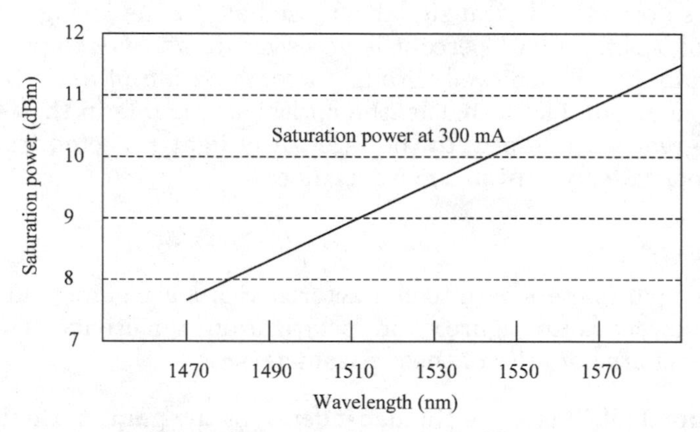

Figure 11.5. Saturation power of a particular SOA at 300 mA as a function of wavelength.

power is about 10 dBm. The saturation power varies with wavelength, as Fig. 11.5 illustrates.

- *Gain bandwidth.* The wavelength at which the SOA has a maximum gain can be tailored to occur anywhere between about 1100 and 1700 nm by changing the composition of the active InGaAsP material. As an example, Fig. 11.6 shows a typical gain versus wavelength characteristic for a device with a peak gain of 25 dB at 1530 nm. The wavelength span over which the gain decreases by less than 3 dB with respect to the maximum gain is known as the *gain bandwidth* or the *3-dB optical bandwidth*. In the example shown in Fig. 11.6, the 3-dB optical bandwidth is 85 nm. Values of up to 100 nm can be achieved.

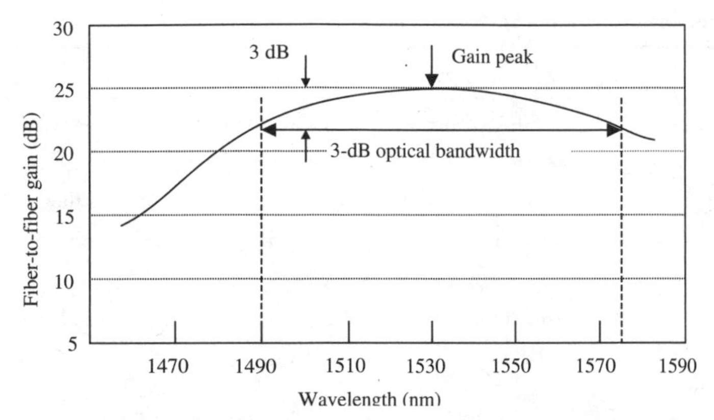

Figure 11.6. Typical gain versus wavelength characteristic for an SOA with a peak gain of 25 dB at 1530 nm. The definition of the 3-dB optical bandwidth is illustrated.

- *Noise figure.* When there is a population inversion in an optical amplifier, some of the electrons sitting in an excited state may *spontaneously* drop back down to the ground state without any external stimulus. This process generates nonsignal photons across a broad spectral range. The SOA then amplifies these photons as they continue through the device, thereby producing a background noise. This noise, which is known as *amplified spontaneous emission (ASE) noise*, degrades the signal-to-noise ratio (SNR) of the amplified signal. The *noise figure* (abbreviated as NF or F) of an SOA is a measure of this degradation. It is equal to the electrical SNR at the input divided by the electrical SNR at the output, where the SNR is measured by a photodetector, that is, $F = \mathrm{SNR_{in}}/\mathrm{SNR_{out}}$. For an ideal photodetector in which the performance is limited by shot noise only, the SNR of the amplified signal is degraded by a factor of 2 (or 3 dB), which is the theoretical lower limit of the noise figure. However, in reality the SOA exhibits a higher noise level because of low coupling efficiency on its input side. For most practical SOAs the value of F is typically in the range of 7 to 11 dB. For optical communication systems an optical amplifier should have a value of F that is as low as possible.
- *Polarization sensitivity.* Since the optical properties of semiconductor materials depend on the polarization state of an incoming light signal, the SOA gain can vary as the polarization state of the signal changes. The maximum gain change at a fixed wavelength resulting from variations in the polarization state is called the *polarization-dependent gain* (PDG). By careful design of the SOA, the polarization sensitivity can be reduced to less than 1 dB.

Table 11.1 lists typical performance values of SOA parameters for devices that can be housed in a standard 14-pin butterfly package. Since there is some coupling loss when one attaches input and output fibers to an SOA, the gain is measured from the input to the output fibers.

TABLE 11.1 Typical Performance Values for SOA Parameters

Parameter	O-band device	C-band device	Conditions
Operating peak wavelength range	1280–1340 nm	1530–1570 nm	—
3-dB bandwidth	60 nm	40 nm	—
Small-signal gain	25 dB	20 dB	Fiber-to-fiber gain
Gain ripple	<0.5 dB	<0.5 dB	Across 3-dB bandwidth
Operating current	250–300 mA	250–300 mA	—
Saturation power	10 dBm	8 dBm	250-mA drive current
Noise figure	9 dB	9 dB	−25-dBm input power
Polarization-dependent gain (PDG)	1 dB	1 dB	—

11.4. Erbium-Doped Fiber Amplifier (EDFA)

The active medium in an optical fiber amplifier consists of a nominally 10- to 30-m length of optical fiber that has been lightly doped (say, 1000 parts per million weight) with a rare-earth element, such as erbium (Er), ytterbium (Yb), neodymium (Nd), or praseodymium (Pr). The host fiber material can be standard silica, a fluoride-based glass, or a multicomponent glass. The operating regions of these devices depend on the host material and the doping elements.

The most common material for long-haul telecommunication applications is a silica fiber doped with erbium, which is known as an *erbium-doped fiber amplifier* or EDFA. Originally the operation of an EDFA by itself was limited to the C-band (1530- to 1560-nm region), since the gain coefficient for erbium atoms is high in this region. This fact actually is the origin of the designation *conventional band* or C-band. Outside of this region the erbium gain peak drops off rapidly, and in the L-band it is only 20 percent of that in the C-band. However, recent improvements in erbium-doped fiber designs and the use of high-power pump lasers operating at wavelengths that are different from those used by C-band pump lasers have allowed the extension of EDFAs into the L-band.

In addition, a combined operation of an EDFA together with Raman amplification techniques for the L-band (see Sec. 11.5) has resulted in a hybrid amplifier that can boost the gain over the 1531- to 1616-nm region with a 3-dB gain bandwidth of 75 nm.

11.4.1. Amplification mechanism

Whereas semiconductor optical amplifiers use external current injection to excite electrons to higher energy levels, optical fiber amplifiers use *optical pumping*. In this process, one uses photons to directly raise electrons into excited states.

The optical pumping process requires the use of three energy levels. The top energy level to which the electron is elevated must lie energetically above the desired lasing level. After reaching its excited state, the electron must release some of its energy and drop to the desired lasing level. From this level a signal photon can then trigger the excited electron into stimulated emission, whereby the electron releases its remaining energy in the form of a new photon with an identical wavelength as the signal photon. Since the pump photon must have a higher energy than the signal photon, the pump wavelength is shorter than the signal wavelength.

Erbium Energy Bands To get a phenomenological understanding of how an EDFA works, we need to look at the energy-level structure of erbium. The erbium atoms in silica are actually Er^{3+} ions, which are erbium atoms that have lost three of their outer electrons. In describing the transitions of the outer electrons in these ions to higher energy states, it is common to refer to the process as "raising the ions to higher energy levels." Figure 11.7 shows a simplified energy-level diagram and various energy-level transition processes of these Er^{3+} ions in silica glass. The two principal levels for telecommunication applications are a *metastable level* (the so-called $^4I_{13/2}$ level) and the $^4I_{11/2}$ *pump level*. The term *metastable* means that the lifetimes for transitions from this state to the ground state are very long compared to the lifetimes of the states that led to this level.

The metastable band is separated from the bottom of the $^4I_{15/2}$ ground-state level by an energy gap ranging from about 0.814 eV at the bottom of the band (corresponding to a 1527-nm photon) to 0.841 eV at the top of the band (corresponding to a 1477-nm photon). The energy band for the pump level exists at a 1.27-eV separation (corresponding to a 980-nm wavelength) from the ground state. The pump band is fairly narrow, so that the pump wavelength must be exact to within a few nanometers.

In normal operation, a pump laser emitting 980-nm photons is used to excite ions from the ground state to the pump level, as shown by transition process

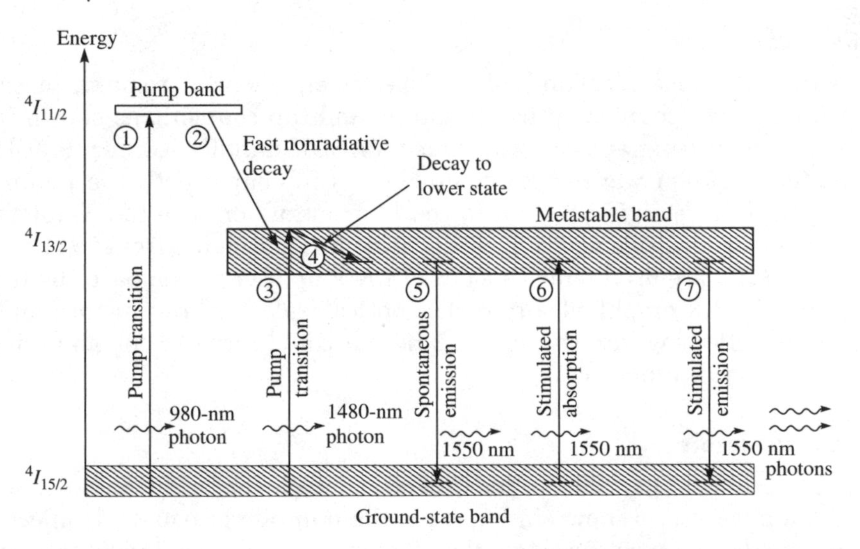

Figure 11.7. Simplified energy-level diagrams and various transition processes of Er^{3+} ions in silica.

1 in Fig. 11.7. These excited ions decay (relax) very quickly (in about 1 μs) from the pump band to the metastable band, shown as transition process 2. During this decay, the excess energy is released as phonons or, equivalently, mechanical vibrations in the fiber. Within the metastable band, the electrons of the excited ions tend to populate the lower end of the band.

Another possible pump wavelength is 1480 nm. The energy of these pump photons is very similar to the signal-photon energy, but slightly higher. The absorption of a 1480-nm pump photon excites an electron from the ground state directly to the lightly populated top of the metastable level, as indicated by transition process 3 in Fig. 11.7. These electrons then tend to move down to the more populated lower end of the metastable level (transition 4). Some of the ions sitting at the metastable level can decay back to the ground state in the absence of an externally stimulating photon flux, as shown by transition process 5. This decay phenomenon is known as *spontaneous emission* and adds to the amplifier noise.

Two more types of transitions occur when a flux of signal photons that have energies corresponding to the bandgap energy between the ground state and the metastable level passes through the device. First, a small portion of the external photons will be absorbed by ions in the ground state, which raises these ions to the metastable level, as shown by transition process 6. Second, in the stimulated emission process (transition process 7) a signal photon triggers an excited ion to drop to the ground state, thereby emitting a new photon of the same energy, wave vector (direction of travel), and polarization as the incoming signal photon. The widths of the metastable and ground-state levels allow high levels of stimulated emissions to occur in the 1530- to 1560-nm range. Beyond 1560 nm, the gain decreases steadily until it reaches 0 dB (unity gain) at 1616 nm.

11.4.2. EDFA configurations

An EDFA consists of an erbium-doped fiber, one or more pump lasers, a passive wavelength coupler, optical isolators, and tap couplers, as shown in Fig. 11.8. The *wavelength-selective coupler* (WSC) handles either 980/1550-nm or 1480/1550-nm wavelength combinations to couple both the pump and signal optical powers efficiently into the fiber amplifier. The tap couplers are wavelength-insensitive with typical splitting ratios ranging from 99 : 1 to 95 : 5. They generally are used on both sides of the amplifier to compare the incoming signal with the amplified output. The optical isolators prevent the amplified signal from reflecting into the device, where it could increase the amplifier noise and decrease its efficiency.

11.4.3. EDFA pump lasers

The erbium-doped fiber in the C-band is pumped optically by 980- and/or 1480-nm pump lasers. As shown in Fig. 11.8, the pump light usually is injected from the same direction as the signal flow. This is known as *codirectional pumping*. It is also possible to inject the pump power in the opposite direction to the signal flow,

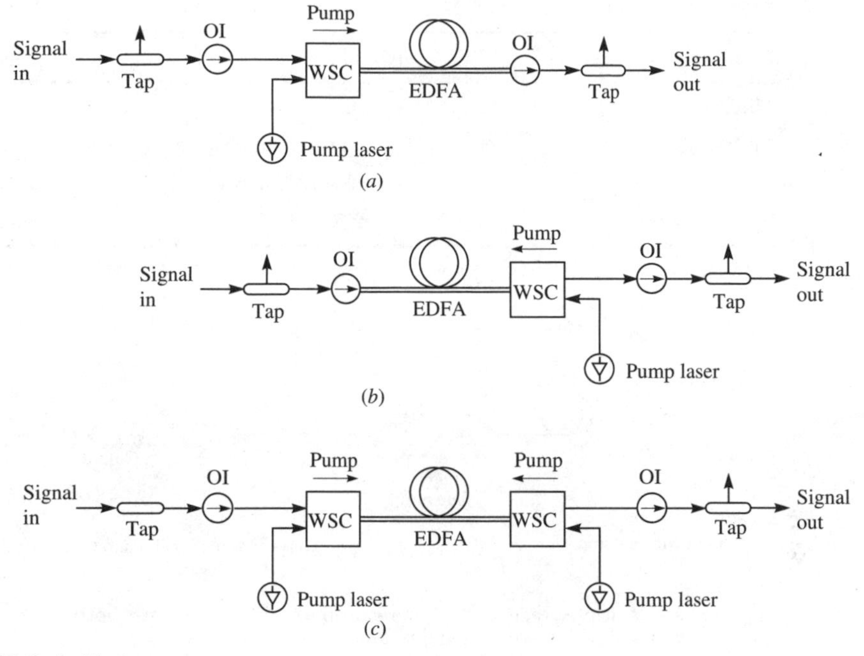

Figure 11.8. Configuration of an EDFA and possible pumping methods. (*a*) Codirectional pumping; (*b*) counterdirectoinal pumping; (*c*) dual pumping.

which is known as *counterdirectional pumping*. One can employ either a single pump source or use *dual-pump schemes*, with the resultant gains typically being +17 and +35 dB, respectively. Counterdirectional pumping allows higher gains, but codirectional pumping gives better noise performance. In addition, for operation in the C-band pumping at 980 nm is preferred, since it produces less noise and achieves larger population inversions than pumping at 1480 nm.

When a single pump is used, a copropagating 980-nm laser minimizes the EDFA noise, which is advantageous in preamplifier applications. On the other hand, use of a counterdirectional pumping 1480-nm laser optimizes the output power but results in higher noise, which is a suitable application for a booster amplifier. Table 11.2 compares the characteristic 980- and 1480-nm pump lasers.

Since the erbium gain spectrum varies by several decibels as a function of wavelength over the C-band, gain-flattening filters typically are used to attenuate wavelengths with higher gain, thereby equalizing the gain across the band. To maintain a reasonable gain in an EDFA, gain flattening generally involves the use of multiple stages of EDFA amplification combined with both codirectional and counterdirectional pumping, as shown in Fig. 11.9. Here the first stage uses a 980-nm codirectional pump whereas the second stage uses a 1480-nm counterdirectional pump laser.

TABLE 11.2. Comparison of EDFA Pump Lasers

Parameter	980-nm laser	1480-nm laser
Minimum noise figure	<4 dB	<5.5 dB
Fiber-coupled power	• 300 mW (standard)	• 250 mW (standard)
	• >500 mW (high-power)	• 310 mW (high-power)
Spectral width	5 nm @ 250 mW	8 nm @ 250 mW

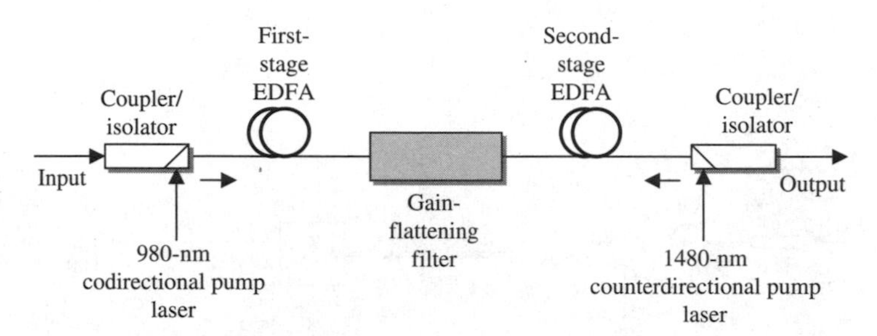

Figure 11.9. Multiple-stage EDFA plus gain-flattening filter with both codirectional and counterdirectional pumping.

Figure 11.10 illustrates the gain spectrum for this configuration with and without gain-flattening filters. Note that in a real device the illustrative curves shown in Fig. 11.10 have a slight amount of ripple in them. Table 11.3 lists some typical performance parameters for fused-fiber and thin-film filter-based couplers that combine a 980-nm pump wavelength with a C-band signal.

11.4.4. EDFA noise

As noted earlier, the dominant noise generated in an optical amplifier results from *amplified spontaneous emission* (ASE). The origin of this is the spontaneous recombination of electrons and holes in the amplifier medium. This recombination gives rise to a broad spectral background of photons that get amplified along with the optical signal. This is shown in Fig. 11.11 for a 1480-nm pump and an EDFA amplifying a signal at 1540 nm.

Receiver Noises Since ASE originates ahead of the photodiode, it gives rise to three different noise components in an optical receiver in addition to the normal thermal noise of the photodetector and the shot noise generated by the signal in the photodetector. This occurs because the photocurrent in the receiver is produced by a number of mixing components (called *beat signals*) between the signal and the optical noise fields, in addition to the currents that arise purely from the signal and the optical noise fields. Several of these terms are small compared to the main noise effect and therefore can be ignored. The following is a brief overview of the various noise effects and their magnitudes.

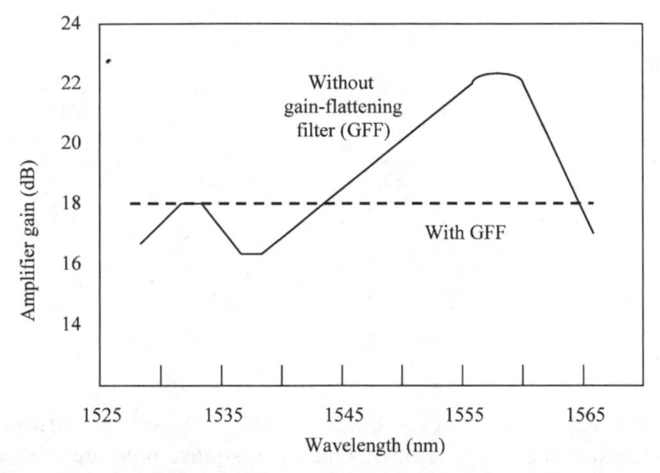

Figure 11.10. Gain spectrum for a multiple-stage EDFA configuration with and without gain-flattening filters.

TABLE 11.3. Typical Performance Values of Different Coupler Types That Combine a 980-nm Pump Wavelength with a C-Band Signal

Parameter	Specification	
Device technology	Fused fiber	Thin-film filter
Pump channel λ range	970–990 nm	965–995 nm
Pass channel λ range	1535–1565 nm	1520–1610 nm
Pump channel insertion loss	0.2 dB	0.6 dB
Pass channel insertion loss	0.2 dB	1.0 dB
Polarization-dependent loss	<0.1 dB	0.1 dB
Polarization mode dispersion	0.05 ps	0.05 ps
Optical power capability	500 mW	500 mW

- There is the shot noise in the photodetector that originates from the signal photons. This is a major noise term, and its associated mean-square noise current is given by

$$\langle i_{\text{shot}}^2 \rangle = \sigma_{\text{shot-s}}^2 = 2q\mathcal{R}GP_{\text{in}}B \tag{11.3}$$

  where q is the electron charge, $\mathcal{R}$ is the photodiode responsivity, G is the amplifier gain, P_{in} is the input power to the amplifier, and B is the front-end receiver electrical bandwidth.
- There is the extra photodetector shot noise due to the addition of ASE-generated photons to the mean optical signal power. However, by using a narrow optical filter at the receiver, most of this noise can be filtered out with the remainder being negligible.
- The thermal noise of the photodetector generally can be neglected when the amplifier gain is large enough.

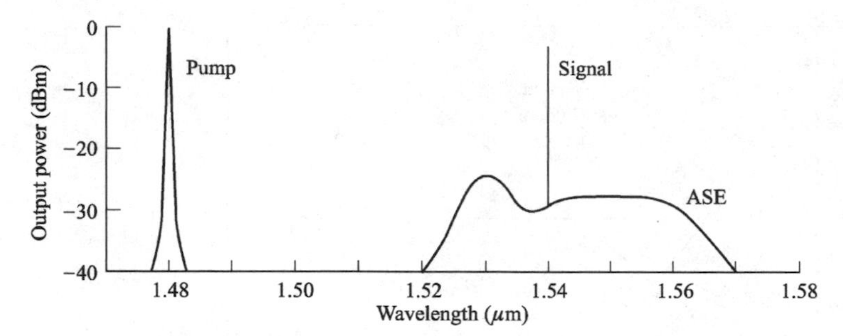

Figure 11.11. Representative 1480-nm pump spectrum and a typical output signal at 1540 nm with the associated ASE noise.

- Another noise term arises from the mixing of the different optical frequencies contained in the light signal and the ASE. The mean-square noise current for this effect is given by

$$\langle i^2_{\text{s-ASE}} \rangle = \sigma^2_{\text{s-ASE}} = 2q\mathcal{R}GP_{\text{in}}B[2\eta n_{\text{sp}}(1-G)] \tag{11.4}$$

  where η is the quantum efficiency of the photodetector and n_{sp} is the *spontaneous emission* or *population inversion factor*. Typical values for n_{sp} are around 2.
- Finally, since the ASE spans a wide optical frequency range, it can beat against itself which gives rise to a broadband ASE-ASE beat noise current. Since the amplified signal power is much larger than the ASE noise power, this term is significantly smaller than the signal-ASE beat noise and can be ignored.

Using these results then yields the following approximate signal-to-noise ratio at the photodetector output

$$\left(\frac{S}{N}\right)_{\text{out}} = \frac{\sigma^2_{\text{ph}}}{\sigma^2_{\text{total}}} = \frac{\mathcal{R}^2G^2P^2_{s,\text{in}}}{\sigma^2_{\text{total}}} \approx \frac{\mathcal{R}P_{s,\text{in}}}{2qB}\frac{G}{1+2\eta n_{\text{sp}}(G-1)} \tag{11.5}$$

From Eq. (11.5) we can then find the noise figure of the optical amplifier, which is a measure of the S/N degradation experienced by a signal after passing through the amplifier. Using the standard definition of *noise figure* as the ratio between the S/N at the input and the S/N at the amplifier output, we have

$$\text{Noise figure} = F = \frac{(S/N)_{\text{in}}}{(S/N)_{\text{out}}} = \frac{1+2\eta n_{\text{sp}}(G-1)}{G} \tag{11.6}$$

When G is large, this becomes $2\eta n_{\text{sp}}$. A perfect amplifier would have $n_{\text{sp}} = 1$, yielding a noise figure of 2 (or 3 dB), assuming $\eta = 1$. That is, using an ideal receiver with a perfect amplifier would degrade S/N by a factor of 2. In a real EDFA, for example, n_{sp} is around 2, so the input S/N gets reduced by a factor of about 4.

11.4.5. Operation in the L-band

Since the erbium ion-emission cross section is lower in the L-band, greater pump powers and longer amplification fibers with higher erbium-ion concentrations are required compared to the C-band. Whereas pumps with wavelengths of 980 and 1480 nm are used for the C-band, high-power 960-nm pump lasers are more advantageous for the L-band. These lasers are constructed from AlGaInAs and have optical output powers greater than 200 mW in uncooled devices while dissipating less than 1 W of electric power. Typical gains range from 27 dB (a factor of 500) at −25-dBm (3-μW) input powers to 20 dB (a factor of 100) at −6-dBm (250-μW) input powers. Noise figures are nominally less than 5.5 dB.

11.5. Raman Amplification

Whereas an EDFA requires a specially constructed optical fiber for its operation, a *Raman amplifier* makes use of the transmission fiber itself as the amplification medium. A Raman amplifier is based on an effect called *stimulated Raman scattering* (SRS). This effect is due to an interaction between an optical energy field and the vibration modes of the lattice structure in a material. Basically what happens here is that an atom first absorbs a photon at one energy and then releases another photon at a lower energy, that is, at a wavelength longer than that of the absorbed photon. The energy difference between the absorbed and the released photons is transformed to a *phonon*, which is a vibration mode of the material. The power transfer results in an upward wavelength shift of 80 to 100 nm, and the shift to a longer wavelength is referred to as the *Stokes shift*. Figure 11.12 illustrates the Stokes shift and the resulting Raman gain spectrum from a pump laser operating at 1445 nm. Here a signal at 1535 nm, which is 90 nm away from the pump wavelength, is amplified.

In a Raman amplifier this process transfers optical energy from a strong laser pump beam to a weaker transmission signal that has a wavelength which is 80 to 100 nm higher than the pumping wavelength. For example, pumping at 1450 nm will lead to a signal gain at approximately 1530 to 1550 nm. Owing to the molecular structure of glass, a number of vibration modes exist so that the optical gain region is about 30 nm wide. In practice one uses several pump lasers

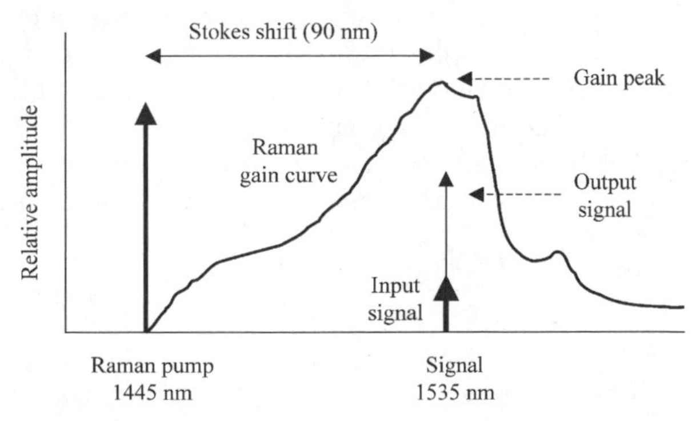

Figure 11.12. Stokes shift and the resulting Raman gain spectrum from a pump laser operating at 1445 nm.

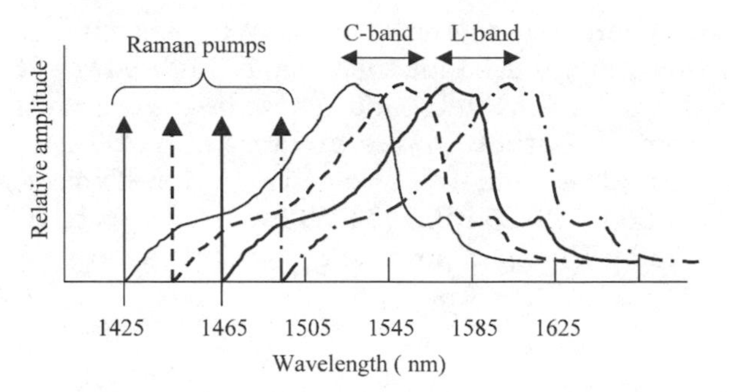

Figure 11.13. Raman gain spectrum for four pump lasers operating at different wavelengths.

to generate a flat wideband gain spectrum. Figure 11.13 shows the Raman gain spectrum for four pump lasers at different wavelengths. This illustrates that by using an appropriate combination of pump wavelengths, it is possible to achieve gain over any desired spectral range.

Traditionally the signal and pump beams travel in opposite directions, but the pumping also can be done codirectionally with the signal. By using counterdirectional pumping, the amplification is distributed within the last 20 to 40 km of transmission fiber. For this reason Raman amplification often is referred to as *distributed amplification*. In general, noise factors limit the practical gain of a distributed Raman amplifier to less than 20 dB.

Pump lasers with high output powers in the 1400- to 1500-nm region are required for Raman amplification of C- and L-band signals. Lasers that provide fiber launch powers of up to 300 mW are available in standard 14-pin butterfly packages.

Figure 11.14 shows the setup for a typical Raman amplification system. Here a pump combiner multiplexes the outputs from four pump lasers operating at different wavelengths (examples might be 1425, 1445, 1465, and 1485 nm) onto a single fiber. These pump power couplers are referred to popularly as *14XX-nm pump-pump combiners*. Table 11.4 lists the performance parameters of a pump combiner based on fused-fiber coupler technology. This combined pump power then is coupled into the transmission fiber in a counterpropagating direction through a broadband WDM coupler, such as those listed in Table 11.5. The difference in the power levels between the two monitoring photodiodes measures the amplification gain. The gain-flattening filter (GFF) is used to equalize the gains at different wavelengths.

11.6. Wavelength Conversion

An optical wavelength converter is a device that can directly translate information on an incoming wavelength to a new wavelength without entering the electrical domain. This is an important component in WDM networks for several

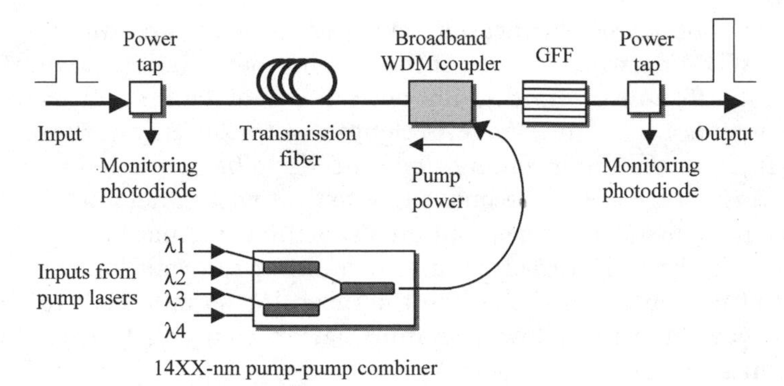

Figure 11.14. Setup for a typical distributed Raman amplification system.

TABLE 11.4. Performance Parameters of a 14XX-nm Pump-Pump Combiner Based on Fused-Fiber Coupler Technology

Parameter	Performance value
Device technology	Fused-fiber coupler
Wavelength range	1420–1500 nm
Channel spacing	Customized: 10–40 nm
	Standard: 10, 15, 20 nm
Insertion loss	<0.8 dB
Polarization-dependent loss	<0.2 dB
Directivity	>55 dB
Optical power capability	3000 mW

TABLE 11.5. Performance Parameters of Broadband WDM Couplers for Combining 14XX-nm Pumps and C-Band or L-Band Signals

Parameter	Performance value	
Device technology	Microoptics	Thin-film filter
Pump channel λ range	1420–1490 nm	1440–1490 nm
Pass channel λ range	1505–1630 nm	1528–1610 nm
Pump channel insertion loss	0.30 dB	0.6 dB
Pass channel insertion loss	0.45 dB	0.8 dB
Polarization-dependent loss	0.05 dB	0.10 dB
Polarization mode dispersion	0.05 ps	0.05 ps
Optical power capability	2000 mW	500 mW

reasons. For notational clarification purposes, suppose we examine what is happening in WDM network 1. First, if there is no coordination of wavelength allocations in different networks, an optical signal coming from an external network may not have the same wavelength as that of any of those used in network 1. In this case the incoming signal needs to be converted to a wavelength that network 1 recognizes. Second, suppose that within network 1 a signal coming into a node needs to be sent out on a specific transmission line. If the wavelength of this signal is already in use by another information channel residing on the destined outgoing path, then the incoming signal needs to be converted to a new wavelength to allow both information channels to traverse the same fiber simultaneously.

Although a number of all-optical techniques have been investigated for achieving wavelength conversion, none of them are commercially mature yet. Therefore, currently the most practical method of wavelength conversion is to change the incoming optical signal to an electrical format and then use this electric signal to modulate a light source operating at a different appropriate wavelength. However, for those readers who are curious about all-optical techniques, the following two subsections briefly discuss two all-optical wavelength conversion methods, one based on optical gating and the other on wave mixing.

11.6.1. Optical-gating wavelength converters

A wide variety of optical-gating techniques using devices such as semiconductor optical amplifiers, semiconductor lasers, or nonlinear optical-loop mirrors have been investigated to achieve wavelength conversion. The use of an SOA in a *cross-gain modulation* (CGM) mode has been one of the most successful techniques for implementing single-wavelength conversion. The configurations for implementing this scheme include the Mach-Zehnder or the Michelson interferometer setups shown in Fig. 11.15.

The CGM scheme relies on the dependency of the refractive index on the carrier density in the active region of the SOA. As depicted in Fig. 11.15, the basic concept is that an incoming information-carrying signal at wavelength λ_s and a continuous-wave (CW) signal at the desired new wavelength λ_c (called the *probe beam*) are simultaneously coupled into the device. The two waves can be either copropagating or counterpropagating. However, the noise in the latter case is higher. The signal beam modulates the gain of the SOA by depleting the carriers, which produces a modulation of the refractive index. When the CW beam encounters the modulated gain and refractive index, its amplitude and phase are changed, so that it now carries the same information as the input signal. As shown in Fig. 11.15, the SOAs are placed in an asymmetric configuration so that the phase change in the two amplifiers is different. Consequently, the CW light is modulated according to the phase difference. A typical splitting ratio is 69/31 percent. These types of converters readily handle data rates of at least 10 Gbps.

A limitation of CGM architecture is that it only converts one wavelength at a time. In addition, the refractive index varies as the carrier density changes

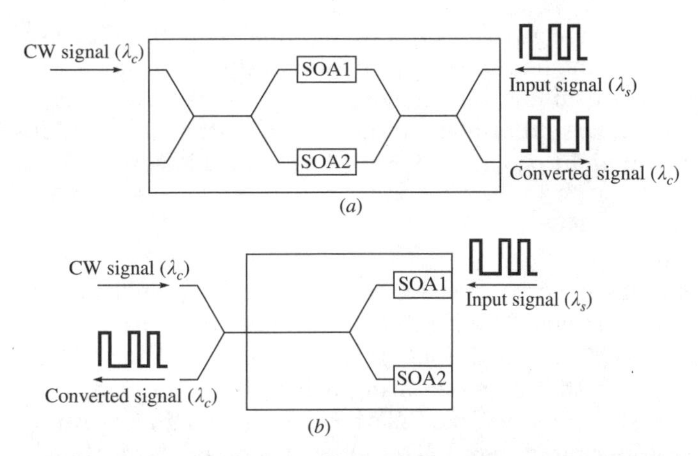

Figure 11.15. (*a*) Mach-Zehnder interferometer and (*b*) Michelson interferometer setups using a pair of SOAs for implementing the cross-phase modulation wavelength conversion scheme.

inside the SOA. This affects the phase of the probe and creates significant pulse distortion.

11.6.2. Wave-mixing wavelength converters

Wavelength conversion based on nonlinear optical wave mixing offers important advantages compared to other methods. This includes a multiwavelength conversion capability and transparency to the modulation format. The mixing results from nonlinear interactions among optical waves traversing a nonlinear material. The outcome is the generation of another wave whose intensity is proportional to the product of the intensities of the interacting waves. The phase and frequency of the generated wave are a linear combination of those of the interacting waves. Therefore the wave mixing preserves both amplitude and phase information, and consequently is the only wavelength conversion category that offers strict transparency to the modulation format.

Two successful schemes are four-wave mixing (FWM) in either passive waveguides or SOAs and difference-frequency generation in waveguides. For wavelength conversion, the *four-wave mixing* scheme employs the mixing of three distinct input waves to generate a fourth distinct output wave. In this method, an intensity pattern resulting from two input waves interacting in a nonlinear material forms a grating. For example, in SOAs there are three physical mechanisms that can form a grating: carrier-density modulation, dynamic carrier heating, and spectral hole burning. The third input wave in the material gets scattered by this grating, thereby generating an output wave. The frequency of the generated output wave is offset from that of the third wave by the frequency difference between the first two waves. If one of the three incident waves contains amplitude, phase, or frequency information and the other two waves are constant, then the generated wave will contain the same information.

Difference-frequency generation in waveguides is based on the mixing of two input waves. Here the nonlinear interaction of the material is with a pump and a signal wave. For example, one demonstration of this technique has shown the simultaneous conversion of eight input wavelengths in the 1546- to 1560-nm region to a set of eight output wavelengths in the 1524- to 1538-nm region.

11.7. Summary

Optical amplifiers have become an essential component in modern lightwave communication systems. The three basic technologies are semiconductor optical amplifiers (SOAs), doped-fiber amplifiers (DFAs), and Raman amplifiers. SOAs are based on the same operating principles as laser diodes, whereas the other two types employ a fiber as the gain mechanism. Among the DFAs, erbium-doped fiber amplifiers (EDFAs) are used widely in the C- and L-bands for optical communication networks. In contrast to an EDFA which uses a specially constructed fiber for the amplification medium, a Raman amplifier makes use of the transmission fiber itself.

The attractiveness of an SOA is that since it is essentially an InGaAsP laser that is operating below its threshold point, it can operate in every fiber wavelength band extending from 1280 nm in the O-band to 1650 nm in the U-band. SOAs are particularly useful in the O-band where other optical amplifier technologies are more difficult to implement. However, the limitation is that the SOA is practical only for single-wavelength amplification. Table 11.1 lists typical values of SOA performance parameters for devices that can be housed in a standard 14-pin butterfly package.

The most common optical fiber amplifier for long-haul telecommunication applications is a silica fiber doped with erbium, which is known as an *erbium-doped fiber amplifier*, or EDFA. Originally the operation of an EDFA by itself was limited to the 1530- to 1560-nm region, which is the origin of the designation *conventional band* or C-band. However, improvements in erbium-doped fiber designs and the use of pump lasers operating at wavelengths different from those used in the C-band have allowed the extension of EDFAs into the L-band. An EDFA needs to be pumped optically, which is done with 980- and 1480-nm lasers for C-band operation and by 960-nm lasers for L-band use. Table 11.2 compares the performances of these two laser types, and Table 11.3 lists the characteristics of couplers used to combine the pump power wavelengths with the C-band and/or L-band transmission signals.

Whereas an EDFA requires a specially constructed optical fiber for its operation, a *Raman amplifier* makes use of the transmission fiber itself as the amplification medium. A Raman amplifier is based on an effect called *stimulated Raman scattering* (SRS). This process transfers optical energy from a strong laser pump beam to a weaker transmission signal that has a wavelength which is 80 to 100 nm higher than the pumping wavelength. For example, pumping at 1450 nm will lead to a signal gain at approximately 1530 to 1550 nm. By using counterdirectional pumping, the amplification is distributed within the last

20 to 40 km of transmission fiber. For this reason Raman amplification often is referred to as *distributed amplification*. Pump lasers with high output powers in the 1400- to 1500-nm region are required for Raman amplification of C- and L-band signals. Lasers that provide fiber launch powers of up to 300 mW are available in standard 14-pin butterfly packages. To realize gain over a wide spectral band, several pump lasers operating at different wavelengths normally are used. These pumps are referred to popularly as *14XX-nm lasers*. Tables 11.4 and 11.5 list the characteristics of couplers used to combine the pump powers and to insert the aggregate pump output onto the fiber along with the C-band and/or L-band transmission signals.

Further Reading

1. M. Young, "Next-generation networks may benefit from SOAs," *Laser Focus World*, vol. 37, pp. 73–79, September 2001.
2. J.-J. Bernard and M. Renaud, "Semiconductor optical amplifiers," *SPIE OE Magazine*, vol. 1, pp. 36–38, September 2001.
3. B. Verbeek, K. Vreeburg, H. Naus, W. van den Brink, L. Lunardi, and A. Turukhin, "SOAs provide multiple network design options," *WDM Solutions*, vol. 4, February 2002 (www.wdm-solutions.com).
4. ITU-T Recommendation G.661, "Definition and Test Methods for the Relevant Generic Parameters of Optical Amplifiers."
5. ITU-T Recommendation G.662, "Generic Characterization of Optical Amplifier Devices and Subsystems."
6. ITU-T Recommendation G.663, "Application-Related Aspects of Optical Amplifier Devices and Subsystems and Comprehensive Appendix on Transmission-Related Aspects."
7. K. Wang, T. Strite, and B. Fidric, "Pump lasers at 960 nm for the L-band," *Lightwave*, vol. 19, pp. 72–76, March 2002 (www.lightwaveonline.com).
8. E. Desurvire, D. Bayart, B. Desthieux, and S. Bigo, *Erbium-Doped Fiber Amplifiers: Devices and System Developments*, Wiley, New York, 2002.
9. J. Connolly, "New pump lasers power Raman amplifier," *WDM Solutions*, vol. 3, pp. 57–62, June 2001 (www.wdm-solutions.com).
10. G. Keiser, *Optical Fiber Communications*, 3d ed., McGraw-Hill, Burr Ridge, Ill., 2000, Chap. 11.

Chapter

12

Wavelength Division Multiplexing

A powerful aspect of an optical communication link is that many different wavelengths selected from the spectral regions ranging from the O-band through the L-band can be sent along a single fiber simultaneously. The technology of combining a number of wavelengths onto the same fiber is known as *wavelength division multiplexing*, or WDM. Conceptually, the WDM scheme is the same as frequency division multiplexing (FDM) used in microwave radio and satellite systems. Just as in FDM, the wavelengths (or optical frequencies) in WDM must be spaced properly to avoid interference between adjacent channels. The key system features of WDM are the following:

- *Capacity upgrade*. The classical application of WDM has been to upgrade the capacity of existing point-to-point fiber optic transmission links. If each wavelength supports an independent network channel of a few gigabits per second, then WDM can increase the capacity of a fiber system dramatically with each additional wavelength channel.
- *Transparency*. An important aspect of WDM is that each optical channel can carry any transmission format. Thus, by using different wavelengths, fast or slow asynchronous and synchronous digital data and analog information can be sent simultaneously, and independently, over the same fiber without the need for a common signal structure.
- *Wavelength routing*. Instead of using electronic means to switch optical signals at a node, a wavelength-routing network can provide a pure optical end-to-end connection between users. This is done by means of *lightpaths* that are routed and switched at intermediate nodes in the network. In some cases, lightpaths may be converted from one wavelength to another wavelength along their route. Chapter 17 gives a detailed treatment of wavelength routing and switching.

This chapter begins by addressing the generic operating principles of WDM and the international standards that have evolved for its implementation. The next topic concerns the components needed for WDM realization. These range in complexity from simple, passive optical splitting or combining elements to sophisticated, dynamically tunable devices. The first component category involves wavelength multiplexers, which are used to combine independent signal streams operating at different wavelengths onto the same fiber. The technologies for achieving this include thin-film filters (TFFs), arrayed waveguide gratings (AWGs), Bragg fiber gratings, diffraction gratings, and interleavers. The final topic concerns wavelength lockers which are important devices in WDM transmitters to maintain the output from a laser diode at a predefined ITU-T frequency with a high precision.

Chapter 13 on operational concepts shows how all the different component puzzle pieces fit together to form a variety of metro and long-distance WDM links and networks. Included there are discussions of WDM applications of dynamic devices such as controllers for tunable transmitters, dynamic gain equalizers (DGEs), tunable wavelength filters, and variable optical attenuators (VOAs).

12.1. Operational Principles of WDM

When optical fiber systems were first deployed, they consisted of simple point-to-point links in which a single fiber line has one light source at its transmitting end and one photodetector at the receiving end. In these early systems, signals from different light sources used separate and uniquely assigned optical fibers. In addition to filling up ducts with fibers, these simplex systems represent a tremendous underutilization of the bandwidth capacity of a fiber.

Since the spectral width of a high-quality source occupies only a narrow slice of optical bandwidth, there are many additional operating regions across the entire spectrum ranging from the O-band through the L-band that can be used simultaneously. The original use of WDM was to upgrade the capacity of installed point-to-point transmission links. This was achieved with wavelengths that were separated from several tens up to 200 nm in order not to impose strict wavelength-tolerance requirements on the different laser sources and the receiving wavelength splitters.

With the advent of tunable lasers that have extremely narrow spectral emission widths, one then could space wavelengths by less than a few nanometers. This is the basis of wavelength division multiplexing, which simultaneously uses a number of light sources, each emitting at a slightly different peak wavelength. Each wavelength carries an independent signal, so that the link capacity is increased greatly. The main trick is to ensure that the peak wavelength of a source is spaced sufficiently far from its neighbor so as not to create interference between their spectral extents. Equally important is the requirement that these peak wavelengths not drift into the spectral territory occupied by adjacent channels. In addition to maintaining strict control of the wavelength, typically system designers include an empty guard band between the channels. Thereby the fidelities of the independent messages from each source are maintained for subsequent conversion to electric signals at the receiving end.

12.1.1. WDM operating regions

To see the potential of WDM, let us first examine the characteristics of a high-quality optical source. As an example, a distributed-feedback (DFB) laser has a frequency spectrum on the order of 1 MHz, which is equivalent to a spectral linewidth of 10^{-5} nm. With such spectral-band widths, simplex systems make use of only a tiny portion of the transmission bandwidth capability of a fiber. This can be seen from Fig. 12.1, which depicts the attenuation of light in a silica fiber as a function of wavelength. The curve shows that the two low-loss regions of a standard single-mode fiber extend over the O-band wavelengths ranging from about 1270 to 1350 nm (originally called the second window) and from 1480 to 1600 nm (originally called the third window).

We can view these regions either in terms of *spectral width* (the wavelength band occupied by the light signal) or by means of *optical bandwidth* (the frequency band occupied by the light signal). To find the optical bandwidth corresponding to a particular spectral width in these regions, we use the fundamental relationship $c = \lambda\nu$, which relates the wavelength λ to the carrier frequency ν, where c is the speed of light. Differentiating this, we have

$$\Delta\nu = \frac{c}{\lambda^2}\,\Delta\lambda \tag{12.1}$$

where the deviation in frequency $\Delta\nu$ corresponds to the wavelength deviation $\Delta\lambda$ around λ.

Now suppose we have a fiber that has the attenuation characteristic shown in Fig. 12.1. From Eq. (12.1) the optical bandwidth is $\Delta\nu = 14$ THz for a usable

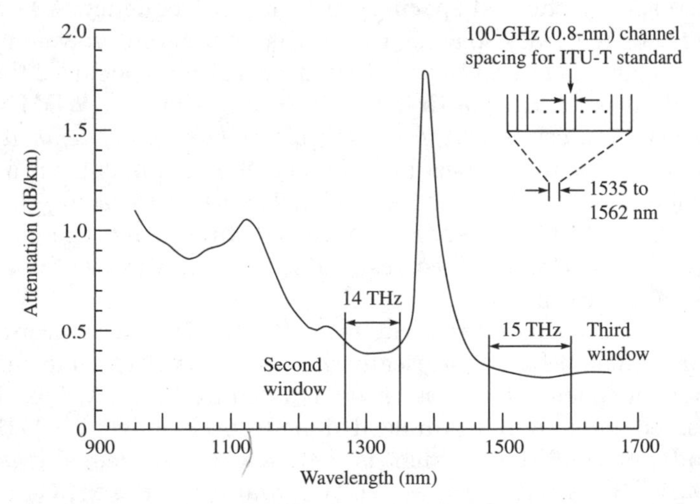

Figure 12.1. The transmission bands of a standard single-mode fiber in the O-band (second window) and from 1480 to 1600 nm (third window) allow the use of many simultaneous channels. The first ITU-T standard for WDM specified channels with 100-GHz spacings.

spectral band $\Delta\lambda = 80$ nm in the center of the O-band. Similarly, $\Delta\nu = 15$ THz for a usable spectral band $\Delta\lambda = 120$ nm in the low-loss region running from near the beginning of the S-band to almost the end of the L-band. This yields a total available fiber bandwidth of about 30 THz in the two low-loss windows. The insert in Fig. 12.1 shows how a series of 100-GHz channel separations fits into the C-band.

Prior to about 2000, the peak wavelengths of adjacent sources typically were restricted to be separated by 0.8 to 1.6 nm (100 to 200 GHz). This was done to take into account possible drifts of the peak wavelength due to aging or temperature effects, and to give both the manufacturer and the user some leeway in specifying and choosing the precise peak emission wavelength. As described in Sec. 12.1.3, the next generation of WDM systems specified both narrower and much wider channel spacings depending on the application and on the wavelength region being used. The much narrower spacings thus require strict wavelength control of the optical source, as discussed in Sec. 12.3. On the other hand, the wider wavelength separations offer inexpensive WDM implementations since wavelength control requirements are relaxed significantly.

Example If one takes a spectral band of 0.8 nm (or, equivalently, a mean frequency spacing of 100 GHz) within which narrow-linewidth lasers are transmitting, then one can send about 36 independent signals in the 1530- to 1560-nm C-band on a single fiber.

12.1.2. WDM standards

Since WDM is essentially frequency division multiplexing at optical carrier frequencies, the WDM standards developed by the International Telecommunication Union (ITU) specify channel spacings in terms of frequency. A key reason for selecting a fixed-frequency spacing, rather than a constant-wavelength spacing, is that when a laser is locked to a particular operating mode, it is the frequency of the laser that is fixed. The first ITU-T specification for WDM was Recommendation G.692, *Optical Interfaces for Multichannel Systems with Optical Amplifiers*. This document specifies selecting the channels from a grid of frequencies referenced to 193.100 THz (1552.524 nm) and spacing them 100 GHz (about 0.8 nm at 1550 nm) apart. Suggested alternative spacings in G.692 include 50 and 200 GHz, which correspond to spectral widths of 0.4 and 1.6 nm, respectively, at 1550 nm.

The literature often uses the term *dense WDM* (DWDM), in contrast to conventional or regular WDM. Historically this term was used somewhat loosely to refer to channel spacings such as those denoted by ITU-T G.692. In 2002 the ITU-T released an updated standard aimed specifically at DWDM. This is Recommendation G.694.1, which is entitled *Dense Wavelength Division Multiplexing (DWDM)*. It specifies WDM operation in the S-, C-, and L-bands for high-quality, high-rate metro-area network (MAN) and wide-area network (WAN) services. It calls out for narrow frequency spacings of 100 to 12.5 GHz (or, equivalently, 0.8 to 0.1 nm at 1550 nm). This implementation requires the use of stable, high-quality, temperature-controlled and wavelength-controlled (frequency-locked) laser diode light sources. For example, the wavelength drift tolerances for 25-GHz channels are ±0.02 nm.

Table 12.1 lists part of the ITU-T G.694.1 DWDM frequency grid for 100- and 50-GHz spacings in the L- and C-bands. The column labeled "50-GHz offset" means that for the 50-GHz grid one uses the 100-GHz spacings with these 50-GHz values interleaved. For example, the 50-GHz channels in the L-band would be at 186.00, 186.05, 186.10 THz, and so on. Appendix C gives a more complete frequency table for the L- and C-bands. Note that when the frequency spacings are uniform, the wavelengths are not spaced uniformly because of the relationship given in Eq. (12.1).

With the production of full-spectrum (low-water-content) fibers, the development of relatively inexpensive VCSEL optical sources, and the desire to have low-cost optical links operating in metro- and local-area networks came the concept of *coarse WDM* (CWDM). In 2002 the ITU-T released a standard aimed specifically at CWDM. This is Recommendation G.694.2, which is entitled *Coarse Wavelength Division Multiplexing (CWDM)*. The CWDM grid is made up of 18 wavelengths defined within the range of 1270 to 1610 nm (O- through L-bands) spaced by 20 nm with wavelength drift tolerances of ±2 nm. This can be achieved with inexpensive VCSEL light sources that are not temperature-controlled. The targeted transmission distance for CWDM is 50 km on single-mode fibers, such as those specified in ITU-T Recommendations G.652, G.653, and G.655. Chapter 13 gives more details on CWDM applications.

TABLE 12.1. Sample Portion of the ITU-T G.694.1 DWDM Grid for 100- and 50-GHz Spacings in the L- and C-Bands

	L-band				C-band			
	100-GHz		50-GHz offset		100-GHz		50-GHz offset	
Unit	THz	nm	THz	nm	THz	nm	THz	nm
1	186.00	1611.79	186.05	1611.35	191.00	1569.59	191.05	1569.18
2	186.10	1610.92	186.15	1610.49	191.10	1568.77	191.15	1568.36
3	186.20	1610.06	186.25	1609.62	191.20	1576.95	191.25	1567.54
4	186.30	1609.19	186.35	1608.76	191.30	1567.13	191.35	1566.72
5	186.40	1608.33	186.45	1607.90	191.40	1566.31	191.45	1565.90
6	186.50	1607.47	186.55	1607.04	191.50	1565.50	191.55	1565.09
7	186.60	1606.60	186.65	1606.17	191.60	1564.68	191.65	1564.27
8	186.70	1605.74	186.75	1605.31	191.70	1563.86	191.75	1563.45
9	186.80	1604.88	186.85	1604.46	191.80	1563.05	191.85	1562.64
10	186.90	1604.03	186.95	1603.60	191.90	1562.23	191.95	1561.83

12.1.3. Generic WDM link

The implementation of WDM networks requires a variety of passive and/or active devices to combine, distribute, isolate, add, drop, attenuate, and amplify optical power at different wavelengths. Passive devices require no external control for their operation, so they have a fixed application in WDM networks. These passive components are used to split and combine or tap off optical signals. The performance of active devices can be controlled electronically, thereby providing a large degree of network flexibility. Active WDM components include tunable optical filters, tunable sources, add/drop multiplexers, VOAs, DGEs, optical switches, and optical amplifiers.

Figure 12.2 shows the implementation of a simple WDM link. The transmitting side has a series of fixed-wavelength or tunable independently modulated light sources, each of which emits signals at a unique wavelength. Here a *multiplexer* (mux) is needed to combine these optical outputs into a continuous spectrum of signals and couple them onto a single fiber. Within the link there may be various types of optical amplifiers and a variety of specialized components (not shown). The spacing between amplifiers is called a *span*.

At the receiving end a *demultiplexer* (demux) is required to separate the optical signals into appropriate detection channels for signal processing. At the transmitter the basic design challenge is to have the multiplexer provide a low-loss path from each optical source to the multiplexer output. Since the optical signals generally are tightly controlled so that they do not emit any significant amount of optical power outside of the designated channel spectral width, interchannel crosstalk factors are relatively unimportant at the transmitting end.

A different requirement exists for the demultiplexer, since photodetectors are usually sensitive over a broad range of wavelengths, which could include all the WDM channels. To prevent spurious signals from entering a receiving channel, that is, to give good channel isolation of the different wavelengths being used, the demultiplexer must exhibit narrow spectral operation, or very stable optical filters with sharp wavelength cutoffs must be used. The tolerable interchannel crosstalk levels can vary widely depending on the application. In general,

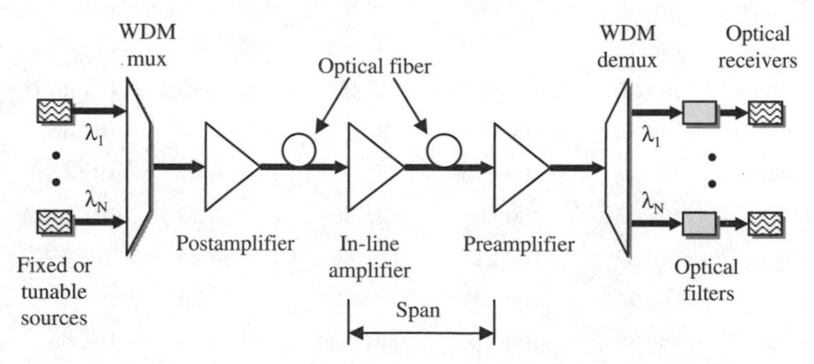

Figure 12.2. Implementation of a simple WDM link showing the use of different amplifiers.

a −10 dB level is not sufficient whereas a level of −30 dB is acceptable. In principle, any optical demultiplexer also can be used as a multiplexer. For simplicity, the word *multiplexer* is used as a general term to refer to both combining and separating functions, except when it is necessary to distinguish the two devices or functions.

12.2. Multiplexers for WDM

A key WDM component is the wavelength multiplexer. The function of this device is to combine independent signal streams operating at different wavelengths onto the same fiber. Many different techniques using specialized components have been devised for combining multiple wavelengths onto the same fiber and separating them at the receiver. Each of these techniques has certain advantages and of course various limitations. These include thin-film filters, arrayed waveguide gratings, Bragg fiber gratings, diffraction gratings, and interleavers. The performance demands on these components are increasing constantly with the desire to support higher channel counts and longer distances between terminals.

Whereas prior to 2000 the standard wavelength spacing was 100 GHz for 2.5-Gbps DWDM links, the current move is toward 10-Gbps ultradense systems operating with channels that are spaced 25 or 12.5 GHz apart. An even further squeezing of the channels is seen in the hyperfine WDM products that have separations down to 3.125 GHz. For 40-Gbps systems the channels nominally are spaced 50 or 100 GHz apart because of the greater impact from nonlinear dispersion effects at these higher data rates. The expansion of WDM channels beyond the C-band into the S- and L-bands has allowed the possibility of sending 320 wavelengths spaced 25 GHz apart in the combined C- and L-band with 10-Gbps transmission rates per channel. This is in contrast to the earlier 96 maximum 10-Gbps channels that were separated by 50 GHz in the C-band.

Multiplexers for CWDM applications have less stringent performance demands for certain parameters such as center wavelength tolerance, its change with temperature, and the passband sharpness. However, they still need to have a good reflection isolation, a small polarization-dependent loss, and low insertion losses. These CWDM devices can be made with thin-film filter technology.

12.2.1. Thin-film filters

Section 9.3 discusses the operational principles of thin-film optical filters. Here we will look at how to use them in a WDM system to create a multiplexer. As Sec. 9.3 describes, a *thin-film filter* allows only a very narrow slice of spectral width to pass through it and reflects all other light outside this band. To create a wavelength multiplexing device for combining or separating N wavelength channels, one needs to cascade $N - 1$ thin-film filters.

Figure 12.3 illustrates a multiplexing function for the four wavelengths λ_1, λ_2, λ_3, and λ_4. Here the filters labeled TFF_2, TFF_3, and TFF_4 pass wavelengths λ_2,

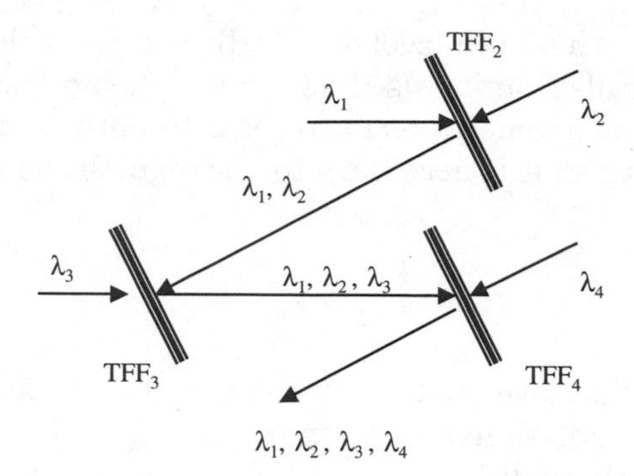

Figure 12.3. Illustration of multiplexing four wavelengths using thin-film filters.

λ_3, and λ_4, respectively, and reflect all others. The filters are set at a slight angle in order to direct light from one TFF to another. First filter TFF$_2$ reflects wavelength λ_1 and allows wavelength λ_2 to pass through. These two signals then are reflected from filter TFF$_3$ where they are joined by wavelength λ_3. After a similar process at filter TFF$_4$ the four wavelengths can be coupled into a fiber by means of a lens mechanism.

To separate the four wavelengths, the directions of the arrows in Fig. 12.3 are reversed. Since a light beam loses some of its power at each TFF because the filters are not perfect, this multiplexing architecture works for only a limited number of channels. This usually is specified as being 16 channels or less.

Table 12.2 lists typical performance parameters for commercially available wavelength multiplexers based on thin-film filter technology. The parameters address 8-channel DWDM devices with 50- and 100-GHz channel spacings and an 8-channel CWDM module.

12.2.2. Fiber Bragg gratings

Section 9.4 discusses the operational principles and performance characteristics of fiber Bragg gratings. This section shows how to use them in conjunction with an optical circulator (see Sec. 9.2) to form a wavelength multiplexer. A *fiber Bragg grating* (FBG) allows optical channel spacings as narrow as 25 GHz. By using special packaging techniques, Bragg gratings can be made to have a very low thermal drift of less than one-half of a picometer (pm) per degree celsius, and they exhibit very low interchannel crosstalk.

In contrast to a thin-film filter, an FBG reflects a narrow spectral slice and allows all other wavelengths to pass through it. To create a device for combining or separating N wavelengths, one needs to cascade $N - 1$ FBGs and $N - 1$ circulators. Figure 12.4 illustrates a multiplexing function for the four wavelengths λ_1, λ_2, λ_3, and λ_4 using three FBGs and three circulators (labeled C$_2$, C$_3$, and C$_4$). The fiber grating filters labeled FBG$_2$, FBG$_3$, and FBG$_4$ are constructed to reflect wavelengths λ_2, λ_3, and λ_4, respectively, and to pass all others.

TABLE 12.2. Typical Performance Parameters for 8-Channel DWDM and CWDM Multiplexers Based on Thin-Film-Filter Technology

Parameter	50-GHz DWDM	100-GHz DWDM	20-nm CWDM
Center wavelength accuracy	±0.1 nm	±0.1 nm	±0.3 nm
Channel passband @ 0.5-dB bandwidth	±0.20 nm	±0.11 nm	±6.5 nm
Insertion loss	≤4.0 dB	≤4.0 dB	≤4.5 dB
Ripple in passband	≤0.5 dB	≤0.5 dB	≥0.5 dB
Adjacent channel isolation	≥23 dB	≥20 dB	≥15 dB
Directivity	≥50 dB	≥55 dB	≥50 dB
Optical return loss	≥40 dB	≥50 dB	≥45 dB
Polarization-dependent loss	≤0.1 dB	≤0.1 dB	≤0.1 dB
Thermal wavelength drift	<0.001 nm/°C	<0.001 nm/°C	<0.003 nm/°C
Optical power capability	500 mW	500 mW	500 mW

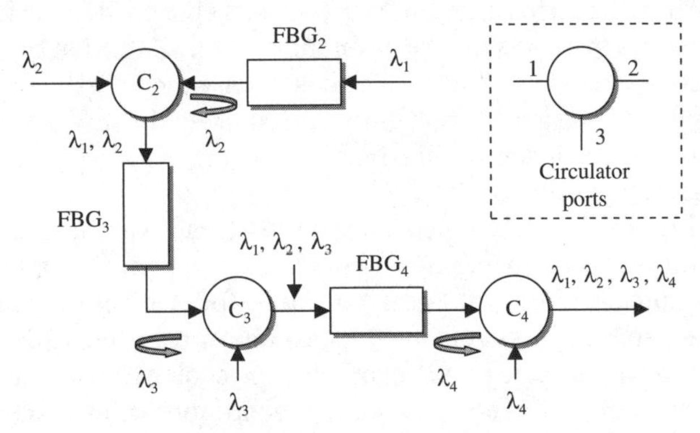

Figure 12.4. Multiplexing of four wavelengths using three FBG devices and three circulators.

To see how the multiplexer functions, first consider the combination of circulator C_2 and fiber filter FBG_2. Here filter FBG_2 reflects wavelength λ_2 and allows wavelength λ_1 to pass through. After wavelength λ_1 passes through FBG_2 it enters port 2 of circulator C_2 and exits from port 3. Wavelength λ_2 enters port 1 of circulator C_2 and exits from port 2. After being reflected from FBG_2 it enters port 2 of circulator C_2 and exits from port 3 together with wavelength λ_1. Next at circulator C_3 wavelength λ_3 enters port 3 of circulator C_3 and exits from port 1 and travels toward FBG_3. After being reflected from FBG_3 it enters port 1 of circulator C_3 and exits from port 2 together with wavelengths λ_1 and λ_2. After a similar process at circulator C_4 and filter FBG_4 to insert wavelength λ_4, the four wavelengths all exit together from port 2 of circulator C_4 and can be coupled easily into a fiber.

Similar to the use of thin-film filters to form multiplexers, the size limitation when using fiber Bragg gratings is that one filter is needed for each wavelength and normally the operation is sequential with wavelengths being transmitted by one filter after another. Therefore the losses are not uniform from channel to channel, since each wavelength goes through a different number of circulators and fiber gratings, each of which adds loss to that channel. This may be acceptable for a small number of channels, but the loss differential between the first and last inserted wavelengths is a restriction for large channel counts.

12.2.3. Arrayed waveguide gratings

An *arrayed waveguide grating* is a third DWDM device category. As shown in Fig. 12.5, an AWG consists of input and output slab waveguide arrays, two identical focusing star couplers, and an interconnection of uncoupled waveguides called a *grating array*. In the grating array region the path length of each waveguide differs by a very precise amount ΔL from the length in adjacent arms. These path length differences introduce precisely spaced delays in the signal phase in each adjacent arm, so the array forms a Mach-Zehnder type of grating. As a result, the second lens (region 5) focuses each wavelength into a different exit port in the output slab array in region 6. Reciprocally, if N wavelengths are inserted into N separate fibers (going from right to left in Fig. 12.5), they all emerge out the same port on the left.

These devices are used widely since they have attractive characteristics such as 25-GHz (0.4-nm at 1550 nm) channel spacings, are compact and can be easily fabricated on silica wafers, and can be made for a large number of WDM channels. Devices ranging in size from 8 to 40 channels are available commercially. Some device designs require a thermoelectric cooler to prevent wavelength drift ($\pm$5 pm achievable), but other packaging techniques offer athermal designs (see Sec. 20.2).

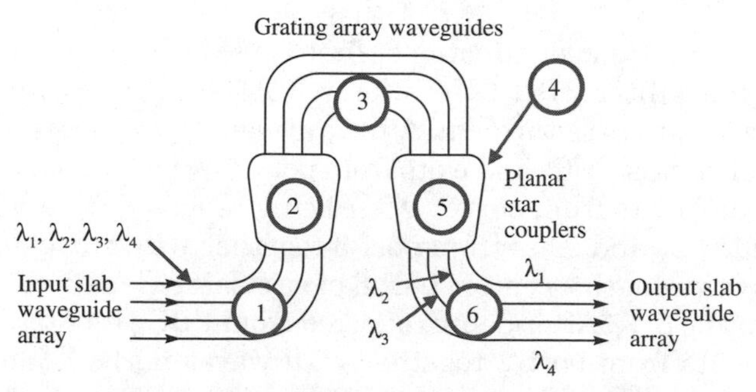

Figure 12.5. Top view of a typical arrayed waveguide grating and designation of its various key operating regions.

AWG Operation The AWG shown in Fig. 12.5 works as follows:

- Starting from the left, the input slab waveguides in region 1 are connected to the planar star coupler (region 2) which acts as a lens.
- The lens distributes the entering optical power among the different waveguides in the grating array in region 3.
- Adjacent waveguides of the grating array in region 3 differ in path length by a precise length ΔL. The path length differences ΔL can be chosen such that all input wavelengths emerge at point 4 with different phase delays

$$\Delta\Phi = 2\pi n_{\text{eff}} \frac{\Delta L}{\lambda_c} \tag{12.2}$$

Here n_{eff} is the effective refractive index of the waveguides, and λ_c is the center wavelength.

- The second lens in region 5 refocuses the light from all the grating array waveguides onto the output slab waveguide array in region 6.
- Thus each wavelength is focused into a different output waveguide in region 6.

Note that the demultiplexing function of the AWG is *periodic*. Thus an important property of the AWG is the *free spectral range* (FSR), which also is known as the *demultiplexer periodicity*. This periodicity is due to the fact that constructive interference at the output star coupler can occur for a number of wavelengths. Basically the FSR specifies the extent of a spectral width that will be separated across the output waveguides. The next chunk of higher or lower spectral width having an equal width will be separated across the same output waveguides.

Example of Free Spectral Range For example, as shown in Fig. 12.6, suppose an AWG is designed to separate light in the 4-THz-wide frequency range in the C-band running from 195.00 THz (1537.40 nm) to 191.00 THz (1569.59 nm) into forty 100-GHz channels. Then it also will separate the next-higher-frequency S-band and lower-frequency

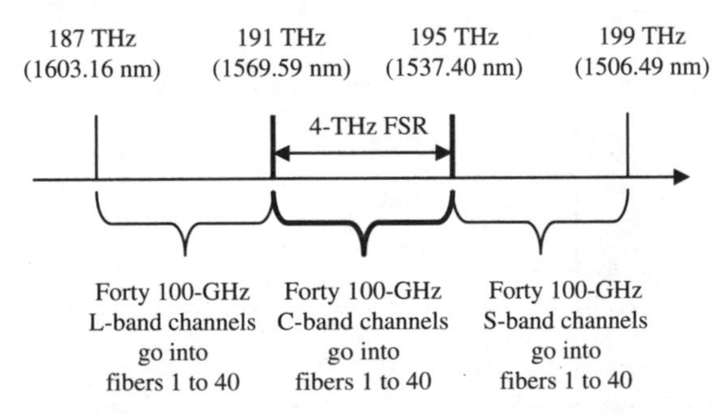

Figure 12.6. The FSR specifies the spectral width that will be separated across the output waveguides of an AWG.

L-band 4-THz-wide spectral chunks into the same 40 output fibers. The free spectral range $\Delta\lambda_{FSR}$ is determined from the relationship

$$FSR = \Delta\lambda_{FSR} = \frac{\lambda_c^2}{\Delta L n_{eff}} \tag{12.3}$$

For example, for the 4-THz frequency range denoted here, the center wavelength λ_c is 1550.5 nm, the free spectral range $\Delta\lambda_{FSR}$ should be at least 32.2 nm in order to separate all the wavelengths into distinct fibers, and the effective refractive index n_{eff} is nominally 1.45 in silica. Then the length difference between adjacent array waveguides is $\Delta L = 51.49\,\mu m$.

The passband shape of the AWG filter versus wavelength can be altered by the design of the input and output slab waveguides. Two common passband shapes are shown in Fig. 12.7. On the left is the *normal* or *gaussian* passband. This passband shape exhibits the lowest loss at the peak, but the fact that it rolls off quickly on either side of the peak means that it requires a high stabilization of the laser wavelength. Furthermore, for applications where the light passes through several AWGs, the accumulative effect of the filtering function reduces the passband to an extremely small value. An alternative to the gaussian passband shape is the *flattop* or *wideband* shape, as shown on the right in Fig. 12.7. This wideband device has a uniform insertion loss across the passband and is therefore not as sensitive to laser drift or the sensitivity of cascaded filters as is the gaussian passband. However, the loss in a flattop device is usually 2 to 3 dB higher than that in a gaussian AWG. Table 12.3 compares the main operating characteristics of these two designs for a typical 40-channel AWG.

12.2.4. Diffraction gratings

A fourth DWDM technology is based on diffraction gratings. A *diffraction grating* is a conventional optical device that spatially separates the different wavelengths contained in a beam of light. The device consists of a set of diffracting

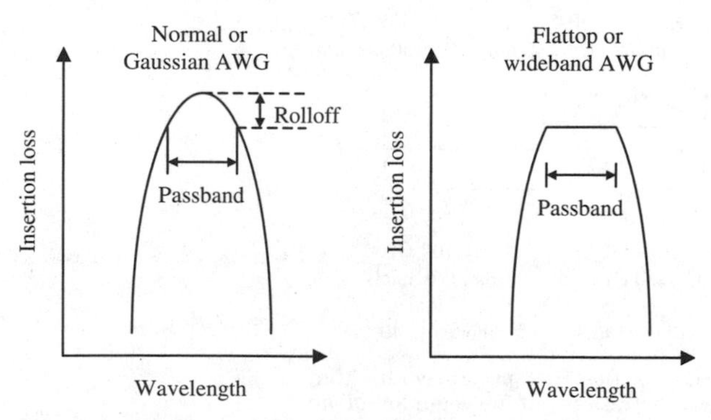

Figure 12.7. Two common optical-filter passband shapes: normal or gaussian and flattop or wideband.

TABLE 12.3. Performance Characteristics of Typical 40-Channel Arrayed Waveguide Gratings

Parameter	Gaussian	Wideband
Channel spacing	100 GHz	100 GHz
1-dB bandwidth	>0.2 nm	>0.4 nm
3-dB bandwidth	>0.4 nm	>0.6 nm
Insertion loss	<5 dB	<7 dB
Polarization-dependent loss	<0.25 dB	<0.15 dB
Adjacent channel crosstalk	30 dB	30 dB
Passband ripple	1.5 dB	0.5 dB
Optical return loss	45 dB	45 dB
Size ($L \times W \times H$)	130 × 65 × 15 (mm)	130 × 65 × 15 (mm)

elements, such as narrow parallel slits or grooves, separated by a distance comparable to the wavelength of light. These diffracting elements can be either reflective or transmitting, thereby forming a reflection grating or a transmission grating, respectively. With diffraction gratings, separating and combining wavelengths is a parallel process, as opposed to the serial process that is used with the fiber-based Bragg gratings.

Adjacent-channel crosstalk in a diffraction grating is very low, usually less than 30 dB. Insertion loss is also low (typically less than 3 dB) and is uniform to within 1 dB over a large number of channels. A passband of 30 GHz at 1-dB ripple is standard. As is the case with other WDM schemes, packaging designs can make the device be athermal, so no active temperature control is needed.

Reflection gratings are fine ruled or etched parallel lines on some type of reflective surface. With these gratings, light will bounce off the grating at an angle. The angle at which the light leaves the grating depends on its wavelength, so the reflected light fans out in a spectrum. For DWDM applications, the lines are spaced equally and each individual wavelength will be reflected at a slightly different angle, as shown in Fig. 12.8. There can be a reception fiber at each of the positions where the reflected light gets focused. Thus, individual wavelengths will be directed to separate fibers. The reflective diffraction grating works reciprocally; that is, if different wavelengths come into the device on the individual input fibers, all the wavelengths will be focused back into one fiber after traveling through the device. One also could have a photodiode array in place of the receiving fibers for functions such as power-per-wavelength monitoring.

One type of *transmission grating*, which is known as a *phase grating*, consists of a periodic variation of the refractive index of the grating. These may be characterized by a Q parameter which is defined as

$$Q = \frac{2\pi\lambda d}{n_g \Lambda^2 \cos\alpha} \tag{12.4}$$

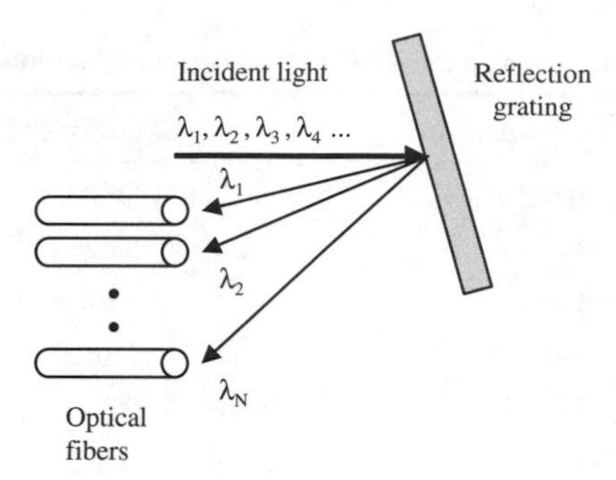

Figure 12.8. The angle at which reflected light leaves a reflection grating depends on its wavelength.

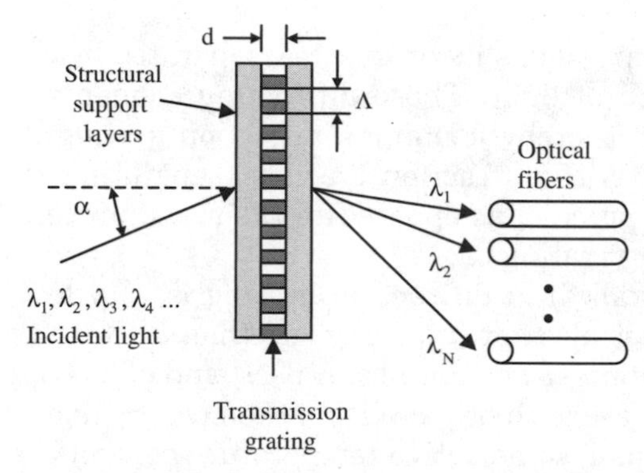

Figure 12.9. Each wavelength emerges at a slightly different angle after passing through a transmission grating.

where λ is the wavelength, d is the thickness of the grating, n_g is the refractive index of the material, Λ is the grating period, and α is the incident angle, as shown in Fig. 12.9. The phase grating is called *thin* for $Q < 1$ and *thick* for $Q > 10$. After a spectrum of wavelength channels passes through the grating, each wavelength emerges at a slightly different angle and can be focused into a receiving fiber.

12.2.5. Interleavers

Another wavelength multiplexing component is an *interleaver*, which is a passive, low-dispersion device that can increase the channel density in a WDM system. This device can combine or separate very high-density channels with a spacing as low as 3.125 GHz. A unique feature is that it can be custom-designed to route or drop a group of channels while allowing all other wavelengths to pass straight through (which commonly are referred to as the *express channels*).

Interleavers are bidirectional, so they can be used as either a multiplexer or a demultiplexer. For simplicity of discussion here we will consider the demultiplexing function. Interleavers are characterized by the designation $1 \times N$, which indicates one input and N output ports, as shown in Fig. 12.10. For example, consider a series of wavelengths separated by 25 GHz entering a 1×4 interleaver. The interleaver splits the incoming channels into four sets of 100-GHz spaced channels. This greatly simplifies further demultiplexing by allowing optical filtering at 100 GHz instead of at the initial 25 GHz. Note that for a $1 \times N$ demultiplexer, every Nth wavelength is selected to exit a particular port. For example, in the 1×4 interleaver, wavelengths λ_1, λ_5, λ_9, . . . exit port 1.

Figure 12.11 shows an application of a demultiplexing function for a combination of 160 C-band and L-band channels spaced by 50 GHz. First a wideband demultiplexer separates the incoming spectrum of wavelengths into 80 C-band and 80 L-band channels, which still are separated by 50 GHz. Next the channels of each band pass through 1×2 interleavers that separate the wavelengths into sets of 40 channels with 100-GHz spacings. Further demultiplexing with less stringent optical filtering then may be carried out.

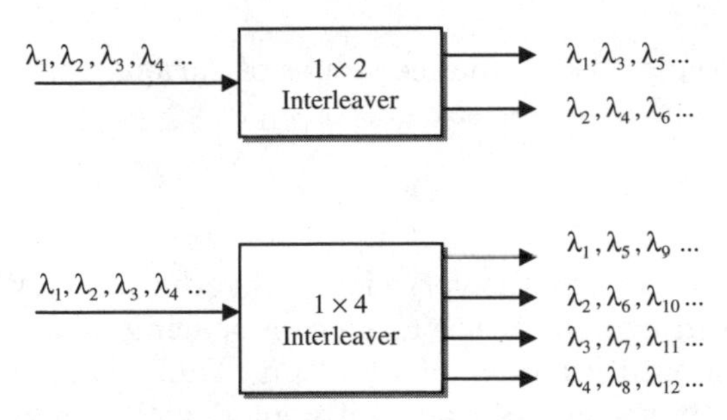

Figure 12.10. Examples of wavelength separations by two different interleavers.

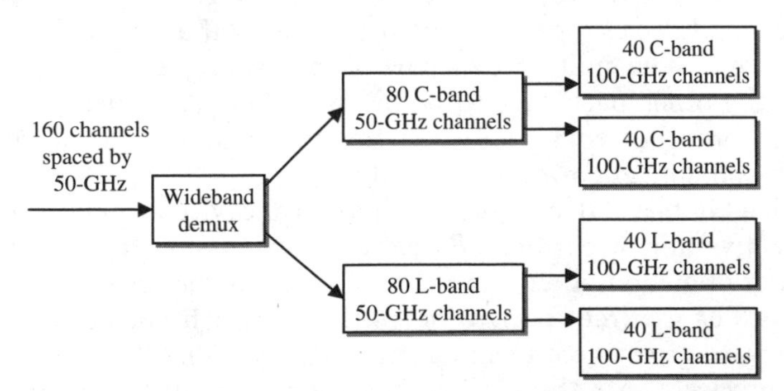

Figure 12.11. An application of an interleaver demultiplexing function for a combination of 160 C- and L-band channels spaced by 50 GHz.

TABLE 12.4. Typical Performance Values for Commercially Available Interleavers

Parameter	Unit	50-GHz interleaver		100-GHz interleaver	
Function		2 × 1 mux	1 × 2 demux	2 × 1 mux	1 × 2 demux
Channel count		80	80	40	40
Wavelength range	nm	C-band or L-band		C-band or L-band	
Insertion loss (IL)	dB	<0.4	<0.8	<0.4	<0.8
IL uniformity	dB	<±0.3	<±0.5	<±0.3	<±0.5
0.5-dB bandwidth	GHz	>11	>10	>22	>20
3-dB bandwidth	GHz	>20	>17	>40	>35
Optical return loss	dB	>50	>50	>50	>50
PDL	dB	<0.1	<0.1	<0.1	<0.1
Dispersion @ ±10 GHz	ps/nm	≤10	≤10	≤10	≤10
Warm-up time	min	10	10	10	10
Package size	cm	15 × 8 × 3		15 × 8 × 3	

Table 12.4 lists some typical performance values of various operating parameters for commercially available interleavers.

12.3. Wavelength Lockers

The move toward spacing wavelengths very closely together in a DWDM system calls for strict wavelength control of lasers since a spacing of 25 GHz, for example, requires a wavelength accuracy of ±0.02 nm. Fabry-Perot etalon-based *wavelength lockers* can offer such accuracy with one device providing multiple wavelength locking across the S-, C-, and L-bands. Since they are very small solid-state devices, they can be integrated into the laser diode package.

Figure 12.12 shows a top-level function of a wavelength locker assembly. Normally a small percentage of the light is tapped off after the laser modulator and is fed into a beam splitter. One part of the beam goes to a reference photodiode, and the other part goes through an etalon. The microprocessor-based transmitter controller then compares the two signals and adjusts the laser wavelength and optical power accordingly.

As described in Sec. 9.3, an *etalon* is an optical cavity formed by two parallel, highly reflective mirror surfaces. Since the transmission through the etalon is a periodic Airy function, it acts as a comb filter. The distance between the maxima is defined as the free spectral range (FSR), which normally is designed to be equivalent to the system channel spacing, say, 100, 50, or 25 GHz. To tune the device precisely onto the ITU channels, one can tilt the etalon to vary the optical path length d, shown in Fig. 12.13, where d' is the physical path length.

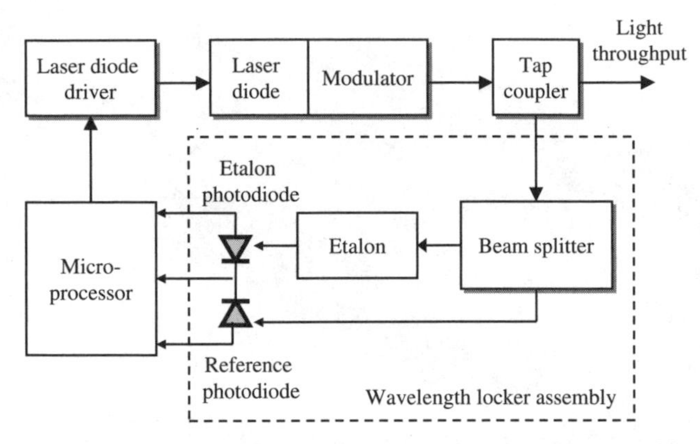

Figure 12.12. Top-level function of a wavelength locker assembly.

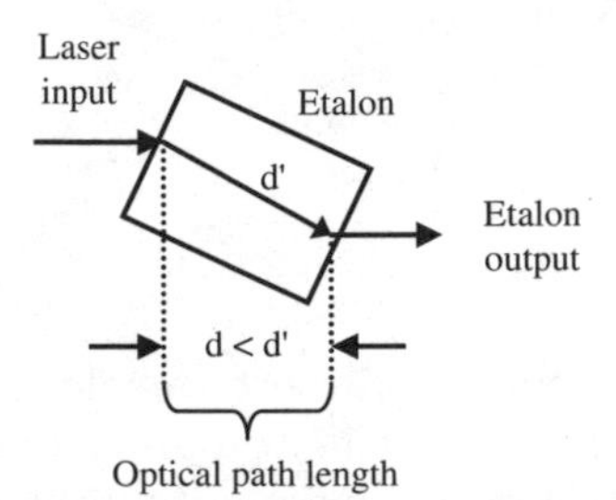

Figure 12.13. Optical path length change when tilting an etalon.

This variation in the optical path length will change the spacing between the peaks to exactly match the channel spacing. The technique of tilting the etalon to achieve a precise wavelength spacing is tedious and costly and cannot be changed once the package is sealed. Therefore a more common and simpler method is to first establish a coarse tuning using the tilting method and then to seal the package. After this one can use a temperature-tuning method to alter the refractive index, which induces a fine-tuning of the optical path length.

Figure 12.14 shows a commercially available frequency control device. Its operation is based on the use of a calibrated optical resonator locked onto a molecular filter, thereby providing an optical frequency grid reference. This particular device provides an absolute multifrequency grid over a 200-nm wavelength range with an FSR as low as 12.5 GHz, which covers the full telecommunication spectrum. The module offers different types of referencing peaks enhanced by various internal filters. Its applications include calibration of optical spectrum analyzers, tunable lasers, wavelength meters, and other instruments, and it also can be used as an absolute wavelength locker.

Figure 12.14. Example of a module that provides an absolute multi-frequency grid over a broad spectrum. (*Photo courtesy of DiCOS Technologies; www.dicostech.com.*)

12.4. Summary

Wavelength division multiplexing allows many different wavelengths selected from the spectral regions ranging from the O-band through the L-band to be sent along a single fiber simultaneously. A wide variety of passive and active components are used for WDM. The technologies include thin-film filters (TFFs), arrayed waveguide gratings (AWGs), Bragg fiber gratings, diffraction gratings, interleavers, wavelength lockers, controllers for tunable transmitters, dynamic gain equalizers (DGEs), tunable wavelength filters, and variable optical attenuators (VOAs).

The first ITU-T specification for WDM was Recommendation G.692, *Optical Interfaces for Multichannel Systems with Optical Amplifiers*. This document specifies selecting the channels from a grid of frequencies referenced to 193.100 THz (1552.524 nm) and spacing them 100 GHz (about 0.8 nm at 1550 nm) apart. Suggested alternative spacings in G.692 include 50 and 200 GHz. In 2002 the ITU-T released an updated standard for dense WDM (DWDM). This is Recommendation G.694.1, which is entitled *Dense Wavelength Division Multiplexing (DWDM)*. It specifies WDM operation in the S-, C-, and L-bands and specifies narrow frequency spacings of 100 to 12.5 GHz (or, equivalently, 0.8 to 0.1 nm at 1550 nm).

With the production of full-spectrum (low-water-content) fibers, the development of relatively inexpensive VCSEL optical sources, and the desire to have low-cost optical links operating in metro- and local-area networks came the concept of *coarse WDM* (CWDM). In 2002 the ITU-T released a standard aimed specifically at CWDM. This is Recommendation G.694.2, which is entitled *Coarse Wavelength Division Multiplexing (CWDM)*. The CWDM grid is made up of 18 wavelengths defined within the range of 1270 to 1610 nm (O- through L-bands) spaced by 20 nm with wavelength drift tolerances of ±2 nm.

Table 12.2 presents some typical performance parameters for 8-channel DWDM and CWDM multiplexers based on thin-film-filter technology. Table 12.3 lists some performance characteristics of typical 40-channel arrayed waveguide gratings. Table 12.4 gives typical performance values for commercially available interleavers.

The implementation of DWDM using very closely spaced channels is not possible without having strict wavelength stability control in the transmitter. As Sec. 12.3 describes, this is the function of commercially available wavelength lockers.

Further Reading

1. G. E. Keiser, "A review of WDM technology and applications," *Optical Fiber Tech.*, vol. 5, pp. 3–39, January 1999.
2. ITU-T Recommendation G.692, *Optical Interfaces for Multichannel Systems with Optical Amplifiers*, October 1998.
3. J. Prieur, G. Pandraud, M. Colin, and B. Vida, "High-port-count interleavers provide network design options," *WDM Solutions*, vol. 4, pp. 27–32, September 2002.
4. J.-P. Laude, *DWDM Fundamentals, Components, and Applications*, Artech House, Boston, 2002.
5. W. T. Boord, T. L. Vanderwert, and R. DeSalvo, "Bulk diffraction gratings play increasing role in optical networking," *Lightwave*, vol. 18, pp. 172–178, March 2001.
6. A. Ashmead, "Electronically switchable Bragg gratings provide versatility," *Lightwave*, vol. 18, pp. 180–184, March 2001.
7. F. Chatain, "Fiber Bragg grating technology passes light to new passive components," *Lightwave*, vol. 18, pp. 186, 190–191, March 2001.
8. H. Shakouri, "Wavelength lockers make fixed and tunable lasers precise," *WDM Solutions*, vol. 4, pp. 23–32, January 2002.

Chapter

13

Constructing the WDM Network Puzzle

In the early days of optical fiber communications, optical link design engineers only needed to be concerned with the operation of a few component types. Mainly these included light sources, optical fibers, photodetectors, connectors, splices, and couplers. The links carried a single wavelength, and the data rates were low enough that the design did not require a great deal of special signal processing to compensate for distortion effects. However, the push to increase the data rate, provide longer transmission distances, and send many wavelengths simultaneously over the same fiber has resulted in the development of numerous, highly sophisticated passive and active optical components to meet the new and ever-increasing link performance demands. The design, installation, and operation of WDM links now have become more complex with the use of these new components.

This chapter describes the major modules of a typical WDM link and explains their functions. The first part outlines performance requirements for wideband long-distance networks (known as *long-haul networks*), for metro networks, and for local-area networks (LANs). Section 13.2 gives a top-level view of how various optical devices and modular components fit into a WDM system and what their functions and impacts are on system operation. Next, Sec. 13.3 gives some examples of coarse WDM (CWDM) and dense WDM (DWDM) networks. Finally, Sec. 13.4 introduces the concepts of monitoring the performance and health of WDM links. Chapter 18 presents further details on this last topic.

13.1. Network Requirements

Figure 13.1 shows the major parts of a generic WDM link. The start of a link (shown on the left) has a series of fixed or tunable laser sources and a multiplexing device for combining the independent light signals from the sources

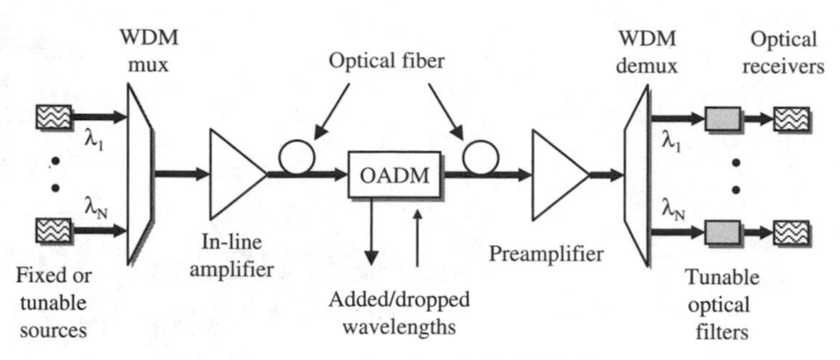

Figure 13.1. The major parts of a generic WDM link.

onto a single fiber. Within the link there may be optical amplifiers, add/drop multiplexers for inserting or subtracting individual wavelengths along the path, and other devices to enhance the link performance. At the end of the link there is a demultiplexing device for separating the wavelengths into independent signal streams and an array of tunable optical receivers.

A major application of WDM technology is to increase the capacity of long-haul links. Owing to the large amount of traffic carried on these long links, high-performance wideband components are required. Metro WDM links have a different set of applications which point to the need for lower-cost narrowband components. One important point is that WDM-based networks are bit-rate- and protocol-independent, so they can carry various types of traffic at different speeds concurrently.

13.1.1. Wideband long-haul WDM network

Wideband long-haul networks are essentially a collection of point-to-point trunk lines with one or more optical add/drop multiplexers (OADMs) for inserting and extracting traffic at intermediate points. Standard transmission distances in long-haul terrestrial WDM links are 600 km with 80 km between optical amplifiers. Since a primary desire is to have a high link capacity, modern systems can carry 160 channels running at 2.5 or 10 Gbps each (OC-48/STM-16 or OC-192/STM-64, respectively). By boosting the transmission capacity, long-haul DWDM networks lower the cost per bit for high-rate traffic.

If the 160 channels are separated by 50 GHz, then the frequency span is 8 THz (8000 GHz or a wavelength band of about 65 nm). This shows that operation is required over both the C- and L-bands simultaneously. As a result, the various active and passive components must meet high performance requirements, such as the following:

- The optical amplifiers need to operate over a wide spectral band (e.g., over both the C- and L-bands).
- High-power pump lasers are needed for the optical amplifiers in order to amplify a large number of channels.

- Each wavelength must emerge from an optical amplifier with the same power level, to prevent an increasing skew in power levels from one wavelength to another as the signals pass through successive amplifiers.
- Strict device temperature and light frequency controls are required of laser transmitters to prevent crosstalk between channels.
- High-rate transmission over long distances requires fast modulators and receivers, forward-error-correction (FEC) schemes, and optical signal-conditioning techniques such as chromatic dispersion and polarization-mode dispersion compensations.

Although these performance requirements result in expensive components, the cost is distributed over many information channels.

13.1.2. Narrowband metro WDM network

Metro topologies can be viewed as consisting of core networks and access networks, as illustrated in Fig. 13.2. Nominally a *metro core network* consists of point-to-point connections between central offices of carriers that are spaced 10 to 20 km apart. These connections typically are configured as SONET/SDH rings. The core ring usually contains from six to eight nodes and nominally is from 80 to 150 km in circumference. The *metro access network* consists of links between end users and a central office. The ring configurations in this case range from 10 to 40 km in circumference and typically contain three or four nodes. Optical add/drop multiplexers provide the capability to add or drop multiple wavelengths to other locations or networks. A router or other switching equipment allows interconnections to a long-haul network.

In contrast to the stringent performance specifications imposed on wideband long-haul WDM systems, the shorter transmission spans in metro and LAN applications relax some of the requirements. In particular, if coarse WDM is

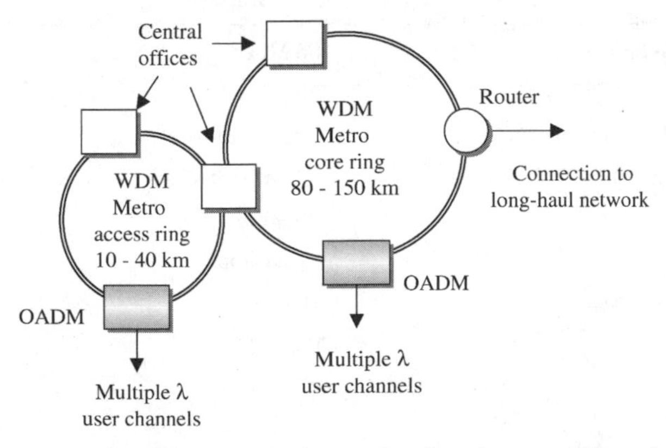

Figure 13.2. Metro topologies can be viewed as consisting of core networks and access networks.

employed, the broad frequency tolerances allow the use of devices that are not temperature-controlled. However, other requirements unique to metro applications arise, such as the following:

- A high degree of connectivity is required to support the meshed traffic in which various wavelengths are put on and taken off the fiber at different points along the path.
- A modular and flexible transmission platform is needed since the wavelength add/drop patterns and link capacities vary dynamically with traffic demands from different nodes.
- Since the add/drop functions can change dynamically from node to node, special components such as variable optical attenuators (VOAs) are needed to equalize the power of newly added wavelengths with those that already are on the fiber.
- Transmission loss in metro applications is fairly high because of interconnection losses and cascaded optical passthrough in successive nodes. Metro-optimized optical amplifiers thus are essential.
- Metro WDM networks must support a wide variety of transmission formats, protocols, and bit rates. As listed in Table 13.1, these include SONET/SDH traffic ranging from OC-3/STM-1 to OC-192/STM-64, ESCON (Enterprise Systems Connection from IBM), FICON (Fibre Channel connection from IBM), Fast Ethernet, Gigabit Ethernet, 10GigE, Fibre Channel, and digital video.

TABLE 13.1. Data Formats and Protocols That Need to Be Accommodated by a Metro WDM Network

Format or protocol	Data rate
OC-3/OC-3c and STM-1/STM-1c	155 Mbps
OC-12/OC-12c and STM-3/STM-3c	622 Mbps
OC-48/OC-48c and STM-16/STM-16c	2.488 Gbps
OC-192/OC-192c and STM-64/STM-64c	9.953 Gbps
Fast Ethernet	125 Mbps
Gigabit Ethernet (GigE)	1.25 Gbps
10-Gigabit Ethernet (10GigE)	10 Gbps
ESCON	200 Mbps
FICON	1 or 2 Gbps
Fibre Channel	133 Mbps to 1.06 Gbps
Digital Video	270 Mbps

13.2. Component Performance in WDM Links

Figure 13.3 shows the major modules contained within a WDM link. These include optical transmitters, wavelength multiplexers, interleavers, optical amplifiers, optical add/drop multiplexers, dispersion compensators, and receivers. The following sections discuss the component constituents of each of these modules and their functions in a WDM link.

13.2.1. DWDM optical transmitters

Chapter 6 gives details of optical sources and their associated transmitters. This section examines the application of transmitters in DWDM and CWDM links.

Figure 13.4 shows the various components of a high-speed DWDM optical transmitter. Although it is possible to use a fixed-wavelength laser in the transmitters, tunable lasers are a key component for simplifying and reducing the costs of DWDM operations. If a laser is tunable over a 4-nm spectral range (or a frequency range of around 500 GHz), then only 8 different laser types are required for 25-GHz channels in the 30-nm-wide C-band as opposed to using 160 different fixed-wavelength lasers. This allows a significant cost savings in stocking spares and in operational versatility for channel configuration.

The tight tolerances on wavelength stability require stringent operational control of laser transmitters in DWDM systems. A commercially available transmission controller might perform such functions as

- Efficient and high-precision laser temperature control that enables a miniature thermoelectric cooler to maintain the laser temperature to within ±0.02°C, since the wavelength changes by approximately 0.09 nm/°C.
- A dynamic wavelength-selection capability over the laser tuning range.
- Stabilization circuitry for wavelength lockers to restrict the laser output wavelength to drift less than a few picometers (e.g., ±5 pm = ±0.005 nm or equivalently a frequency accuracy of ±0.6 GHz).

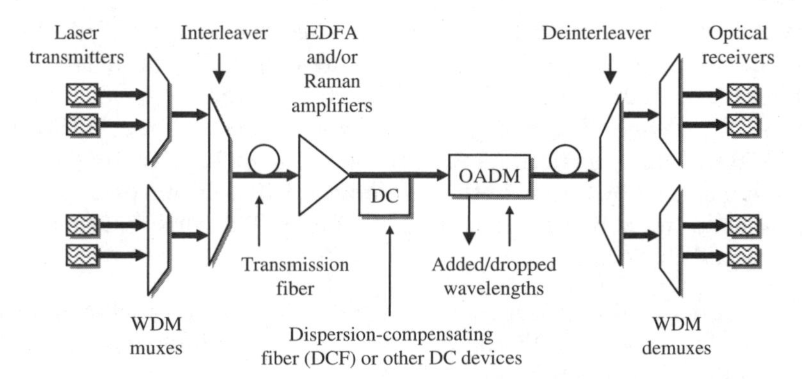

Figure 13.3. The major modules contained within a WDM link.

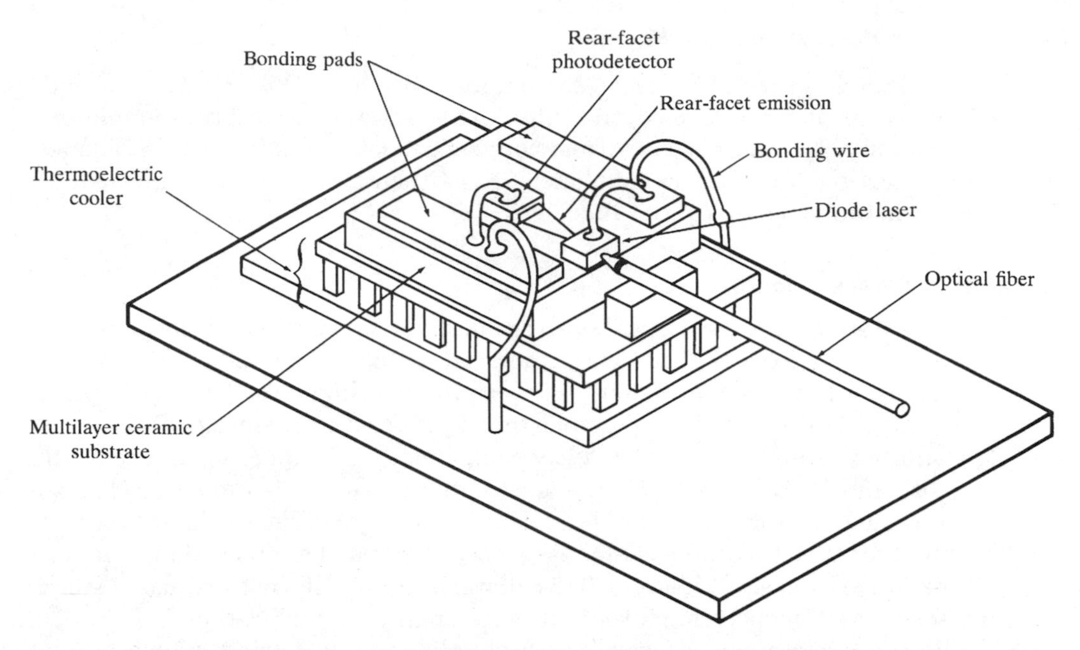

Figure 13.4. Construction of a laser transmitter that has a thermoelectric cooler and a monitoring photodiode.

- Continuous adjustment of the laser bias current (e.g., to within 7 mV).
- Stable optical output-power monitoring in order to have a reference for output-level control. The monitoring is done by built-in low-profile photodiodes.
- Constant optical-output control (e.g., within ±0.2 dB) provided by a transmitter-power controller that works in conjunction with a variable optical attenuator.
- Monitoring and alarm notification of abnormal operating conditions such as instability in temperature, output power, or drive current.

13.2.2. CWDM optical transmitters

The concept of coarse WDM arose from the production of full-spectrum fibers and the desire to have low-cost optical links operating in metro- and local-area networks. In 2002 the ITU-T released Recommendation G.694.2, entitled *Coarse Wavelength Division Multiplexing* (*CWDM*). The CWDM grid is made up of 18 wavelengths defined within the range of 1270 to 1610 nm (O- through L-bands) spaced by 20 nm with wavelength drift tolerances of ±2 nm.

The wider wavelength spacing in CWDM applications results in the following operational characteristics:

- Distributed-feedback (DFB) lasers or vertical cavity surface emitting lasers, popularly known as VCSELs (see Chap. 6), which are not temperature-controlled. These can be used since it is not necessary to maintain the lasers at a precise wavelength.

- Since a thermoelectric cooler is not needed, nominal power requirements are 0.5 W compared to 4 W for temperature-controlled DWDM lasers.
- Whereas typical wavelength tolerances for DWDM lasers are on the order of ±0.1 nm, the manufacturing tolerances for CWDM lasers are around ±3 nm. This significantly reduces yield costs.
- The elimination of thermoelectric cooler requirements and the higher manufacturing yields reduce both the cost and the size of CWDM transmitters by a factor of 4 or 5 compared to DWDM devices.
- Wideband thin-film-filter technology can be used for wavelength multiplexing at about one-half the cost of DWDM components. Figure 13.5 shows that a TFF with a 3-dB passband of typically 13 nm can accommodate the wavelength drift of 0.09 nm/°C of a typical DFB laser between the temperature ranges of 0 to 50°C.

13.2.3. Wavelength multiplexing devices

As described in Chap. 12, four of the basic device types for wavelength multiplexing are thin-film filters, fiber Bragg gratings, arrayed waveguide gratings, and bulk diffraction gratings. The performance demands on these components are increasing with the desire to support higher channel counts and longer distances between terminals.

Whereas prior to 2000 the standard wavelength spacing was 100 GHz for 2.5-Gbps DWDM links, subsequently service providers started deploying 10-Gbps ultradense systems operating with channels that are spaced 25 or 12.5 GHz apart. An even further squeezing of the channels is seen in the hyperfine WDM products that have separations down to 3.125 GHz. For 40-Gbps systems the channels nominally are spaced 50 or 100 GHz apart because of the greater impact from nonlinear dispersion effects at these higher data rates. The expansion of WDM channels beyond the C-band into the S- and L-bands has demonstrated the possibility of sending at least 320 wavelengths spaced 25 GHz apart with 10-Gbps transmission rates per channel in the combined

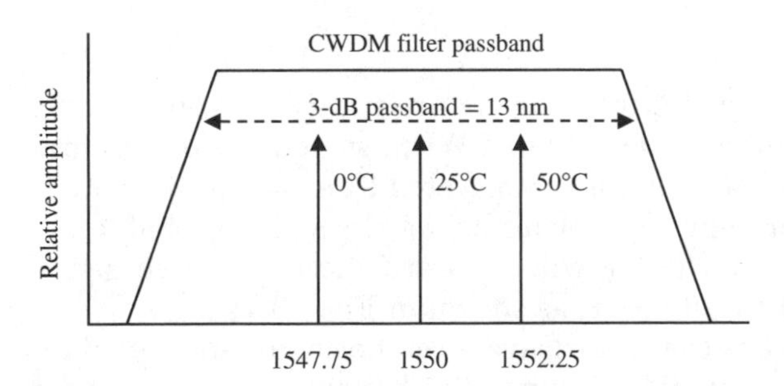

Figure 13.5. A CWDM thin-film filter with a 3-dB passband of 13 nm allows a wavelength drift of 0.09 nm/°C in the temperature range of 0 to 50°C.

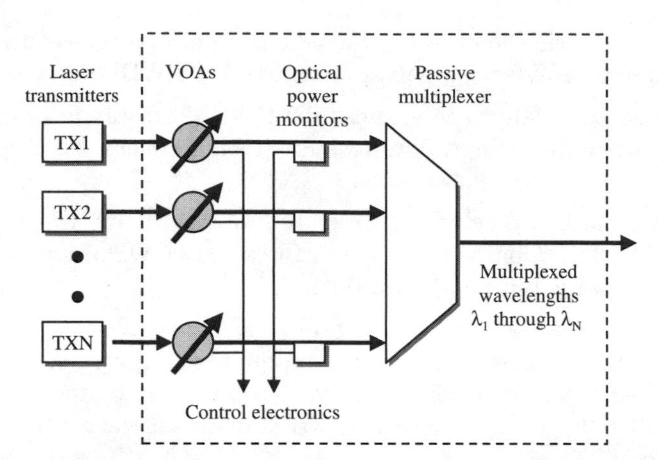

Figure 13.6. DWDM multiplexers use power monitors and VOAs to balance and control light levels across the ITU grid.

C- and L-band. This is in contrast to the earlier 72 maximum 10-Gbps channels that were separated by 50 GHz in the C-band.

Multiplexers for high-channel-count DWDM systems require a means to balance and control optical powers across the ITU grid. This can be achieved with a device that combines a passive wavelength multiplexer (e.g., an AWG- or a TFF-based device) with miniature power monitors, variable optical attenuators, and built-in control electronics. Figure 13.6 illustrates this concept. Such devices are available commercially for channel numbers ranging from 4 to 40 with 50- or 100-GHz spacing in the C- or L-band. The output from the power monitor is fed into a microprocessor which is user-configurable to maintain a uniform power level across all channels to within ± 0.5 dB over an input range of -50 to $+10$ dBm.

Multiplexers for CWDM applications have less stringent performance demands for certain parameters such as center wavelength tolerance, its change with temperature, and the passband sharpness. However, they still need to have a good reflection isolation, a small polarization-dependent loss, and low insertion losses. Passive CWDM devices can be made with thin-film-filter technology.

13.2.4. Interleavers

As described in Chap. 12, interleavers are passive, low-dispersion devices that can increase the channel density in a WDM system. They can multiplex or separate very high-density channels separated by as low as 3.125 GHz. As a simple example, consider eight wavelengths λ_1 through λ_8 that are separated by 50 GHz. A 1×2 interleaver will separate these into two sets of four wavelengths separated by 100 GHz, as shown in Fig. 13.7.

A unique feature is that interleavers can be custom-designed to route or drop a group of channels while allowing all other wavelengths to pass through the device. Alternatively a specific set of wavelengths which are not grouped can be chosen as the add/drop channels that can be demultiplexed at a certain node.

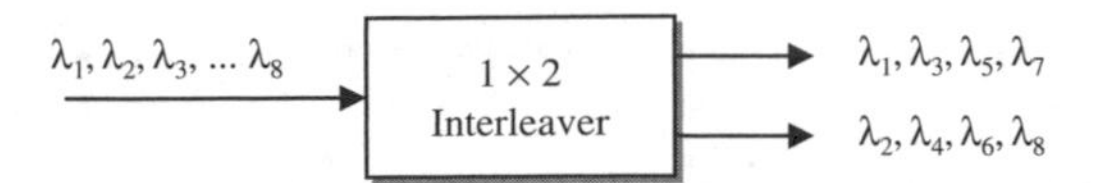

Figure 13.7. A 1 × 2 interleaver separates eight wavelengths into two sets of four wavelengths.

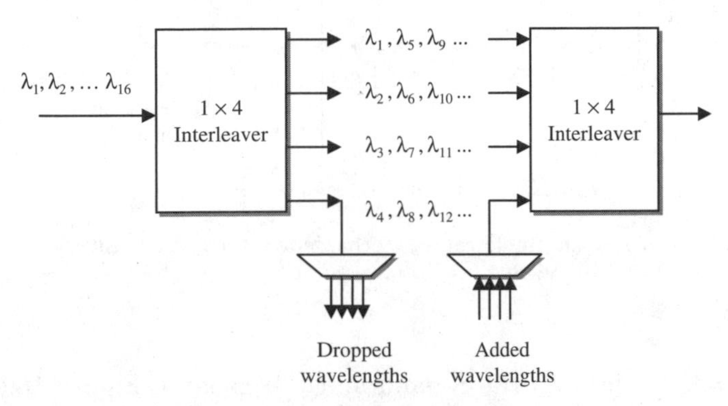

Figure 13.8. An interleaver can route or drop a group of channels while allowing all other wavelengths to pass through the device.

The remainder of the wavelengths then pass through and travel to the next node. These through-going wavelengths are known as *express traffic*. Figure 13.8 shows an example of this. Here 16 wavelengths enter a 1 × 4 deinterleaver that separates them into four sets. A second identical 1 × 4 interleaver then can recombine them onto a single fiber. Wavelengths λ_4, λ_8, λ_{12}, and λ_{16} are chosen as the add/drop channels. After being dropped, separated, and processed, these four wavelengths can be multiplexed again and inserted into the second interleaver to rejoin the express wavelengths. Although the majority of the traffic is express, the transmission through an interleaver/deinterleaver pair can place high demands on the power budget and the performance requirements to minimize channel mixing and crosstalk.

13.2.5. Wideband optical amplifiers

An amazingly rapid development of various configurations of optical amplifiers for use in WDM systems occurred since 1990. Whereas EDFA devices originally dominated for operation in the C-band (1530 to 1560 nm), Raman amplifiers for operation in any of the optical fiber transmission bands also are in use now. The use of optical amplifiers has spawned a whole new area of both passive and active optical components. Most importantly are the pump lasers that are required for signal amplification.

When EDFAs first started being used, their operation was limited to the 1530- to 1560-nm C-band in which a standard erbium-doped fiber has a high gain response. By adding a Raman amplification mechanism, the gain response

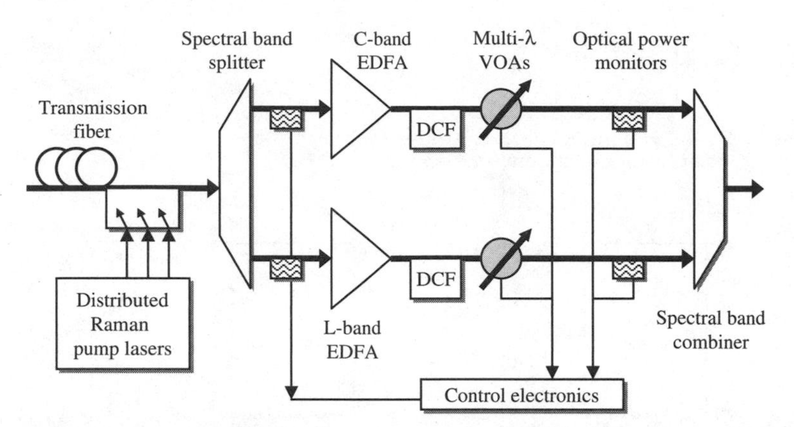

Figure 13.9. Adding a Raman amplification mechanism extends the gain response of an EDFA into the L-band.

can be extended into both the S-band and the L-band. Figure 13.9 illustrates the concept for operation in the C- and L-bands. Here a multiple-wavelength distributed Raman amplifier pump unit is added ahead of a band-splitting device. The Raman amplification boosts the power level in the L-band. After passing through the band splitter, the gains of the L-band wavelengths are further enhanced with an L-band EDFA. Following the amplification process, the wavelengths are recombined with another wideband multiplexing unit.

Other components used in this amplification system include dispersion-compensating fibers or DCFs (see Sec. 13.2.8) and gain equalization units. Gain equalization is accomplished by monitoring the optical power levels of each wavelength entering and leaving the C- and L-band EDFAs and then using a series of VOAs to adjust the wavelength power levels individually.

13.2.6. Metro optical amplifiers

A high-performance EDFA for long-haul wideband links requires peripheral components such as gain-flattening filters, cooled high-power (greater than 300-mW or +25-dBm) pump lasers, and sophisticated frequency and temperature control electronics. On the other hand, an EDFA for metro applications can be a narrowband device that typically amplifies from one to eight wavelengths. Since this covers a range of only a few nanometers, gain-flattening filters are not needed and uncooled pump lasers with a lower output power (around 80 mW or +19 dBm) can be used in metro EDFAs.

Figure 13.10 shows a schematic of a narrowband EDFA. A C-band/980-nm coupler combines the output from a 980-nm pump laser with metro-based wavelengths in the C-band. A pump reflection filter following the erbium-doped fiber prevents the 980-nm light from being coupled into the transmission fiber. The optical isolator prevents reflected C-band light from coupling back into the

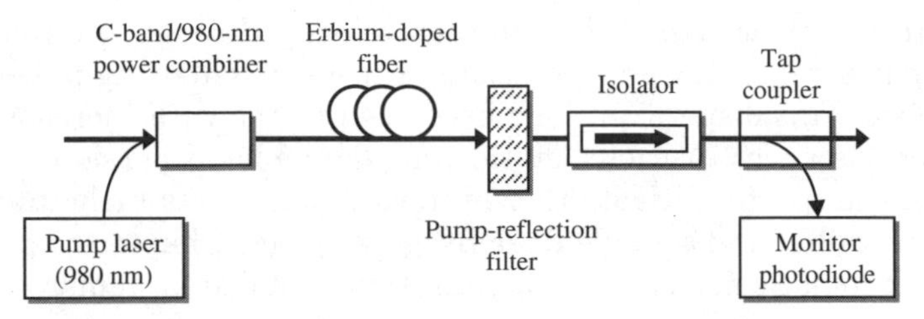

Figure 13.10. Schematic of a narrowband EDFA for use in metro networks.

EDFA where it would create output instabilities. The monitor photodiode verifies that the pump laser output is providing the desired EDFA gain.

13.2.7. Optical add/drop multiplexer

An *optical add/drop multiplexer* (OADM) allows the insertion or extraction of a wavelength from a fiber at a point between terminals. An OADM can operate either statically or dynamically. Some vendors call a dynamic device a *reconfigurable* OADM (R-OADM). A static version obviously is not as flexible and may require a hardware change if a different wavelength needs to be dropped or added. For example, a static OADM might use two optical circulators in conjunction with a series of fixed-wavelength fiber Bragg gratings. A dynamic or reconfigurable OADM results if the gratings are tunable. Although a dynamic feature adds greater flexibility to a network, this versatility also requires more careful system design. In particular, tunable (wavelength-selectable) optical filters may be needed at the drop receivers, and the optical signal-to-noise ratio for each wavelength must be analyzed more exactly.

Depending on whether an engineer is designing a metropolitan-area network (MAN) or a long-haul network, different performance specifications need to be addressed when implementing an OADM capability in the network. In general, because of the nature of the services provided, changes in the add/drop configuration for a long-haul network tend to occur less frequently than in a MAN. In addition, the channel spacing is much narrower in a long-haul network, and the optical amplifiers which are used must cover a wider spectral band. For an interesting analysis of the EDFA performance requirements and the link power budgets used for an OADM capability in a MAN environment, the reader is referred to the paper by Pan et al.

13.2.8. Chromatic dispersion compensation

Chromatic dispersion and polarization-mode dispersion are the two principal signal-distorting mechanisms in optical fiber links. To mitigate these effects, different techniques usually are implemented in separate modules for each dispersion type at the end of a fiber or following an EDFA. *Chromatic dispersion* occurs because any optical pulse contains a spectrum of wavelengths. Since each wavelength travels at a slightly different velocity through the fiber, the pulse progressively spreads out and eventually causes neighboring pulses to overlap and interfere. This dispersion issue becomes greater with increasing data rate and longer distances. Various static and dynamic dispersion compensation methods are in use to mitigate these interference effects including dispersion-compensating fiber and tunable fiber Bragg gratings. Chapter 15 presents more details on chromatic dispersion compensation in WDM systems.

13.2.9. Polarization mode dispersion compensation

Polarization mode dispersion (PMD) results from the fact that light signal energy at a given wavelength in a single-mode fiber actually occupies two orthogonal polarization states or modes (see Fig. 4.10). At the start of the fiber the two polarization states are aligned. However, fiber material is not perfectly uniform throughout its length. In particular, the refractive index varies slightly across any given cross-sectional area, which is known as the *birefringence* of the material. Consequently, each polarization mode will encounter a slightly different refractive index, so that each will travel at a slightly different velocity. The resulting difference in propagation times between the two orthogonal polarization modes will cause pulse spreading. This is the basis of polarization mode dispersion. PMD is not a fixed quantity but fluctuates with time due to factors such as temperature variations and stress changes on the fiber. It varies as the square root of distance and thus is specified as a maximum value in units of $\text{ps}/\sqrt{\text{km}}$. A typical value is $D_{\text{PMD}} = 0.05\,\text{ps}/\sqrt{\text{km}}$. Since PMD is a statistically varying parameter, it is more difficult to control than chromatic dispersion, which has a fixed value. Chapter 15 presents more details on PMD compensation.

13.3. WDM Network Applications

WDM systems are the traditional commercial choice to alleviate traffic congestion by increasing the bandwidth of existing fiber optic backbones. An important point is that WDM networks are bit-rate- and protocol-independent, which means they can carry various types of traffic at different speeds concurrently.

This section gives some examples of DWDM and CWDM networks.

13.3.1. DWDM networks

Dense WDM enables large channel counts within a limited spectral band, such as the C-band, but can be expensive to implement. However, DWDM is cost-effective in long-haul transport networks and large metro rings. In these cases the cost is justified since it is distributed over many high-capacity long-distance channels.

Figure 13.11 shows a generic long-haul DWDM network. Such networks typically are configured as large rings in order to offer reliability and survivability features. For example, if there is cable cut somewhere, the traffic that was supposed to pass through that fault can be routed in the opposite direction on the

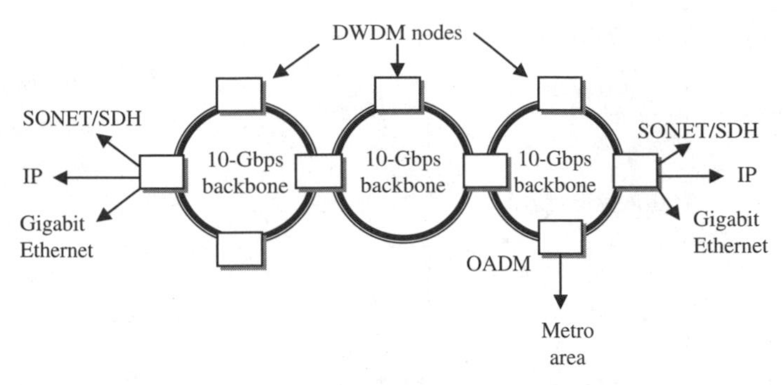

Figure 13.11. A generic long-haul DWDM network which is configured as a set of large rings.

ring and still reach its intended destination. Shown in Fig. 13.11 are three 10-Gbps DWDM rings and the major switching centers where wavelengths can be regenerated, routed, added, or dropped. The links between DWDM nodes have optical amplifiers every 80 km to boost the optical signal amplitude and regenerators every 600 km to overcome degradation in the quality of the optical signals. Extended-reach long-haul networks allow path lengths without regenerators of several thousand kilometers. Also illustrated are typical services between two end users, such as SONET/SDH, Gigabit Ethernet, or IP traffic.

13.3.2. CWDM networks

Coarse WDM applications include enterprise networks, metropolitan networks, storage area networks, and access rings. For example, within the facilities of a business organization, CWDM easily can increase the bandwidth of an existing Gigabit Ethernet optical infrastructure without adding new fiber strands. The simplest update is in a point-to-point configuration in which two user endpoints are connected directly via a fiber link. When implementing a major capacity upgrade of telecommunication campus links, CWDM enables enterprises to add or drop up to eight channels into a pair of single-mode fibers, as shown in Fig. 13.12, therefore minimizing or even negating the need for additional fiber. Since CWDM is protocol-independent, such an upgrade allows the transport of various traffic such as SONET, Gigabit Ethernet, multiplexed voice, video, or Fibre Channel on any of the wavelengths.

A more complex network is the hub-and-spoke configuration, as shown in Fig. 13.13. Here multiple nodes (or *spokes*) are connected with a central location (called a *hub*). The hubs are interconnected by means of a ring of single-mode fiber. Each hub-node connection can consist of a one or several wavelengths, each carrying a full Gigabit Ethernet channel or other protocol. Protection from fiber cuts in the ring (e.g., from cable ruptures by an errant backhoe) is achieved by connecting the hubs and nodes through bidirectional links in the optical ring. One popular application for this architecture is Gigabit Ethernet metro access rings used by telecommunication service providers.

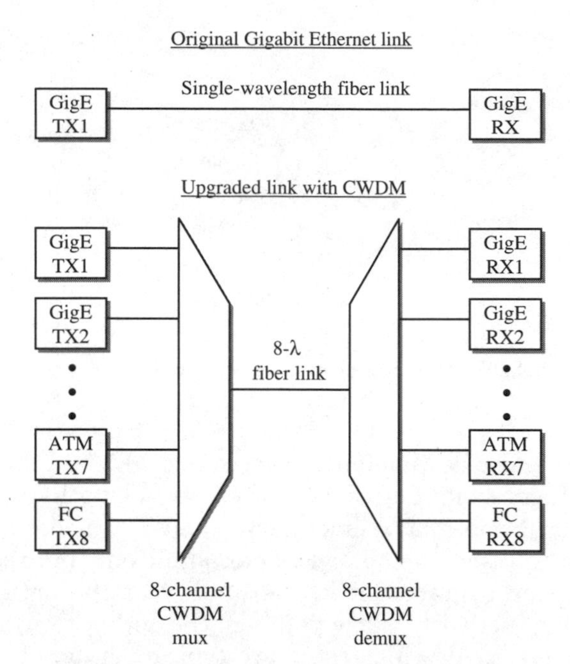

Figure 13.12. Implementing a major capacity upgrade of telecommunication campus links by means of CWDM.

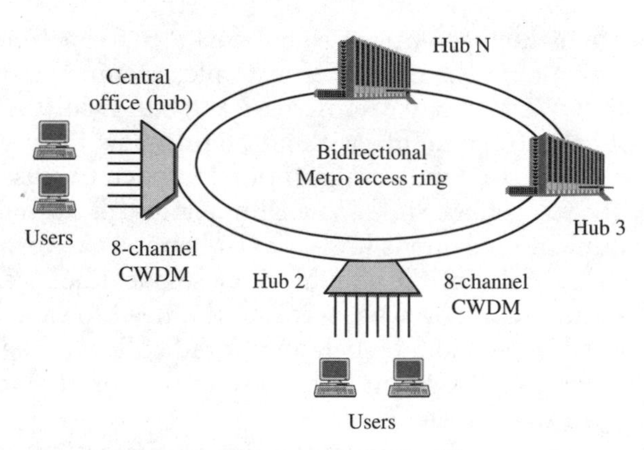

Figure 13.13. A hub-and-spoke LAN and metro network configuration.

13.4. Performance Monitoring Needs

To ensure that information is sent reliably, electric networks monitor signal integrity at strategic points in the transmission path. Traditionally this is done by monitoring the bit error rate (BER) of the data. Although BER testing can be done on an end-to-end basis in a WDM network, a different mechanism based on examining the characteristics of the optical signal is needed to verify that an optical link is operating properly.

One means is to use a miniature spectrum analyzer to monitor power versus wavelength at given points along an optical link. The resultant information of signal power versus wavelength is useful for determining the optical signal-to-noise ratio (OSNR), finding the absolute wavelength of a specific channel, and verifying that the channel spacing is adhering to the ITU wavelength specification.

The points along a link where this monitoring may be done include at the laser transmitters, at the output of optical amplifiers, at an add/drop multiplexer, and immediately following a demultiplexing device. Chapter 18 presents further details on these functions.

13.5. Summary

The discipline of wavelength division multiplexing has undergone dramatic changes in the past several years. New passive and active components are appearing steadily, and different wavelength combining methods are being examined. The technology has moved from multiplexing a few widely spaced wavelengths onto a fiber to dense WDM in multiple bands, coarse WDM ranging from the O- through the L-band, and recently hyperfine WDM where 320 wavelengths spaced 12.5 GHz apart are packed into a narrow spectral band. The future holds more excitement as researchers examine components and signal processing techniques needed for implementing wavelength channels spaced 6.25 GHz apart.

Further Reading

The following are some general sources for further study on basic aspects of optical fiber communications and WDM applications.

1. G. Keiser, *Optical Fiber Communications*, 3d ed., McGraw-Hill, Burr Ridge, Ill., 2000.
2. R. Ramaswami and K. N. Sivarajan, *Optical Networks*, 2d ed., Morgan Kaufmann, San Francisco, 2002.
3. A. Gumaste and T. Antony, *DWDM Network Designs and Engineering Solutions*, Prentice Hall, Upper Saddle River, N.J., 2003.
4. J. Hecht, *Understanding Fiber Optics*, 4th ed., Prentice Hall, Upper Saddle River, N.J., 2002.
5. I. P. Kaminow and T. Li, *Optical Fiber Telecommunications, IVA and IVB*, Academic Press, New York, 2002.
6. K. Liu and J. Ryan, "All the animals in the zoo: The expanding menagerie of optical components," *IEEE Communications Mag.*, vol. 39, pp. 110–115, July 2001.
7. H. Shakouri, Sr., "Temperature tuning the etalon," *Fiberoptic Product News*, vol. 17, pp. 38–40, February 2002.
8. C. Duvall, "VCSELs make metro networks dynamic," *WDM Solutions*, vol. 2, pp. 35–38, November 2000.
9. J. Chon, A. Zeng, P. Peters, B. Jian, A. Luo, and K. Sullivan, "Integrated interleaver technology enables high performance in DWDM systems," *Proc. National Fiber Optic Engineers Conference (NFOEC)*, pp. 1410–1421, September 2001.

10. M. Islam and M. Nietubyc, “Raman reaches for ultralong-haul WDM,” *WDM Solutions*, vol. 4, pp. 51–56, March 2002.
11. J. J. Pan, J. Jiang, X. Qiu, and K. Guan, “Scalable EDFAs simplify metro design,” *WDM Solutions*, vol. 4, pp. 15–19, July 2002.
12. L. Grüner-Nielsen, S. N. Knudsen, B. Edvold, T. Veng, D. Magnussen, C. C. Larsen, and H. Damsgaard, “Dispersion compensating fibers,” *Optical Fiber Technology*, vol. 6, pp. 164–180, 2000.

Chapter

14

Performance Measures

A major challenge in the operation of WDM networks is how to verify that the system is functioning properly. Thus there is a crucial need to monitor each wavelength intelligently in order to meet network reliability requirements and to guarantee a specific *quality of service* (QoS) to the end customer, as spelled out in a *service-level agreement* (SLA). The key performance parameters to monitor are wavelength, optical power, and optical signal-to-noise ratio (OSNR).

This chapter examines the parameters needed to evaluate the performance of a network. Section 14.1 discusses the concept of bit error rate (BER) for measuring the performance of digital systems, which encompass the predominant application of fiber optic links. Since the BER depends on the OSNR, Sec. 14.2 addresses that topic. Next, Sec. 14.3 discusses the performance of analog links, which is given in terms of a carrier-to-noise ratio. How to measure these parameters is the theme of Sec. 14.4. Chapter 18 gives more details on performance monitoring and control of optical communication links.

14.1. Digital Link Performance

In the operation of a digital optical fiber link, first a light source launches a certain amount of optical power into a fiber. As the optical signal travels along the link, it becomes attenuated due to loss mechanisms in the fiber, at connectors, and in other components. There also may be signal degradation mechanisms due to factors such as chromatic dispersion, polarization mode dispersion, nonlinear effects in the fiber, and various electrical and optical noises. As described in Chap. 7, there is a lower limit as to how weak an optical signal a receiver can detect in the presence of noise and interference. The optical power level at the end of a link defines the *signal-to-noise ratio* at the receiver, which commonly is used to measure the performance of both analog and digital communication systems. Note that the signal-to-noise ratio is designated as either *S/N* or SNR. Here we will use the abbreviation SNR.

In a digital system a photodetector in the receiver produces an output voltage proportional to the incident optical power level. This voltage is compared once per bit time to a threshold voltage level to determine whether a pulse is present at the photodetector in that time slot. Ideally the measured voltage would always exceed the threshold voltage when a 1 is present and would be less than the threshold voltage when no pulse (a 0) was sent. In an actual system, various noises and interference effects cause deviations from the expected output voltage, which leads to errors in the interpretation of the received signal.

14.1.1. Bit error rate

The most common figure of merit for digital links is the *bit error rate*, which commonly is abbreviated as BER. This is defined as the number of bit errors N_E occurring over a specific time interval, divided by the total number of bits N_T sent during that interval; that is, BER = N_E/N_T. The error rate is expressed by a number, such as 10^{-9}, which states that on the average one error occurs for every billion pulses sent. Typical error rates specified for optical fiber telecommunication systems range from 10^{-9} to 10^{-15}. The BER also is known as the *error probability*, which commonly is abbreviated as P_e.

The SNR is related to the BER through the expression

$$\text{BER} = \frac{1}{\sqrt{2\pi}} \int_0^{\infty} \exp\left(-\frac{x^2}{2}\right) dx \approx \frac{1}{\sqrt{2\pi}} \frac{e^{-Q^2/2}}{Q} \tag{14.1}$$

where the symbol Q traditionally represents the SNR for simplicity of notation. The approximation on the right-hand side holds for BER $<10^{-3}$, which means it is accurate for all cases of interest in optical fiber communications. Figure 14.1 shows how the BER or P_e varies with Q. Some commonly quoted values are $Q = 6$ for BER $= 10^{-9}$, $Q = 7$ for BER $= 10^{-12}$, and $Q = 8$ for BER $= 10^{-15}$.

The Q value is defined as

$$Q = \frac{I_1 - I_0}{\sigma_1 + \sigma_0} \tag{14.2}$$

where I_1 and I_0 are the average detected signal currents for 1 and 0 bits, respectively, and σ_1 and σ_0 are the corresponding detected *root-mean-square* (rms) *noise* values, assuming a non-return-to-zero (NRZ) code and an equal number of 1 and 0 pulses.

Certain signal degradation effects such as fiber attenuation are linear processes that can be overcome by increasing the received optical power. Other types of noise sources, such as laser relative intensity noise (RIN), are independent of signal strength and can create a noise floor that limits the system performance. Section 14.3 describes the effects of RIN.

14.1.2. Eye diagrams

The eye diagram technique is a simple but powerful measurement method for assessing the data-handling ability of a digital transmission system. This

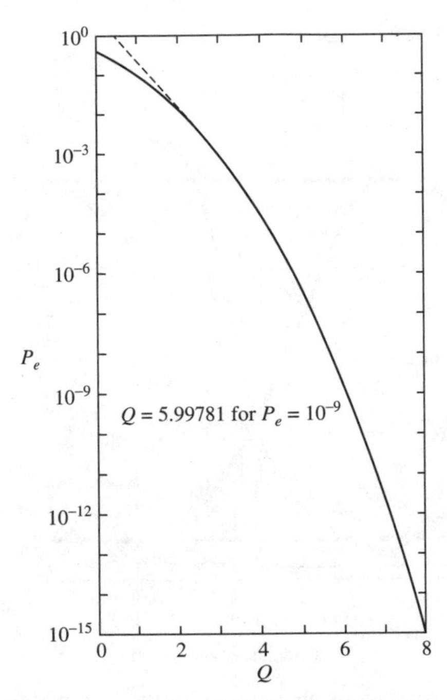

Figure 14.1. Plot of the BER versus the factor Q. The approximation from Eq. (14.1) is shown by the dashed line.

method has been used extensively for evaluating the performance of wire systems and also can be applied to optical fiber data links. The eye pattern measurements are made in the time domain and allow the effects of waveform distortion to be shown immediately on the display screen of standard BER test equipment. Figure 14.2 shows a typical display pattern, which is known as an *eye pattern* or an *eye diagram*. The basic upper and lower bounds are determined by the logic 1 and 0 levels, shown by b_{on} and b_{off}, respectively.

A great deal of system performance information can be deduced from the eye pattern display. To interpret the eye pattern, consider Fig. 14.2 and the simplified drawing shown in Fig. 14.3. The following information regarding the signal amplitude distortion, timing jitter, and system rise time can be derived:

- The *width of the eye opening* defines the time interval over which the received signal can be sampled without error due to interference from adjacent pulses (known as *intersymbol interference*).
- The best time to sample the received waveform is when the *height of the eye opening* is largest. This height is reduced as a result of amplitude distortion in the data signal. The vertical distance between the top of the eye opening and the maximum signal level gives the degree of distortion. The more the eye closes, the more difficult it is to distinguish between 1s and 0s in the signal.

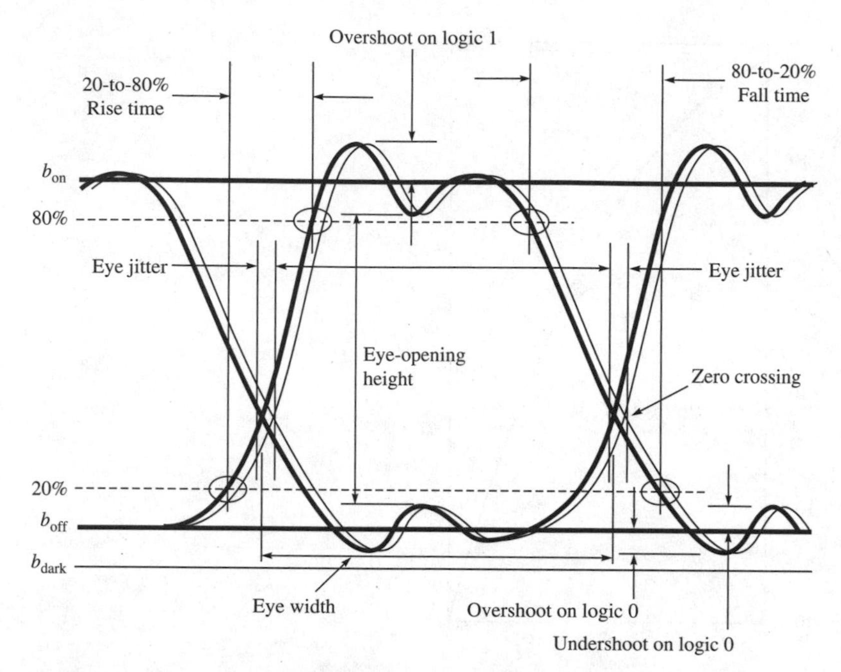

Figure 14.2. General configuration of an eye diagram showing definitions of fundamental measurement parameters.

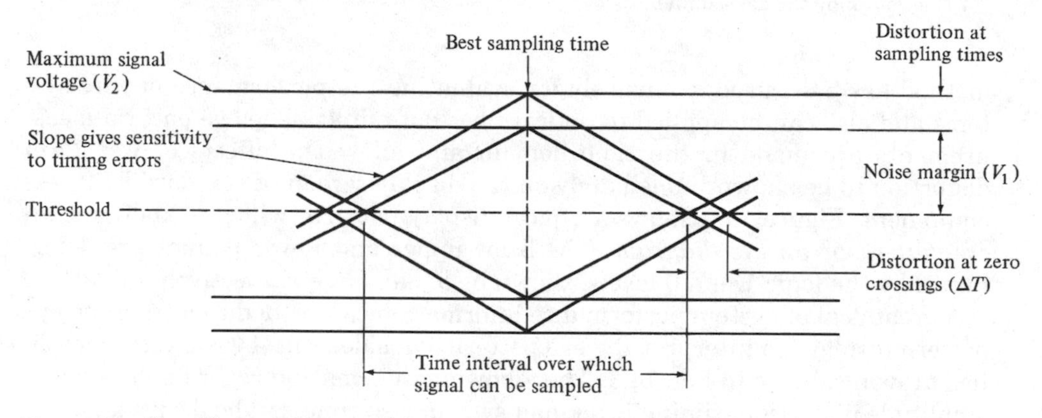

Figure 14.3. Simplified eye diagram showing key performance parameters.

- The height of the eye opening at the specified sampling time shows the noise margin or immunity to noise. *Noise margin* is the percentage ratio of the peak signal voltage V_1 for an alternating bit sequence (defined by the height of the eye opening) to the maximum signal voltage V_2 as measured from the threshold level, as shown in Fig. 14.3. That is,

$$\text{Noise margin (percent)} = \frac{V_1}{V_2} \times 100 \text{ percent} \tag{14.3}$$

- The rate at which the eye closes as the sampling time is varied (i.e., the slope of the eye-pattern sides) determines the sensitivity of the system to *timing errors*. The possibility of timing errors increases as the slope becomes more horizontal.
- *Timing jitter* (also referred to as *eye jitter* or *phase distortion*) in an optical fiber system arises from noise in the receiver and pulse distortion in the optical fiber. If the signal is sampled in the middle of the time interval (i.e., midway between the times when the signal crosses the threshold level), then the amount of distortion ΔT at the threshold level indicates the amount of jitter. Timing jitter is thus given by

$$\text{Timing jitter (percent)} = \frac{\Delta T}{T_b} \times 100 \text{ percent} \tag{14.4}$$

where T_b is the bit interval.
- Traditionally, the *rise time* is defined as the time interval between the point where the rising edge of the signal reaches 10 percent of its final amplitude to the time where it reaches 90 percent of its final amplitude. However, in measuring optical signals, these points are often obscured by noise and jitter effects. Thus, the more distinct values at the 20 percent and 80 percent threshold points normally are measured. To convert from the 20 to 80 percent rise time to a 10 to 90 percent rise time, one can use the approximate relationship

$$T_{10\text{–}90} = 1.25 T_{20\text{–}80} \tag{14.5}$$

- Any nonlinearities of the channel transfer characteristics will create an asymmetry in the eye pattern. If a purely random data stream is passed through a purely linear system, all the eye openings will be identical and symmetric.

14.2. Optical Signal-to-Noise Ratio

An important point about the BER is that it is determined principally by the *optical signal-to-noise ratio*, or OSNR. Therefore it is the OSNR that is measured when a WDM link is installed and when it is in operation. The OSNR does not depend on factors such as the data format, pulse shape, or optical filter bandwidth, but only on the average optical signal power P_{signal} and the average optical noise power P_{noise}. The TIA/EIA-526-19 standard defines OSNR over a given reference spectral bandwidth B_{ref} (normally 0.1 nm) as (in decibels)

$$\text{OSNR(dB)} = 10 \log \frac{P_{\text{signal}}}{P_{\text{noise}}} + 10 \log \frac{B_m}{B_{\text{ref}}} \tag{14.6}$$

where B_m is the noise-equivalent measurement bandwidth of the instrument being used, for example, an optical spectrum analyzer. The noise may be from sources such as the transmitter, crosstalk, or amplified spontaneous emission (ASE) from an erbium-doped fiber amplifier (EDFA). OSNR is a metric that can be used in the design and installation of networks as well as to check the health and status of individual optical channels.

In a transmission link consisting of a chain of optical amplifiers, normally the ASE noise from the optical amplifier dominates, so that one can neglect the receiver thermal noise and the shot noise. For that case the parameter Q is related to the OSNR by the expression

$$Q = \frac{2\sqrt{B_o/B_e}\,\mathrm{OSNR}}{1 + \sqrt{1 + 4(\mathrm{OSNR})}} \tag{14.7}$$

Here B_o is the bandwidth of an optical bandpass filter in front of the receiver and B_e is the electrical noise-equivalent bandwidth in the receiver.

As an example, consider a 2.5-Gbps system where one needs $Q = 7$ for BER $= 10^{-12}$. For this system let the electrical bandwidth $B_e = 2\,\mathrm{GHz}$, since for a bit rate of B bps a receiver bandwidth of at least $B_e = B/2\,\mathrm{Hz}$ is required. Furthermore, let the optical filter that is placed at the receiver have a bandwidth $B_o = 32\,\mathrm{GHz}$. In this case the required OSNR is 4.81 or 6.8 dB. However, there are still a number of other signal degradation effects such as chromatic dispersion, polarization mode dispersion, and various nonlinear processes that impose a higher OSNR requirement on the system. For practical purposes, in the design of an optical system containing a chain of amplifiers, engineers assume that an OSNR of at least 20 dB is needed at the receiver to compensate for these signal impairments.

Now let us see how this applies to the design of a system with a chain of N erbium-doped fiber optical amplifiers, as shown in Fig. 14.4. Let each EDFA have a noise figure $F = 5\,\mathrm{dB}$. Assume that the receiver will need to see an OSNR of better than 20 dB/0.1 nm of signal spectral width to maintain a BER $<10^{-12}$, take the transmitter launch power to be 1 mW (0 dBm) per DWDM channel, and let the link have an average loss of 0.25 dB/km. Now suppose we want to determine the maximum transmission distance without regenerators in a link using either 100-km or 50-km amplifier spacing. This is an important question, since the cost of the system will depend greatly on the number of amplifiers being used. For an optical link consisting of N amplified spans, let each span of length

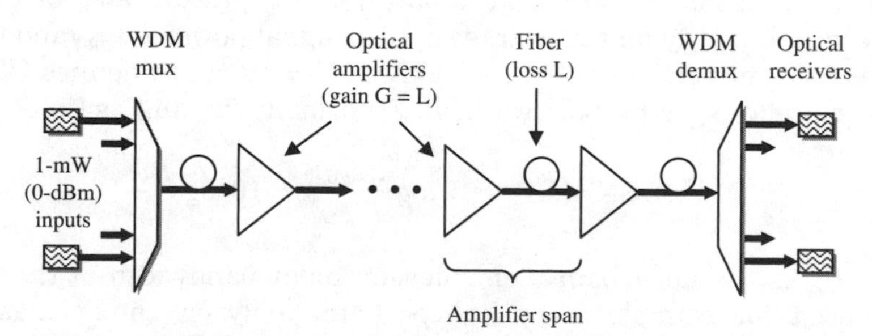

Figure 14.4. System with a chain of N erbium-doped fiber optical amplifiers in which the amplifier gain equals the span loss.

L have a loss αL (where α is the fiber attenuation) and an in-line amplifier with a gain equal to the span loss. The OSNR then is

$$\text{OSNR} = \frac{P_{\text{signal}}}{N \, \alpha L \, F \, h\nu \, \Delta\nu} \tag{14.8}$$

Here, $h\nu$ is the photon energy, and $\Delta\nu$ is the optical frequency range in which the OSNR is measured, which typically is 12.5 GHz (a 0.1-nm width at 1550 nm). At 1550 nm, we have $(h\nu)(\Delta\nu) = 1.58 \times 10^{-6}$ mW, so that $10 \log[(h\nu)(\Delta\nu)] = -58$ dBm. Then Eq. (14.8) may be expressed in decibels as

$$\begin{aligned} \text{OSNR(dB)} &= P_{\text{signal}}(\text{dBm}) - 10 \log N - \alpha L(\text{dB}) - F(\text{dB}) - 10 \log[(h\nu)(\Delta\nu)] \\ &= P_{\text{signal}}(\text{dBm}) - 10 \log N - \alpha L(\text{dB}) - F(\text{dB}) + 58 \text{ dBm} \end{aligned} \tag{14.9}$$

For a 50-km spacing between amplifiers, the span loss $\alpha L = 12.5$ dB, and for a 100-km spacing we have $\alpha L = 25$ dB. Then with OSNR $= 20$ dB, it is seen that for a 100-km spacing the limit on N is 6 (or a link length of 700 km), whereas for a 50-km spacing the limit on N is 112 in-line amplifiers (or a link length of 5650 km).

This illustrates the strong effect of in-line amplifier spacing on the OSNR. For short-haul systems the OSNR can be improved by merely launching more power into the fiber. On the other hand, in long-haul systems, nonlinear effects in the fiber will limit the maximum allowable launch power. Chapter 15 discusses these nonlinear effects and the limits they place on system performance. Note that in either the long-haul or the short-haul case, usually there are maximum power limitations that are allowed based on laser power-level safety considerations.

14.3. Analog Link Performance

Although most fiber optic systems are implemented digitally, there are certain applications where it is desirable to transmit analog signals directly over the fiber without first converting them to a digital form. These applications include cable television (CATV) distribution and microwave links such as connections between remotely located antennas and base stations.

14.3.1. Carrier-to-noise ratio (CNR)

Traditionally, in an analog system a *carrier-to-noise ratio analysis* is used instead of a signal-to-noise ratio analysis, since the information signal normally is superimposed on a radio-frequency (RF) carrier through an optical intensity modulation scheme. To find the carrier power, consider first the generated analog signal. As shown in Fig. 14.5, the drive current through the optical source is the sum of a fixed bias current and a time-varying sinusoid. If the time-varying analog drive signal is $s(t)$, then the output optical power $P(t)$ is

$$P(t) = P_t[1 + ms(t)] \tag{14.10}$$

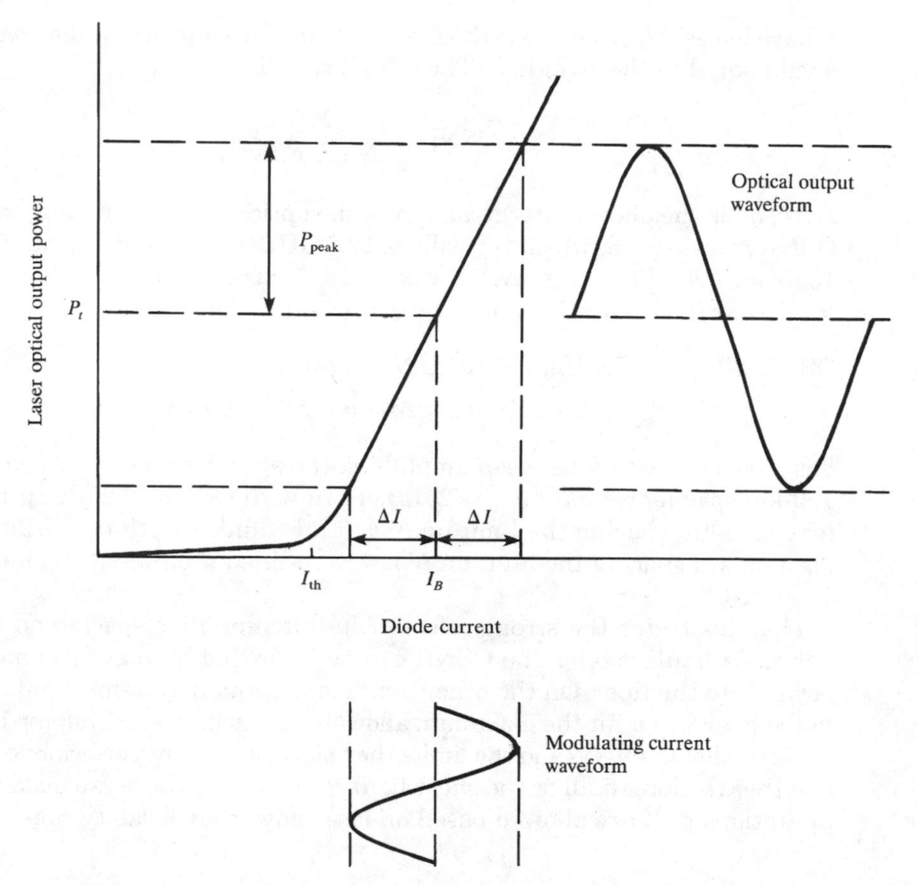

Figure 14.5. Biasing conditions of a laser diode and its response to analog signal modulation.

Here P_t is optical power level at the bias current point, and the *modulation index m* is given by

$$m = \frac{P_{\text{peak}}}{P_t} \tag{14.11}$$

where P_{peak} is defined in Fig. 14.5. Typical values of m for analog applications range from 0.25 to 0.50.

In an analog system the main noise sources are due to thermal noise, shot noise in the photodetector, and *relative intensity noise* (RIN) from the laser. RIN arises from random intensity fluctuations within a semiconductor laser, which produce optical intensity noise. These fluctuations could arise from temperature variations or from spontaneous emission contained in the laser output. RIN is measured in decibels per hertz with typical values ranging from −135 to −150 dB/Hz. Further details are given below.

For a sinusoidal received signal, the carrier power C at the output of the receiver (in units of A^2, where A designates amperes) is

$$C = \frac{1}{2}(m\mathcal{R}MP_0)^2 \tag{14.12}$$

where $\mathcal{R}$ is the responsivity of the photodetector, M is the photodetector gain, and P_0 is the average received optical power. The carrier-to-noise ratio then is given by

$$\text{CNR} = \frac{{}^1\!/_2\,(m\mathcal{R}MP_0)^2}{\text{RIN}(\mathcal{R}P_0)^2 B_e + 2q\mathcal{R}M^2F(M)P_0B_e + (4k_B T/R_{\text{eq}})B_eF_t} \tag{14.13}$$

where the noise terms in the denominator are due to RIN, shot, and thermal noises, respectively. For the other terms, $F(M)$ is the photodetector noise figure, k_B is Boltzmann's constant, R_{eq} is the equivalent resistance of the photodetector load and the preamplifier, and F_t is the noise figure of the preamplifier.

Relative Intensity Noise RIN is defined in terms of a noise-to-signal power ratio. This ratio is the mean-square optical intensity variations in a laser diode divided by the average laser light intensity. The mean-square noise current resulting from these variations is given by

$$\langle i_{\text{RIN}}^2 \rangle = \sigma_{\text{RIN}}^2 = \text{RIN}(\mathcal{R}P_0)B_e \tag{14.14}$$

The CNR due to laser amplitude fluctuations only is $\text{CNR}_{\text{RIN}} = C/\sigma_{\text{RIN}}^2$.

14.3.2. Limiting conditions on CNR

Let us now look at some limiting conditions. When the optical power level at the receiver is low, the preamplifier circuit noise dominates the system noise. For this

$$\text{CNR}_{\text{limit 1}} = \frac{{}^1\!/_2\,(m\mathcal{R}MP_0)^2}{(4k_BT/R_{\text{eq}})B_eF_t} = \frac{C}{\langle i_T^2 \rangle B_e} \tag{14.15}$$

In this case the CNR is directly proportional to the square of the received optical power. This means that for each 1-dB variation in received optical power, CNR will change by 2 dB.

For intermediate power levels the shot noise of the photodetector will dominate the system noise. In this case we have

$$\text{CNR}_{\text{limit 2}} = \frac{{}^1\!/_2\,m^2\mathcal{R}P_0}{2qF(M)B_e} \tag{14.16}$$

Here the CNR is directly proportional to the received optical power, so that for each 1-dB variation in received optical power, the CNR will change by 1 dB.

If the laser has a high RIN value so that its associated noise dominates over the other noise terms, then the carrier-to-noise ratio becomes

$$\mathrm{CNR}_{\mathrm{limit}\ 3} = \frac{\frac{1}{2}\,(mM)^2}{\mathrm{RIN}(B_e)} \tag{14.17}$$

which is a constant. In this case the performance cannot be improved unless the modulation index is increased.

As an example of the limiting conditions, consider a link with a laser transmitter and a *pin* photodiode receiver having the following characteristics:

Transmitter	Receiver
$m = 0.5$	$\mathcal{R} = 0.6\,\mathrm{A/W}$
RIN $= -143\,\mathrm{dB/Hz}$	$B = 10\,\mathrm{MHz}$
$P = 0\,\mathrm{dBm}$	$R_{\mathrm{eq}} = 750\,\Omega$
	$F_t = 3\,\mathrm{dB}$

where P is the optical power coupled into the fiber. To see the effects of the different noise terms on the carrier-to-noise ratio, Fig. 14.6 shows a plot of CNR as a function of the optical power level at the receiver. Note that in this plot the

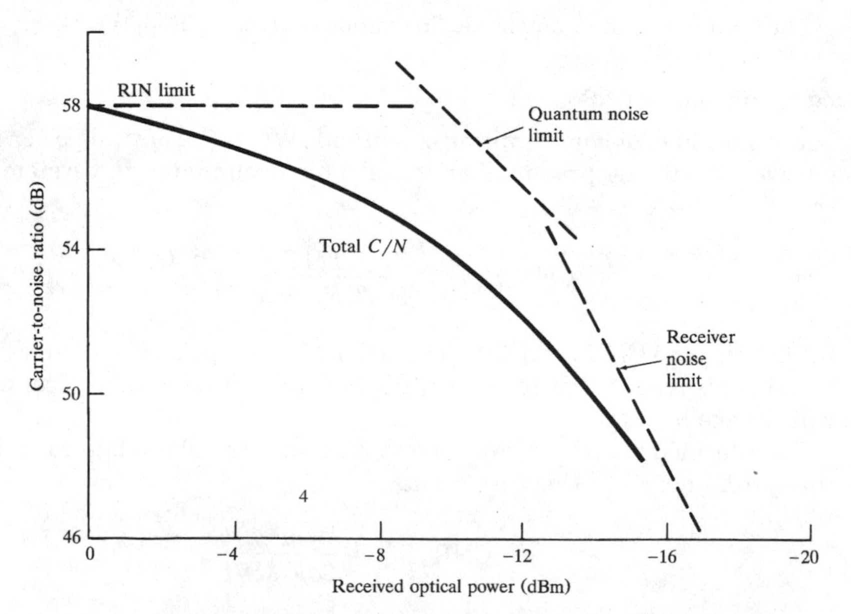

Figure 14.6. CNR as a function of optical power level at the receiver. In this case, RIN dominates at high powers, shot noise gives a 1-dB drop in CNR for each 1-dB power decrease at intermediate levels, and receiver thermal noise yields a 2-dB CNR rolloff per 1-dB drop in received power at low light levels.

power values *decrease* from left to right. At high received powers the source noise arising from RIN effects dominates to give a constant CNR. At intermediate levels the shot noise is the main contributor, with a 1-dB drop in CNR for every 1-dB decrease in received optical power. For low light levels, the thermal noise of the receiver is the limiting noise term, yielding a 2-dB rolloff in CNR for each 1-dB drop in received optical power.

It is important to note that the limiting factors can vary significantly depending on the transmitter and receiver characteristics. For example, for low-impedance amplifiers the thermal noise of the receiver can be the dominating performance limiter for all practical link lengths.

14.4. Measuring Performance Parameters

Knowing what is happening at the optical layer of a DWDM link is a critical issue for network management. The major challenge is how to do real-time dynamic optical monitoring of each channel in order to gather performance information for controlling wavelength drifts and power variations.

Several *optical channel performance monitoring* (OCPM) devices that simultaneously check the operational characteristics of all individual channels are available commercially. They provide rapid channel identification and noninvasive wavelength, power, and OSNR measurements of all DWDM channels. As shown in Fig. 14.7, an OCPM nominally consists of a spectrum-separating element, a photodetection unit, and an electronic processing unit. The spectral element separates the individual wavelengths of the composite DWDM stream. The photodetection unit is usually an array of detectors that converts the optical signal to an electric signal for further processing. Information derived from the measurements includes the central wavelength of each channel, central wavelength shifts with respect to the ITU grid, individual channel powers, channel power distribution, the presence of channels, and the OSNR of each channel.

As Fig. 14.7 shows, a fraction (usually about 1 or 2 percent) of the light power is tapped from the optical signal on the DWDM trunk line for monitoring

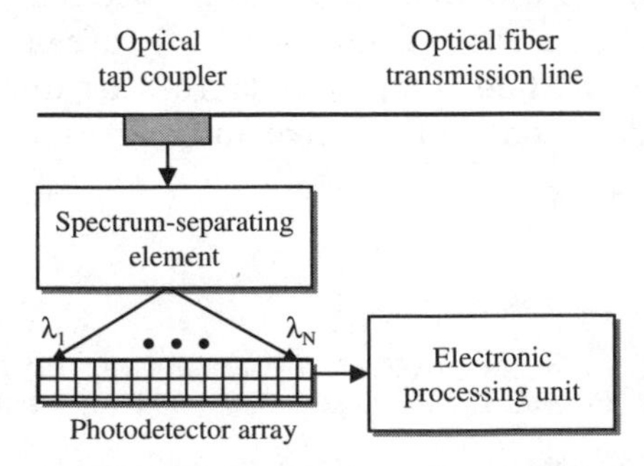

Figure 14.7. Elements and operational concepts of an optical channel performance monitor (OCPM).

purposes. Since the tapped-off signal will not be added back to the main data stream, there are no effects on the properties of the transmitted data and thus OCPM provides a noninvasive measurement process. Chapter 18 presents further details of how OCPM devices are used in an optical network.

14.5. Summary

WDM network operators need to monitor each wavelength intelligently to verify that the system is functioning properly. This is crucial in order to meet network reliability requirements and to guarantee a specific quality of service (QoS) to the end customer as spelled out in a service-level agreement (SLA). The key performance parameters to monitor are wavelength, optical power, and optical signal-to-noise ratio.

The most common figure of merit for digital links is the *bit error rate* or BER. This is the number of bit errors N_E occurring over a specific time interval, divided by the total number of bits N_T sent during that interval. BER is expressed by a number, such as 10^{-9}, which states that on the average one error occurs for every billion pulses sent. Typical error rates specified for optical fiber telecommunication systems range from 10^{-9} to 10^{-15}. The eye diagram technique is a standard measurement method for assessing the data-handling ability of a digital transmission system. The eye pattern measurements allow the effects of waveform distortion to be shown immediately on the display screen of standard BER test equipment.

An important point about the BER is that it is determined principally by the *optical signal-to-noise ratio*. Therefore it is the OSNR that is measured when a WDM link is installed and when it is in operation. The OSNR does not depend on factors such as the data format, pulse shape, or optical filter bandwidth, but only on the average optical signal power and the average optical noise power.

In certain applications it is desirable to transmit analog signals directly over the fiber without first converting them to a digital form. These applications include cable television (CATV) distribution and microwave links such as connections between remotely located antennas and base stations. Traditionally, in an analog system a carrier-to-noise ratio analysis is used, since the information signal normally is superimposed on a radio-frequency (RF) carrier through an optical intensity modulation scheme. The main noise sources are due to thermal noise, shot noise in the photodetector, and relative intensity noise from the laser.

Further Reading

1. The Telecommunications Industry Association: TIA/EIA-526-19 Standard, *OFSTP-19, Optical Signal-to-Noise Ratio Measurement Procedures for Dense Wavelength-Division Multiplexed Systems,* October 2000.
2. G. Keiser, *Optical Fiber Communications*, 3d ed., McGraw-Hill, Burr Ridge, Ill., 2000.
3. B. A. Forouzan, *Introduction to Data Communications and Networking*, 2d ed., McGraw-Hill, Burr Ridge, Ill., 2001.

4. D. Derickson, ed., *Fiber Optic Test and Measurement*, Prentice Hall, Upper Saddle River, N.J., 1998.
5. R. Ramaswami and K. N. Sivarajan, *Optical Networks*, 2d ed., Morgan Kaufmann, San Francisco, 2002.
6. W. I. Way, *Broadband Hybrid Fiber/Coax Access System Technologies*, Academic, New York, 1998.
7. S. Ovadia, *Broadband Cable TV Access Networks: From Technologies to Applications*, Prentice Hall, Upper Saddle River, N.J., 2001.

Chapter

15

Performance Impairments

The information-carrying capacity of a fiber is limited by various internal distortion mechanisms, such as signal dispersion factors and nonlinear effects. The two main distortion categories for high-performance single-mode fibers are chromatic and polarization mode dispersions, which cause optical signal pulses to broaden as they travel along a fiber. Nonlinear effects occur when there are high power densities (optical power per cross-sectional area) in a fiber. Their impact on signal fidelity includes shifting of power between wavelength channels, appearances of spurious signals at other wavelengths, and decreases in signal strength. These nonlinear effects can be especially troublesome in high-rate WDM links. Important challenges in designing such multiple-channel optical networks include

- Transmission of the different wavelength channels at the highest possible bit rate
- Transmission over the longest possible distance with the smallest number of optical amplifiers
- Network architectures that allow simple and efficient network operation, control, and management

15.1. Impairment Effects

The design of a WDM network requires the use of proven design practices when one is selecting optical fibers, specifying component types and the values of their operating parameters, and laying out networks to mitigate performance degradation processes. Signal impairment effects that are inherent in optical fiber transmission links and which can seriously degrade network performance include these:

- *Dispersion*, which refers to any effect that causes different components of a transmitted signal to travel at different velocities in an optical fiber. The result is that a pulse spreads out progressively as it travels along a link. This spreading leads to interference between adjacent pulses (called *intersymbol interference*) and may limit the distance a pulse can travel. The three main types of dispersion are modal dispersion, chromatic dispersion, and polarization mode dispersion.

Modal dispersion occurs only in multimode fibers in which each mode travels at a different velocity. Multimode fibers are used mainly for short-distance communications (such as local-area networks) since modal dispersion limits the link length. See Chap. 4 for more details on this effect.

Chromatic dispersion originates from the fact that each wavelength travels at a slightly different velocity in a fiber. Two factors contribute to chromatic dispersion, as Sec. 15.2 describes. Whether one implements high-speed single-wavelength or WDM networks, this effect can be mitigated by the use of various dispersion compensation schemes.

Polarization mode dispersion (PMD) arises in single-mode fibers because the two fundamental orthogonal polarization modes in a fiber travel at slightly different speeds owing to fiber birefringence. This effect cannot be mitigated easily and can be a very serious impediment for links operating at 10 Gbps and higher.

- *Nonlinear effects*, which arise at high power levels because both the attenuation and the refractive index depend on the optical power in a fiber. The nonlinear processes can be classified into the following two categories:

  Nonlinear inelastic scattering processes, which are interactions between optical signals and molecular or acoustic vibrations in a fiber

  Nonlinear variations of the refractive index in a silica fiber that occur because the refractive index is dependent on intensity changes in the signal
- *Nonuniform gain* across the desired wavelength range of optical amplifiers in WDM links. This characteristic can be equalized over the desired wavelength range by techniques such as the use of grating filters or variable optical attenuators (VOAs), as described in Chap. 10.
- *Reflections* from splices and connectors that can cause instabilities in laser sources. These can be eliminated by the use of optical isolators.

When any of these dispersion or nonlinear effects contribute to signal impairment, there is a reduction in the signal-to-noise ratio (SNR) of the system from the ideal case. This reduction in SNR is known as the *power penalty* for that effect, which generally is expressed in decibels.

15.2. Chromatic Dispersion

The index of refraction of silica varies with wavelength; for example, it ranges from 1.453 at 850 nm to 1.445 at 1550 nm. Furthermore, as described in Chap. 6, a light pulse from an optical source contains a certain slice of wavelength spectrum. For example, a modulated laser diode source may emit pulses that have an 0.1-nm spectral width. Consequently each wavelength within an optical pulse will see a slightly different refractive index and therefore will travel at a slightly different speed through the fiber (recall from Chap. 3 that light speed $s = c/n$). Therefore the range of arrival times at the fiber end of the spectrum of wavelengths will lead to pulse spreading. This effect is known as *material dispersion*.

Another dispersion factor is *waveguide dispersion*. This occurs because the various frequency components of a pulse travel with slightly different group velocities in a fiber, and thus arrive at different times at the fiber end.

The combination of these two factors is called *chromatic dispersion*. This often is referred to simply as *dispersion*, since it has a strong impact on the design of single-mode fiber transmission links. Chromatic dispersion is a fixed quantity at a specific wavelength and is measured in units of picoseconds per kilometer of fiber per nanometer of optical source spectral width, that is, it is measured in ps/(km · nm). For example, a single-mode fiber might have a chromatic dispersion value of $D_{CD} = 2\,\text{ps/(km} \cdot \text{nm)}$ at 1550 nm. Figure 15.1 shows the chromatic dispersion as a function of wavelength for several different fiber types, which are described in Sec. 4.8.

The accumulated dispersion increases with distance along a link. Therefore, either a transmission system has to be designed to tolerate the total dispersion, or some type of dispersion compensation method has to be employed. Although the exact calculation of the effect of dispersion is quite complex, a basic estimate of what limitation dispersion imposes on link performance can be made by specifying that the accumulated dispersion should be less than a fraction of the bit period. For example, the Bellcore/Telcordia standard GR-253-CORE specifies that for a 1-dB performance penalty the accumulated dispersion should be less

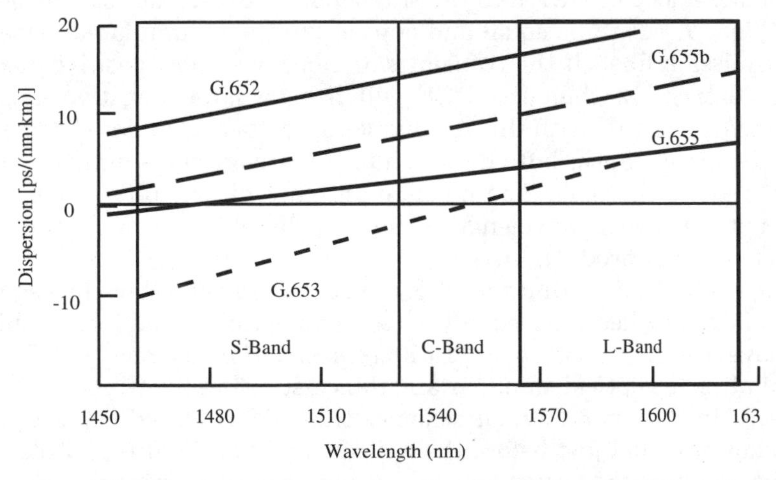

Figure 15.1. Chromatic dispersion as a function of wavelength in various spectral bands for several different fiber types.

than 0.306 of a bit period. If B is the bit rate and $\Delta\lambda$ is the spectral width of the light source, then for a link of length L this limitation can be expressed as

$$|D_{CD}|\, LB\,\Delta\lambda < 0.306 \tag{15.1}$$

As an example, if $D_{CD} = 2\,\text{ps/(km} \cdot \text{nm)}$, $B = 2.5\,\text{Gbps}$, and $\Delta\lambda = 0.1\,\text{nm}$, then the maximum allowed link length is $L = 612\,\text{km}$.

An important factor to remember when you are designing a WDM system is that in order to mitigate some of the nonlinear effects described in Sec. 15.5, the dispersion must be a positive value (or a negative value) across the entire spectral band of the WDM system. As described in Sec. 4.8, this was the motivation for designing the G.655 and G.655b optical fibers.

15.3. Dispersion Compensation

A large base of G.653 dispersion-shifted fiber has been installed throughout the world for use in single-wavelength transmission systems. As described in Sec. 15.5, a nonlinear effect called four-wave mixing (FWM) can be a significant problem for these links when one attempts to upgrade them with high-speed dense WDM technology in which the channel spacings are less than 100 GHz and the bit rates are in excess of 2.5 Gbps. One approach to reducing the effect of FWM is to use *dispersion compensation* techniques that negate the accumulated dispersion of the transmission fiber. Two possible methods are the insertion of a dispersion-compensating fiber into the link or the use of a chirped Bragg grating.

15.3.1. Dispersion-compensating fiber

A *dispersion-compensating fiber* (DCF) has a dispersion characteristic that is opposite that of the transmission fiber. Dispersion compensation is achieved by inserting a loop of DCF into the transmission path. The total dispersion in the DCF loop needs to be equal and opposite to the accumulated dispersion in the transmission fiber. If the transmission fiber has a low positive dispersion [say, $2.3\,\text{ps/(nm} \cdot \text{km)}$], then the DCF will have a large negative dispersion [say, $-90\,\text{ps/(nm} \cdot \text{km)}$]. With this technique, the total accumulated dispersion is zero after some distance, but the absolute dispersion per length is nonzero at all points along the fiber. The nonzero absolute dispersion value causes a phase mismatch between wavelength channels, thereby destroying the possibility of effective FWM production.

Figure 15.2 shows that the DCF can be inserted at either the beginning or the end of an installed fiber span between two optical amplifiers. A third option is to have a DCF at both ends. In *precompensation* schemes the DCF is located right after the optical amplifier and thus just before the transmission fiber. Conversely, in *postcompensation* schemes the DCF is placed right after the transmission fiber and just before the optical amplifier. Figure 15.2 also shows plots of the accumulated dispersion and the power level as functions of distance along the fiber. These plots are called *dispersion maps* and *power maps*, respectively.

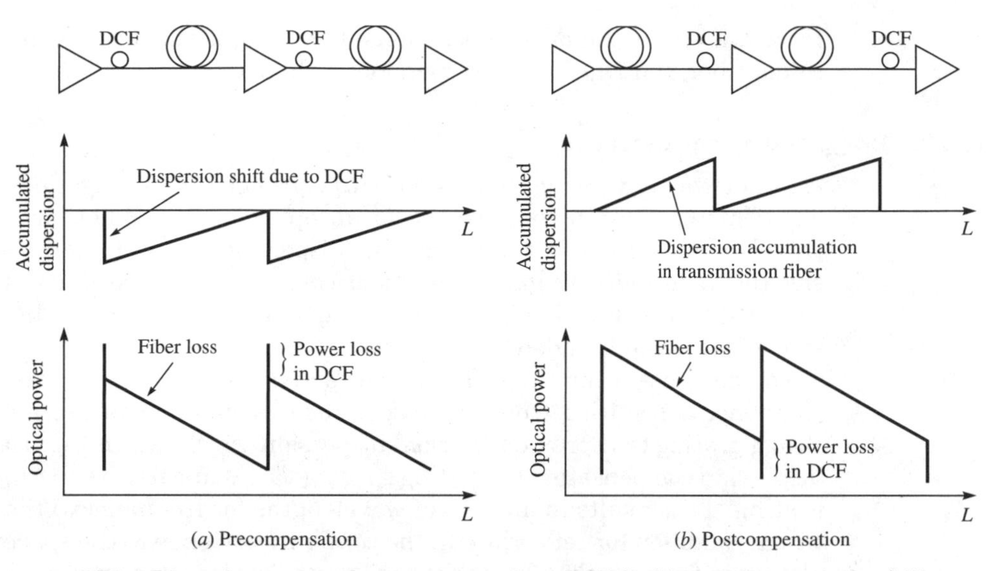

Figure 15.2. Dispersion maps and power maps using a DCF for (*a*) precompensation and (*b*) postcompensation methods.

As Fig. 15.2*a* illustrates, in precompensation the DCF causes the dispersion to drop quickly to a low negative level from which it slowly rises towards zero (at the next optical amplifier) with increasing distance along the trunk fiber. This process repeats itself following amplification. The power map shows that the optical amplifier first boosts the power level to a high value. Since the DCF is a loop of fiber, there is a drop in power level before the signal enters the actual transmission path, in which it decays exponentially before being amplified once more.

Similar processes occur in postcompensation, as shown in Fig. 15.2*b*. In either case the accumulated dispersion is near zero after some distance to minimize the effects of pulse spreading, but the absolute dispersion per length is nonzero at all points, thereby causing a phase mismatch between different wavelengths, which mitigates FWM effects.

In actual systems, both experiments and simulations have shown that a *combination* of postcompensation and precompensation provides the best solution for dispersion compensation. When one is implementing such a compensation technique, the length of the DCF should be as short as possible since the special fiber used has a higher loss than the transmission fiber. The loss is around 0.5 dB/km at 1550 nm compared to 0.21 dB/km for G.655 fiber. Since around 1 km of DCF is needed for every 10 to 12 km of operational fiber, the additional DCF loss needs to be taken into account when one is designing a link. The required length L_{DCF} of the DCF fiber can be calculated by using the expression

$$L_{\mathrm{DCF}} = |D_{\mathrm{TX}}/D_{\mathrm{DCF}}| \times L \tag{15.2}$$

Here L is the length of the operational fiber, D_{TX} is the dispersion of the operational fiber, and D_{DCF} is the dispersion of the DCF.

15.3.2. Bragg grating compensators

Another way of viewing dispersion is to consider the propagation speed of the different wavelength constituents of an optical pulse. When an optical pulse travels along a fiber in the anomalous-dispersion region (where $D_{TX} > 0$), the shorter-wavelength (higher-frequency) components of the pulse travel faster than the longer-wavelength (shorter-frequency) components. This is a dispersive effect which broadens the pulse.

To compensate for the difference in arrival times of the various frequency components resulting from anomolous dispersion, one can use a chirped fiber Bragg grating that provides normal dispersion. As shown in Fig. 15.3, in such a dispersion compensator the grating spacing varies linearly over the length of the grating. This results in a range of wavelengths (or frequencies) that satisfy the Bragg condition for reflection. In the configuration shown, the spacings decrease along the fiber which means that the Bragg wavelength decreases with distance along the grating length. Consequently the shorter-wavelength components of a pulse travel farther into the fiber before being reflected. Thereby they experience greater delay in going through the grating than the longer-wavelength components. The relative delays induced by the grating on the different frequency components of the pulse are the opposite of the delays caused by the fiber. This results in dispersion compensation, since it compresses the pulse.

Prior to the year 2000, manufacturing difficulties limited gratings to lengths of about 10 cm. Since the round-trip time T_R inside the grating of length L_G is given by $T_R = 2n_G L_G c$, where n_G is the refractive index of the grating fiber, the maximum round-trip delay time of light through a 10-cm-long grating is 1 0 μs. The delay per unit length is 500 ps/nm, which corresponds to the product of the dispersion D_G of the grating and the spectral width $\Delta\lambda$ of the light being delayed, that is,

$$\frac{T_R}{L_G} = D_G \Delta\lambda \tag{15.3}$$

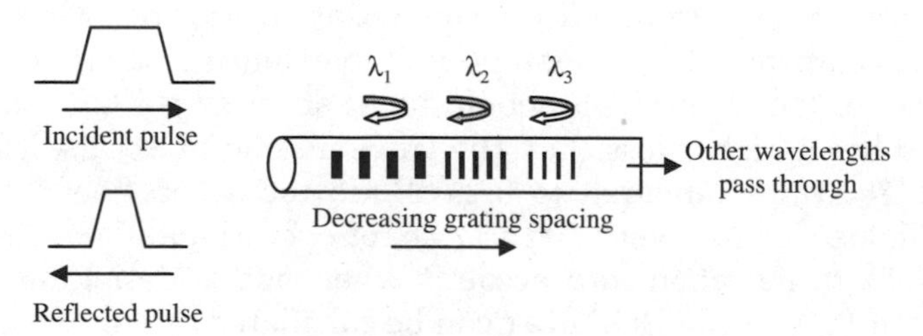

Figure 15.3. Chromatic dispersion compensation can be accomplished through the use of a chirped fiber Bragg grating.

Now one can make some application tradeoffs. Suppose one selects a compensation value of 500 ps/nm. Then, for example, since for G.655 fiber the total dispersion that arises over a 100-km length is about 500 ps/nm, in order to use a 10-cm chirped Bragg grating to compensate for this dispersion, the gratings must be used over a very narrow bandwidth of less than 1 nm. That means the short gratings made prior to 2000 only could be used to compensate for individual wavelengths in a WDM system.

In 2001, 3M Corporation developed a manufacturing technique that can make fiber gratings several meters long. For example, now it is possible to fabricate a chirped fiber Bragg grating that is about 2 m long, has a bandwidth of over 30 nm, an insertion loss of about 1 dB, and a delay slope of $-$ 1.1 ps/nm^2. These types of chirped Bragg gratings are multichannel components that are capable of compensating for chromatic dispersion across the entire C-band or L-band in individual fibers.

15.4. Polarization Mode Dispersion (PMD)

Polarization mode dispersion (PMD) results from the fact that light signal energy at a given wavelength in a single-mode fiber actually occupies two orthogonal polarization states or modes. Figure 4.10 shows this condition. At the start of the fiber the two polarization states are aligned. However, fiber material is not perfectly uniform throughout its length. In particular, the refractive index is not perfectly uniform across any given cross-sectional area, which is known as the *birefringence* of the material. Consequently each polarization mode will encounter a slightly different refractive index, so that each will travel at a slightly different velocity, and the polarization orientation will rotate with distance. The resulting difference in propagation times between the two orthogonal polarization modes will result in pulse spreading. This is the basis of polarization mode dispersion. PMD is not a fixed quantity but fluctuates with time due to factors such as temperature variations and stress changes on the fiber. Since these external stresses vary slowly with time, the resulting PMD also fluctuates slowly. It varies as the square root of distance and thus is specified as a maximum anticipated value in units of ps/$\sqrt{\text{km}}$.

A typical PMD value for a fiber is $D_{\text{PMD}} = 0.05\,\text{ps}/\sqrt{\text{km}}$, but the cabling process can increase this value. The PMD value does not fluctuate widely for cables that are enclosed in underground ducts or in buildings. However it can increase periodically to over 1 ps/$\sqrt{\text{km}}$ for outside cables that are suspended on poles, since such cables are subject to wide variations in temperature, wind-induced stresses, and elongations caused by ice loading.

Pulse spreading Δt_{PMD} resulting from polarization mode dispersion is given by

$$\Delta t_{\text{PMD}} = D_{\text{PMD}} \times \sqrt{\text{fiber length}} \tag{15.4}$$

As an example, consider a 100-km-long fiber for which $D_{\text{PMD}} = 0.5\,\text{ps}/\sqrt{\text{km}}$. Then the pulse spread over this distance is $\Delta t_{\text{PMD}} = 5.0\,\text{ps}$. Suppose one wants to send an NRZ-encoded signal over this distance and the power-penalty requirement is that the pulse spread can be no more than 10 percent of a pulse width T. In this case the maximum possible data rate is

$$\text{Maximum data rate} = \frac{0.1}{\Delta t_{\text{PMD}}} = 20\,\text{Gbps}$$

Whereas several methods exist for mitigating the effects of chromatic dispersion, it is more difficult to compensate for polarization mode dispersion. This is so because PMD varies with wavelength and slowly drifts with time in a random fashion. Since this factor requires any compensation technique to adapt dynamically to polarization state changes while the system is running, no practical PMD compensation method has been implemented as yet (for further details see the paper by Sunnerud et al.).

15.5. Nonlinear Effects

This section addresses the origins of the two nonlinear categories and shows the limitations they place on system performance. The first category encompasses the nonlinear inelastic scattering processes. These are *stimulated Raman scattering* (SRS) and *stimulated Brillouin scattering* (SBS). The second category of nonlinear effects arises from intensity-dependent variations in the refractive index in a silica fiber. This produces effects such as *self-phase modulation* (SPM), *cross-phase modulation* (XPM), and *four-wave mixing* (FWM). In the literature, FWM is also referred to as *four-photon mixing* (FPM), and XPM is sometimes designated by CPM. Table 15.1 gives a summary of these effects.

The SBS, SRS, and FWM processes result in gains or losses in a wavelength channel that are dependent on the optical signal intensity. These nonlinear processes provide gains to some channels while depleting power from others, thereby producing crosstalk between the wavelength channels. In analog video systems, SBS significantly degrades the carrier-to-noise ratio when the scattered power is equivalent to the signal power in the fiber. Both SPM and XPM affect only the phase of signals, which causes chirping in digital pulses. This can worsen pulse broadening due to dispersion, particularly in very high-rate systems (>10 Gbps).

Viewing these nonlinear processes in a little greater detail, Sec. 15.5.1 first shows how to define the distances over which the processes are important. Sections 15.5.2 and 15.5.3 then qualitatively describe the different ways in which the stimulated scattering mechanisms physically affect a lightwave

TABLE 15.1. Summary of Nonlinear Effects in Optical Fibers

Origin	Single-channel	Multiple-channel
Index-related	Self-phase modulation	Cross-phase modulation Four-wave mixing
Scattering-related	Stimulated Brillouin scattering	Stimulated Raman scattering

system. Section 15.5.4 presents the origins of SPM and XPM and shows their degradation effect on system performance. The mechanisms giving rise to FWM and the resultant physical effects on link operation are outlined in Sec. 15.5.5. As noted in Sec. 15.3, FWM can be suppressed through clever arrangements of fibers having different dispersion characteristics.

15.5.1. Effective length and area

Modeling the nonlinear processes can be quite complicated, since they depend on the transmission length, the cross-sectional area of the fiber, and the optical power level in the fiber. The difficulty arises from the fact that the impact of the nonlinearity on signal fidelity increases with distance. However, this is offset by the continuous decrease in signal power along the fiber due to attenuation. In practice, one can use a simple but sufficiently accurate model that assumes the power is constant over a certain fiber length, which is less than or equal to the actual fiber length. This *effective length* L_{eff}, which takes into account power absorption along the length of the fiber (i.e., the optical power decays exponentially with length), is given by

$$L_{\text{eff}} = \frac{1 - e^{-\alpha L}}{\alpha} \tag{15.5}$$

Given a typical attenuation of $\alpha = 0.22\,\text{dB/km}$ (or, equivalently, $\alpha = 5.07 \times 10^{-2}\,\text{km}^{-1}$) at 1550 nm, this yields an effective length of about 20 km when $L_{\text{eff}} \gg 1/\alpha$. When there are optical amplifiers in a link, the signal impairments owing to the nonlinearities do not change as the signal passes through the amplifier. In this case, the effective length is the sum of the effective lengths of the individual spans between optical amplifiers. If the total amplified link length is L_A and the span length between amplifiers is L, the effective length is approximately

$$L_{\text{eff}} = \frac{1 - e^{-\alpha L}}{\alpha} \frac{L_A}{L} \tag{15.6}$$

Figure 15.4 illustrates the effective length as a function of the actual system length. The two curves shown are for a nonamplified link with $\alpha = 0.22\,\text{dB/km}$ and an amplified link with a 75-km spacing between amplifiers. As indicated by Eq. (15.6), the total effective length decreases as the amplifier span increases.

The effects of nonlinearities increase with the light intensity in a fiber. For a given optical power, this intensity is inversely proportional to the cross-sectional area of the fiber core. Since the power is not distributed uniformly over the fiber-core cross section, for convenience one can use an *effective cross-sectional area* A_{eff}. Although this can be calculated accurately from mode-overlap integrals, in general the effective area is close to the actual core area. As a rule of thumb, standard nondispersion-shifted single-mode fibers have effective areas of $80\,\mu\text{m}^2$, dispersion-shifted fibers have effective areas of $55\,\mu\text{m}^2$, and dispersion-compensated fibers have effective areas on the order of $20\,\mu\text{m}^2$.

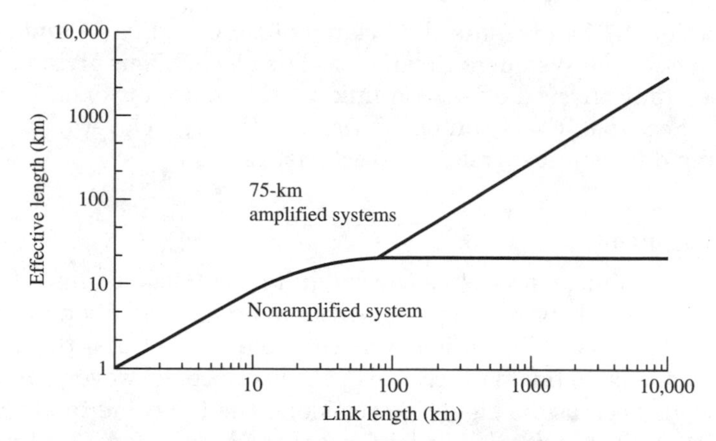

Figure 15.4. Effective length as a function of the actual link length for a link without amplifiers and a link with optical amplifiers spaced 75 km apart.

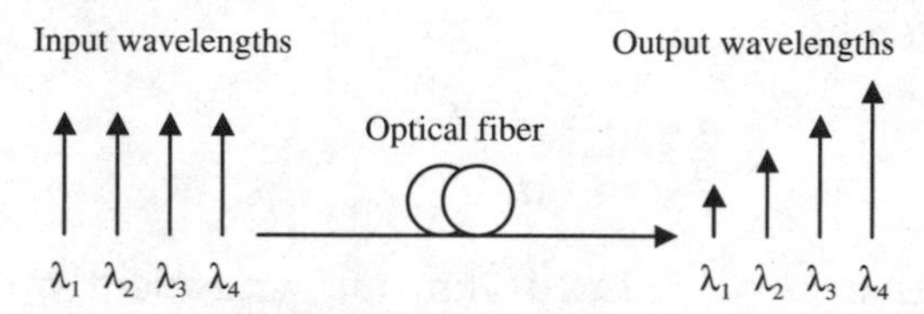

Figure 15.5. SRS generates scattered light at a longer wavelength, thereby decreasing the power in the pump wavelength signal.

15.5.2. Stimulated Raman scattering

Stimulated Raman scattering is an interaction between light waves and the vibrational modes of silica molecules. If a photon with energy $h\nu_1$ is incident on a molecule having a vibrational frequency ν_m, the molecule can absorb some energy from the photon. In this interaction the photon is scattered, thereby attaining a lower frequency ν_2 and a corresponding lower energy $h\nu_2$. The modified photon is called a *Stokes photon*. Because the optical signal wave that is injected into a fiber is the source of the interacting photons, it is often called the *pump wave*, since it supplies power for the newly generated wave.

This process generates scattered light at a wavelength longer than that of the incident light. If another signal is present at this longer wavelength, the SRS light will amplify it and the pump wavelength signal will decrease in power. Figure 15.5 illustrates this effect. Consequently, SRS could limit the performance of a multichannel optical communication system by transferring energy from short-wavelength channels to neighboring higher-wavelength channels. This is a broadband effect that can occur in both directions. Powers in channels separated by up to 16 THz (125 nm) can be coupled through the SRS effect, thereby producing crosstalk between wavelength channels.

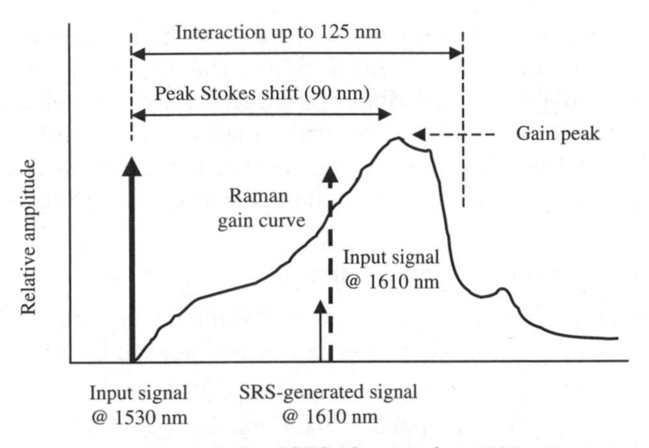

Figure 15.6. As a result of SRS, the signal at 1530 nm acts as a pump for the signal at 1610 nm.

In general, if the optical power per channel is not excessively high (e.g., less than 1 mW each), then the effects of SRS do not contribute significantly to the eye closure penalty as a function of transmission distance. Furthermore, since the generated wavelength is around 90 nm away from the signal wavelength, crosstalk from SRS is not a significant problem unless the system operates over a wide spectral range. As an example, consider a WDM system that operates over both the C-band and the L-band (from 1530 to 1620 nm). As shown in Fig. 15.6 for two specific wavelengths in this range, as a result of SRS the signal at 1530 nm acts as a pump for the signal at 1610 nm.

15.5.3. Stimulated Brillouin scattering

Stimulated Brillouin scattering arises when light waves scatter from acoustic waves. The resultant scattered wave propagates principally in the backward direction in single-mode fibers. This backscattered light experiences gain from the forward-propagating signals, which leads to depletion of the signal power. The frequency of the scattered light experiences a Doppler shift given by

$$\nu_B = \frac{2nV_s}{\lambda} \tag{15.7}$$

where n is the index of refraction and V_s is the velocity of sound in the material. In silica this interaction occurs over a very narrow *Brillouin linewidth* of $\Delta\nu_B = 20\,\text{MHz}$ at 1550 nm. For $V_s = 5760$ m/s in fused silica, the frequency of the backward-propagating light at 1550 nm is downshifted by 11 GHz (0.09 nm) from the original signal. This shows that the SBS effect is confined within a single wavelength channel in a WDM system. Thus, the effects of SBS accumulate individually for each channel, and consequently they occur at the same power level in each channel as occurs in a single-channel system.

System impairment starts when the amplitude of the scattered wave is comparable to the signal power. For typical fibers the threshold power for this process is around 10 mW for single-fiber spans. In a long fiber chain containing optical amplifiers, there are normally optical isolators to prevent backscattered signals from entering the amplifier. Consequently, the impairment due to SBS is limited to the degradation occurring in a single amplifier-to-amplifier span.

15.5.4. Self-phase modulation and cross-phase modulation

The refractive index n of many optical materials has a weak dependence on optical intensity I (equal to the optical power per effective area in the fiber) given by

$$n = n_0 + n_2 I = n_0 + n_2 \frac{P}{A_{\text{eff}}} \tag{15.8}$$

where n_0 is the ordinary refractive index of the material and n_2 is the *nonlinear index* coefficient. In silica, the factor n_2 varies from 2.2 to 3.4×10^{-8} μm²/W. The nonlinearity in the refractive index is known as the *Kerr nonlinearity*. This nonlinearity produces a carrier-induced phase modulation of the propagating signal, which is called the *Kerr effect*. In single-wavelength links, this gives rise to *self-phase modulation* (SPM), which converts optical power fluctuations in a propagating light wave to spurious phase fluctuations in the same wave.

To see the effect of SPM, consider what happens to the optical pulse shown in Fig. 15.7 as it propagates in a fiber. Here the time axis is normalized to the parameter t_0, which is the pulse half-width at the $1/e$ intensity point. The edges of the pulse represent a time-varying intensity, which rises rapidly from zero to a maximum value and then returns to zero. In a medium that has an intensity-dependent refractive index, a time-varying signal intensity will produce a time-varying refractive index. Thus the index at the peak of the pulse will be slightly different from the value in the wings of the pulse. The leading edge will see a positively changing index change (represented by $+dn/dt$), whereas the trailing edge will see a negative change (represented by $-dn/dt$).

This temporally varying index change results in a temporally varying phase change, shown by $d\phi/dt$ in Fig. 15.7. The consequence is that the instantaneous optical frequency differs from its initial value across the pulse. That is, since the phase fluctuations are intensity-dependent, different parts of the pulse undergo different phase shifts. This leads to what is known as *frequency chirping*, in that the rising edge of the pulse experiences a shift toward higher frequencies whereas the trailing edge of the pulse experiences a shift toward lower frequencies. Since the degree of chirping depends on the transmitted power, SPM effects are more pronounced for higher-intensity pulses.

For some types of fibers, the time-varying phase may result in a power penalty owing to a spectral broadening of the pulse as it travels along the fiber. In the wavelength region where chromatic dispersion is negative, the leading edge of the pulse travels faster and thus moves away from the center of the pulse. The

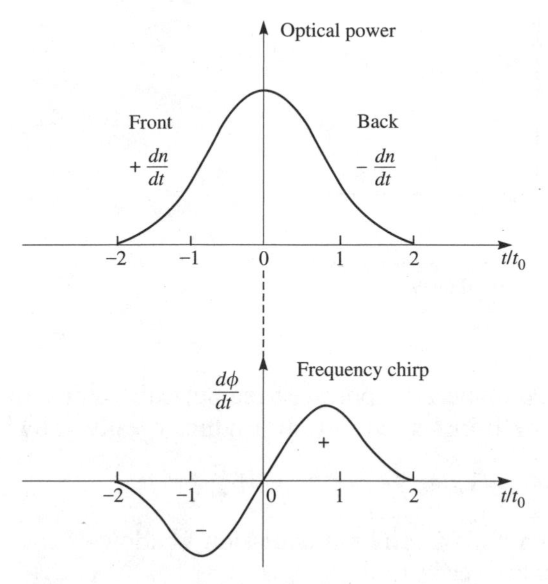

Figure 15.7. Phenomenological description of spectral broadening of a pulse due to self-phase modulation.

trailing edge travels more slowly and thus also moves away from the center of the pulse, but in the opposite direction from the leading edge. Therefore in this case chirping worsens the effects of pulse broadening. On the other hand, in the wavelength region where chromatic dispersion is positive, the leading edge of the pulse travels more slowly and thus moves toward the center of the pulse. Similarly, the trailing edge travels faster and thus also moves toward the center of the pulse from the other direction. In this case, SPM causes the pulse to narrow, thereby partly compensating for chromatic dispersion.

In WDM systems, the refractive index nonlinearity gives rise to *cross-phase modulation* (XPM), which converts power fluctuations in a particular wavelength channel to phase fluctuations in other copropagating channels. This can be mitigated greatly in WDM systems operating over standard non-dispersion-shifted single-mode fiber, but can be a significant problem in WDM links operating at 10 Gbps and higher over dispersion-shifted fiber. When combined with fiber dispersion, the spectral broadening from SPM and XPM can be a significant limitation in very long transmission links, such as cross-country or undersea systems.

15.5.5. Four-wave mixing

Four-wave mixing is a third-order nonlinearity in silica fibers that is analogous to intermodulation distortion in electrical systems. When wavelength channels

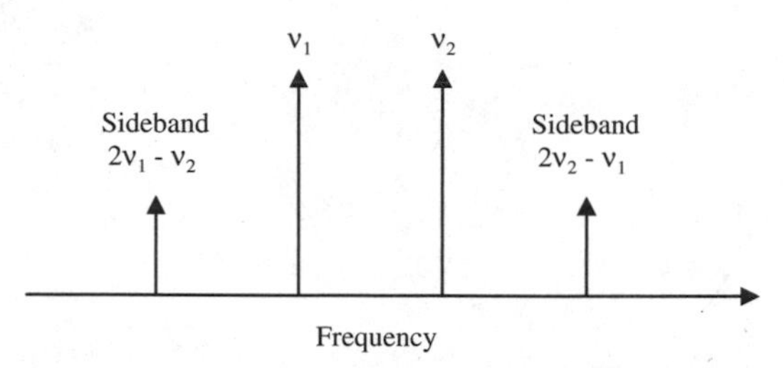

Figure 15.8. Two optical waves at frequencies ν_1 and ν_2 mix to generate two third-order sidebands.

are located near the zero-dispersion point, three optical frequencies (ν_i, ν_j, ν_k) will mix to produce a fourth intermodulation product ν_{ijk} given by

$$\nu_{ijk} = \nu_i + \nu_j - \nu_k \qquad \text{with } i, j \neq k \tag{15.9}$$

When this new frequency falls in the transmission window of the original frequencies, it can cause severe crosstalk.

Figure 15.8 shows a simple example for two waves at frequencies ν_1 and ν_2. As these waves copropagate along a fiber, they mix and generate sidebands at $2\nu_1 - \nu_2$ and $2\nu_2 - \nu_1$. Similarly, three copropagating waves will create nine new optical sideband waves at frequencies given by Eq. (15.9). These sidebands will travel along with the original waves and will grow at the expense of signal strength depletion. In general, for N wavelengths launched into a fiber, the number of generated mixing products M is

$$M = \frac{N^2}{2}(N - 1) \tag{15.10}$$

If the channels are equally spaced, a number of the new waves will have the same frequencies as the injected signals. Thus, the resultant crosstalk interference plus the depletion of the original signal waves can severely degrade multichannel system performance unless steps are taken to diminish it.

The efficiency of four-wave mixing depends on fiber dispersion and the channel spacings. Since the dispersion varies with wavelength, the signal waves and the generated waves have different group velocities. This destroys the phase matching of the interacting waves and lowers the efficiency at which power is transferred to newly generated frequencies. The higher the group velocity mismatches and the wider the channel spacings, the lower the four-wave mixing. The G.655 and G.655b fibers were developed specifically to address the FWM issues.

15.6. Summary

Signal dispersion factors and nonlinear effects in fibers limit the information carrying capacity in optical links. For high-performance single-mode fibers, chromatic and polarization mode dispersions cause optical signal pulses to broaden as they travel along a fiber. Nonlinear effects occur when there are high power densities (optical power per cross-sectional area) in a fiber. Their impact on signal fidelity includes shifting of power between wavelength channels, appearances of spurious signals at other wavelengths, and decreases in signal strength. These nonlinear effects can be especially troublesome in high-rate WDM links.

When any of these dispersion or nonlinear effects contribute to signal impairment, there is a reduction in the signal-to-noise ratio (SNR) of the system from the ideal case. This reduction in SNR is known as the *power penalty* for that effect, which generally is expressed in decibels.

Chromatic dispersion originates from the fact that each wavelength travels at a slightly different velocity in a fiber. Whether one implements high-speed single-wavelength or WDM networks, this effect can be mitigated by the use of dispersion-compensating fiber or a chirped Bragg grating. *Polarization mode dispersion* (PMD) arises in single-mode fibers because the two fundamental orthogonal polarization modes in a fiber travel at slightly different speeds owing to fiber birefringence. This effect cannot be mitigated easily and can be a very serious impediment for links operating at 10 Gbps and higher.

Two categories of nonlinear effects can place limitations on system performance. The first category encompasses the nonlinear inelastic scattering processes. These are stimulated Raman scattering (SRS) and stimulated Brillouin scattering (SBS). The second category of nonlinear effects arises from intensity-dependent variations in the refractive index in a silica fiber. This produces effects such as self-phase modulation (SPM), cross-phase modulation (XPM), and four-wave mixing (FWM). Table 15.1 gives a summary of these effects.

Further Reading

1. Telcordia, *SONET Transport Systems, Common Generic Criteria*, GR-253-CORE, Piscataway, N.J., September 2000.
2. L. Grüner-Nielsen, S. N. Knudsen, B. Edvold, T. Veng, D. Magnussen, C. C. Larsen, and H. Damsgaard, "Dispersion compensating fibers," *Optical Fiber Technology*, vol. 6, pp. 164–180, 2000.
3. M. Karlsson, J. Brentel, and P. A. Andrekson, "Long-term measurement of PMD and polarization drift in installed fibers," *J. Lightwave Technology*, vol. 18, pp. 941–951, July 2000.
4. H. Sunnerud, C. Xie, M. Karlsson, R. Samuelsson, and P. A. Andrekson, "A comparison between different PMD compensation techniques," *J. Lightwave Technology*, vol. 20, pp. 368–378, March 2002.
5. F. Forghieri, R. W. Tkach, and A. R. Chraplyvy, "Fiber nonlinearities and their impact on transmission systems," Chap. 8, pp. 196–264, in I. P. Kaminow and T. L. Koch, eds., *Optical Fiber Telecommunications—III*, vol. A, Academic, New York, 1997.

Chapter

16

Optical Link Design

The preceding chapters have presented the fundamental characteristics of individual building blocks of an optical fiber communication link and various concepts, such as WDM, for implementing links. This chapter describes how these individual parts can be put together to form complete optical fiber transmission links. The main emphasis is on digital systems.

First Sec. 16.1 looks at the design criteria for a point-to-point link. This includes examining the components that are available for a particular application and seeing how they relate to the system performance criteria (such as bit error rate and dispersion). Section 16.2 then shows that for a given set of components and a specified set of system requirements, one can carry out a power budget analysis to determine whether the designed fiber optic link meets the allowed attenuation across the link or if amplifiers are needed periodically to boost the power level. Next Sec. 16.3 describes how to perform a system rise-time analysis to verify that the overall bandwidth requirements are met.

Section 16.4 addresses line-coding schemes that are suitable for digital data transmission over optical fibers. These coding schemes are used to introduce randomness and redundancy into the digital information stream to ensure efficient timing recovery and to facilitate error monitoring at the receiver.

To increase the end-to-end fidelity of an optical transmission line, forward error correction (FEC) can be used if the bit error rate is limited by optical noise and dispersion. This is the topic of Sec. 16.5.

Owing to the complexity of modern optical fiber links, a variety of component, link, and system modeling simulation tools have been developed. Section 16.6 describes the characteristics of some commercially available tools.

16.1. System Considerations

The design of a high-quality transmission link involves a series of tradeoffs among the many interrelated performance variables of each component based on the system operating requirements. Thus the actual link design analyses may require several iterations before they are completed satisfactorily. Since performance and cost constraints are very important factors in a transmission link, the designer must choose the components carefully to ensure that the link meets the operational specifications over the expected system lifetime without overstating the component requirements.

16.1.1. System requirements

The following key system requirements are needed in analyzing a link:

- The transmission distance
- The data rate or channel bandwidth
- The bit error rate (BER)
- The number of WDM channels
- The link margin
- Acceptable power penalties

To fulfill these requirements, the designer has a choice of many components with various levels of associated characteristics, such as those listed in Table 16.1.

TABLE 16.1. Types and Characteristics of Components Used for Optical Link Design

Component	Type	Characteristics
Optical fibers	Single-mode or multimode	Attenuation, dispersion, CWDM or DWDM use
Light source	LED, DFB laser, VCSEL, or other	Modulation rate, output power, wavelength, spectral width, cost
Photodetector	*pin* or APD	Sensitivity, responsivity
Connectors	Single- or multiple-channel	Loss, size, mounting type
Wavelength multiplexers	AWG, TFF, grating-based	Cost, channel width
Optical amplifiers	EDFA, Raman, SOA	Complex long-haul, lower-cost metro, gain, wavelength range
Passive components	Optical filters, OADM, dispersion compensators, optical isolators, couplers	Peak wavelength, spectral range, loss, size, cost, reliability
Active components	VOA, dynamic gain equalizers, tunable optical filters, optical add/drop multiplexers	Tuning speed, peak wavelength, spectral range, loss, size, cost, reliability
Monitoring devices	BER tester, spectrum analyzer, power meter	Optical power, wavelength, OSNR

16.1.2. Link margin

Link margin (also called a *loss margin* or a *system margin*) is an optical-power safety factor for link design. This involves adding extra decibels to the power requirements to compensate for possible unforeseen link degradation factors. These degradations could arise from factors such as a dimming of the light source over time, aging of other components in the link, the possibility that certain splices or connectors in the actual link have a higher loss than anticipated, or additional losses occurring when a cable is repaired.

ITU-T Recommendation G.957 specifies that a link margin ranging from 3.0 to 4.8 dB should be allowed between the transmitter and the receiver to offset possible equipment degradation. In an actual system, designers typically add a link margin of 3 to 10 dB depending on the performance requirements of the application, the number of possible repairs, and the system cost.

16.1.3. Power penalties

Certain operational factors in a link usually contribute to signal impairment. Among these factors are modal noise, chromatic dispersion, polarization mode dispersion, reflection noise in the link, low extinction ratios in the laser, or frequency chirping. When any of these dispersion or nonlinear effects contribute to signal impairment, there is a reduction in the signal-to-noise ratio (SNR) of the system from the ideal case. This reduction in SNR is known as the *power penalty* for that effect, which generally is expressed in decibels.

- Modal noise arises when one is using multimode fiber and a laser source. When one is using a laser diode source, initially the optical output excites only a few lower-order modes at the beginning of a multimode fiber. As the light travels along the fiber, its power gets coupled back and forth among various propagating modes. This leads to a power fluctuation at the fiber end, which degrades the signal-to-noise ratio, thereby leading to a power penalty. The use of a light-emitting diode (LED) greatly reduces modal noise since an LED has a wide spectral output range and therefore couples power into the various possible modes more uniformly.
- Chromatic dispersion is described in Sec. 15.2. The power penalty arising from chromatic dispersion (CD) can be calculated from

  $$P_{\mathrm{CD}} = -5 \log [1-(4BLD\Delta\lambda)^2] \qquad (16.1)$$

  where B is the bit rate in gigabits per second, L is the fiber length in kilometers, D is the chromatic dispersion in ps/(nm·km), and $\Delta\lambda$ is the spectral width of the source in nanometers. To keep the power penalty less than 0.5 dB, a well-designed system should have the quantity $BLD\,\Delta\lambda < 0.1$.
- Polarization mode dispersion (PMD) arises in single-mode fibers because the two fundamental orthogonal polarization modes in a fiber travel at slightly different speeds owing to fiber birefringence, as described in Sec. 15.4. As

described by Poole and Nagel, to avoid having a power penalty of 1 dB or greater for a fractional time of 30 min/yr, the average differential time delay between the two different polarization states must be less than 0.14 of the bit period.

- The *extinction ratio* in a laser is defined as the ratio of the "on" power for a logic 1 to the "off" power for a logic 0. (Note that sometimes in earlier literature the extinction ratio is defined alternatively as the ratio of the "off" power for a logic 0 to the "on" power for a logic 1.) Ideally one would like the extinction ratio to be infinite, so that there would be no power penalty from this condition. However, the extinction ratio must be finite in an actual system in order to reduce the rise time of laser pulses. The power penalty increases significantly for lower extinction ratios. If r_e represents the ratio of the average power in a logic 1 to the average power in a logic 0, then the power penalty P_{ER} arising from a nonideal extinction ratio (ER) is given by

$$P_{\mathrm{ER}} = -10 \log \frac{r_e - 1}{r_e + 1} \tag{16.2}$$

  In practice, optical transmitters have minimum extinction ratios ranging from 7 to 10 (8.5 to 10 dB), for which the power penalties range from 1.25 to 0.87 dB. A minimum extinction ratio of 18 is needed to have a power penalty of less than 0.5 dB.

- Frequency chirping is described in Sec. 15.5. The chirping power penalty arises from the fact that the light output experiences a dynamic spectral broadening (or a frequency chirp) when the laser is directly modulated. The chirping power penalty is reduced for higher bias settings, but this increases the penalty arising from the lower extinction ratio. When analyzed in conjunction with the extinction ratio degradation, the combined power penalty typically is less than 2 dB for an extinction ratio setting of about 10.
- Power penalties for nonlinear effects in optical fibers are described in Sec. 15.5.

16.2. Link Power Budget

In carrying out a *link power budget analysis*, one first determines the *power margin* between the optical transmitter output and the minimum receiver sensitivity needed to establish a specified BER. This margin then can be allocated to fiber, splice, and connector losses, plus any additional margins required for other components, possible device degradations, transmission line impairments, or temperature effects. If the choice of components did not allow the desired transmission distance to be achieved, the components might have to be changed or amplifiers might have to be incorporated into the link.

The first step in evaluating a power budget is to decide at which wavelength to transmit and then select components that operate in this region. If the distance over which the signal is to be transmitted is not far, for example, in a campus network, then operation in the 800- to 900-nm region may be desirable to save on component costs. On the other hand, if the transmission distance is relatively long and the bit rate is high, the lower attenuation and smaller dispersion of the O- or C-bands may be more advantageous.

After a wavelength has been chosen, the next step is to interrelate the system performances of the three major optical link building blocks, that is, the receiver, transmitter, and optical fiber. Normally the designer chooses the characteristics of two of these elements and then computes those of the third to see if the system performance requirements are met. The procedure we shall follow here is to select first the photodetector and then the optical source. Then one can see how far signals can be sent over a particular fiber before an amplifier is needed.

16.2.1. Power budgeting process

Figure 16.1 shows a hypothetical point-to-point link. Here there are connectors on each end of the link and N splices located periodically along the cable length. The optical power arriving at the photodetector depends on the amount of light coupled into the fiber minus the losses incurred along the path. The link loss budget is derived from the sequential loss contributions of each element in the link. Each of these losses is expressed in decibels as

$$\text{Loss} = 10 \log \frac{P_{\text{out}}}{P_{\text{in}}} \tag{16.3}$$

where P_{in} and P_{out} are the optical powers entering and exiting, respectively, a fiber, splice, connector, or other link element.

The link loss budget simply considers the total optical power loss P_T that is allowed between the light source and the photodetector and allocates this loss to factors such as cable attenuation, connector and splice losses, losses in other link components, and system margin. Thus, referring to Fig. 16.1, if P_S is the optical power emerging from the end of a fiber flylead attached to the source and if P_R is the minimum receiver sensitivity needed for a specific BER, then

$$\begin{aligned} P_T &= P_S - P_R \\ &= 2 \times \text{connector loss} + \alpha L + N \times \text{splice loss} \\ &\quad + \text{other losses} + \text{system margin} \end{aligned} \tag{16.4}$$

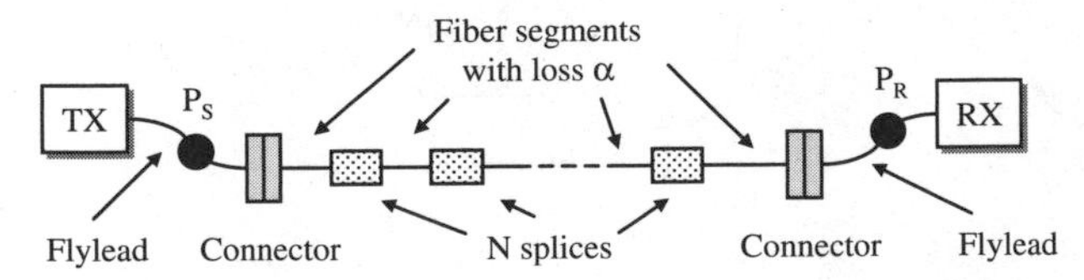

Figure 16.1. A hypothetical point-to-point link that contains N periodic splices along the cable and has connectors on each end.

where α is the fiber attenuation (dB/km) and L is the link length. As noted in Sec. 16.1.2, the link margin normally is selected to be between 3 and 10 dB.

Now let us look at some examples of how to calculate a link loss budget. We will use examples of links operating at 850 nm, at 1310 nm, and in the C-band.

16.2.2. Fast Ethernet LAN example

A *Fast Ethernet* link is known as a 100BASE-T link. Typically such a link is used in a local-area network (LAN) and runs at 100 Mbps. Since the distance is short, we can use components operating at 850 nm, which are less expensive than longer-wavelength devices. Suppose that two computers are to be connected by means of a 150-m fiber link length that has been installed within a building, as shown in Fig. 16.2. The fiber here is multimode fiber with a 50-μm core diameter and has an attenuation of 2.5 dB/km at 850 nm. The fiber ends are terminated in a connector in a patch panel. A 5-m patch cord with connectors on both ends is used to connect each computer to this fiber line. Thus there are four connectors in the transmission path, but there are no splices.

For this system we assume the desired BER is 10^{-9} (i.e., at most one error can occur for every 10^9 bits sent). Figure 16.3 shows that for a silicon *pin* photodiode the required input signal at the receiver at 100 Mbps is -32 dBm (or 630 nW, which is 32 dB below 1 mW) in order to have BER $= 10^{-9}$ at 850 nm. Assume that the 100BASE-T transmitter couples -20 dBm (10 μW) into the fiber and that each connector has a maximum loss of 0.7 dB.

A convenient procedure for calculating the power budget is to use a tabular or spreadsheet form. This is illustrated in Table 16.2 for the case here. This table lists the components in the leftmost column and the associated optical output, sensitivity, or loss in the center column. The rightmost column gives the power margin available after subtracting the component loss from the total

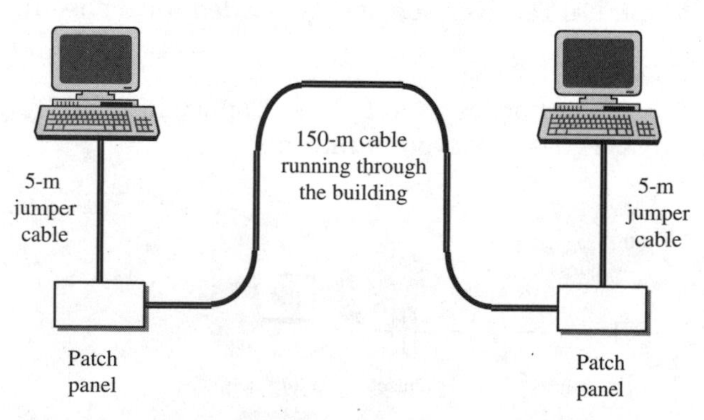

Figure 16.2. Two computers connected by a 150-m fiber link plus two jumper cables within a building.

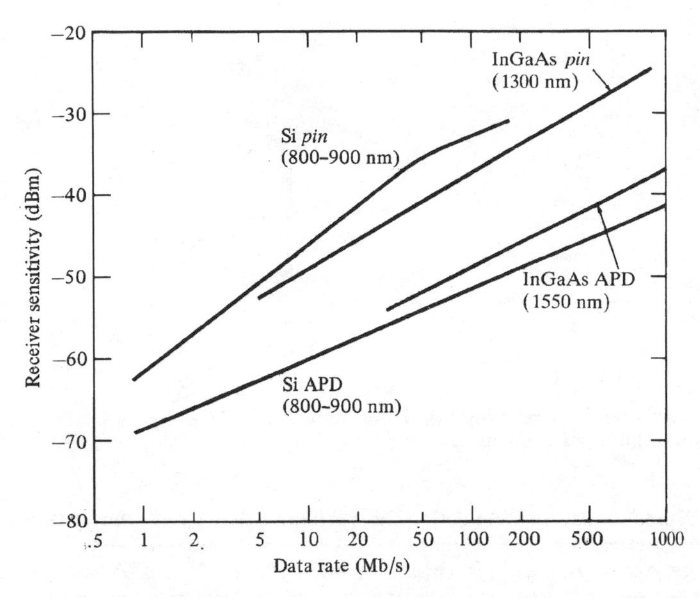

Figure 16.3. Receiver sensitivities as a function of bit rate. The Si *pin*, Si APD, and InGaAs *pin* curves are for a 10^{-9} BER. The InGaAs APD curve is for a 10^{-11} BER.

TABLE 16.2. Spreadsheet for Calculating a LAN Link Power Budget

Component/loss parameter	Output/sensitivity/loss	Power margin, dB
Coupled LED output	−20 dBm	
pin sensitivity at 100 Mbps	−32 dBm	
Allowed loss [−20 − (−32)]		12
Source connector loss	−0.7 dB	11.3
2 × Jumper connector loss	−1.4 dB	9.9
Cable attenuation (160 m)	−0.4 dB	9.5
Receiver connector loss	−0.7 dB	8.8 (final margin)

optical power loss that is allowed between the light source and the photodetector. In this case the allowable loss is

$$P_T = P_S - P_R = -20\,\text{dBm} - (-32\,\text{dBm}) = 12\,\text{dB}$$

The final power margin is 8.8 dB, which is a sufficient margin for this link.

16.2.3. SONET/SDH link example

An engineer plans to design a 2.5-Gbps SONET OC-48 (or equivalently, an SDH STM-16) link over a 30-km path length, as shown in Fig. 16.4. The basic question

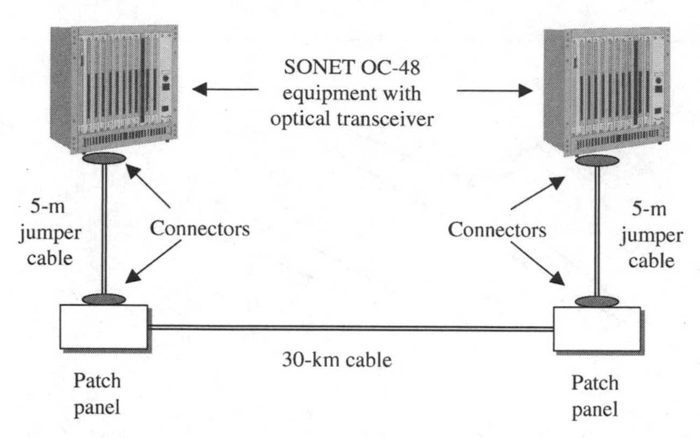

Figure 16.4. A 2.5-Gbps link running over a 30-km path length with a short optical jumper cable at each end.

TABLE 16.3. Spreadsheet for Calculating the 1310-nm SONET Link Power Budget

Component/loss parameter	Output/sensitivity/loss	Power margin, dB
Coupled laser diode output	−2 dBm	
APD sensitivity at 2.5 Gbps	−32 dBm	
Allowed loss [−2 − (−32)]		30.0
Jumper cable loss (2 × 1.5 dB)	−3 dB	27.0
Splice loss (5 × 0.1 dB)	−0.5 dB	26.5
Connector loss (4 × 0.6 dB)	−2.4 dB	24.1
Cable attenuation (30 km)	−18 dB	6.1 (final margin)

is whether to operate at 1310 nm or to use more costly 1550-nm components. Therefore the first step is to calculate the 1310-nm power budget. Suppose the installed fiber for the link meets the G.655 specification and that at 1310 nm the fiber attenuation is 0.6 dB/km versus 0.3 dB/km at 1550 nm. For the 30-km cable span, there is a splice with a loss of 0.1 dB every 5 km (a total of 5 splices). The engineer selects a laser diode that can launch −2 dBm of optical power into the fiber and an InGaAs avalanche photodiode (APD) with a −32-dBm sensitivity at 2.5 Gbps. Assume that here, because of the way the equipment is arranged, a short optical jumper cable is needed at each end between the transmission cable and the SONET equipment rack. Assume that each jumper cable introduces a loss of 1.5 dB. In addition, there is a 0.6-dB connector loss at each fiber joint (two at each end because of the jumper cables for a total of four connectors).

Table 16.3 gives the spreadsheet for calculating the 1310-nm link power budget. The final link margin is 6.1 dB. Therefore operation at 1310 nm is adequate in this case. Note that there still is the question of whether the link meets the bandwidth requirements, which is addressed in Sec. 16.3.

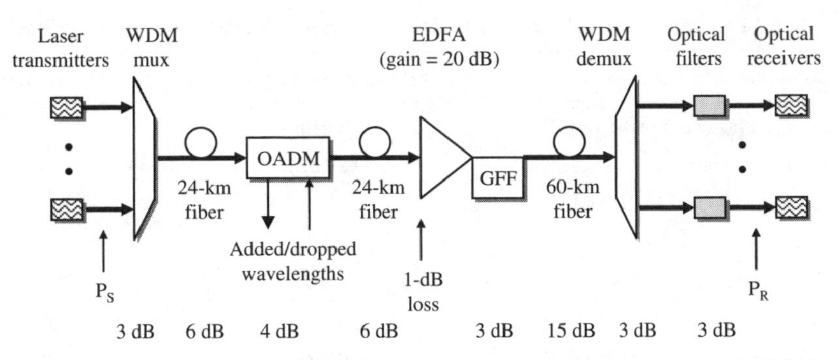

Figure 16.5. Example of optical power losses of various components in a WDM link.

16.2.4. DWDM link example

As a more complex example, consider the WDM link shown in Fig. 16.5. Assume this is a four-channel link with each channel running at 10 Gbps. The system operates in the C-band and contains an optical add/drop multiplexer (OADM), an EDFA with a gain of 20 dB, a gain-flattening filter (GFF) following the EDFA, and optical filters at the receivers. The fiber has an attenuation of 0.25 dB/km. The individual laser diode transmitters have fiber-coupled outputs of $P_S = 2$ dBm, and the individual InGaAs APD receivers need a power level of at least $P_R = -24$ dBm to maintain a 10^{-11} BER.

The task is to make sure there is sufficient power margin in the link. The various power losses are shown at the bottom in Fig. 16.5 and are listed in Table 16.4. The final margin is only 2.0 dB. This is not sufficient, particularly since no allowance was made yet for any possible power penalties. Thus an optical amplifier is needed, for example, just ahead of the wavelength demultiplexer.

16.3. Rise-Time Budget

A *rise-time budget analysis* is a convenient method for determining the dispersion limitation of an optical link. This is particularly useful for a digital link. In this approach the total rise time t_{sys} of the link is the root-sum-square calculation of the rise times from each contributor t_i to the pulse rise-time degradation, that is, if there are N components in a link that affect the rise time then

$$t_{sys} = \left(\sum_{i=1}^{N} t_i^2 \right)^{1/2} \tag{16.5}$$

The five basic elements that may limit the system speed significantly are the transmitter rise time t_{TX}, the modal dispersion rise time t_{mod} of multimode fiber, the chromatic dispersion (CD) rise time t_{CD} of the fiber, the polarization mode dispersion (PMD) rise time t_{PMD} of the fiber, and the receiver rise time t_{RX}.

TABLE 16.4. Spreadsheet for Calculating the C-Band WDM Link Power Budget

Component/loss parameter	Output/sensitivity/loss	Power margin, dB
Coupled laser diode output	+2 dBm	
APD sensitivity at 10 Gbps	−24 dBm	
Allowed loss [+2 − (−24)]		26.0
WDM mux loss	−3 dB	23.0
Cable attenuation (24 km)	−6 dB	17.0
OADM loss	−4 dB	13.0
Cable attenuation (24 km)	−6 dB	7.0
EDFA coupling loss	−1 dB	6.0
EDFA gain	+20 dB	26.0
GFF loss	−3 dB	23.0
Cable attenuation (60 km)	−15 dB	8.0
WDM demux loss	−3 dB	5.0
Optical filter loss	−3 dB	2.0 (final margin)

Substituting these parameters into Eq. (16.5) then yields

$$t_{\text{sys}} = (t_{\text{TX}}^2 + t_{\text{mod}}^2 + t_{\text{CD}}^2 + t_{\text{PMD}}^2 + t_{\text{RX}}^2)^{1/2} \tag{16.6}$$

Single-mode fibers do not experience modal dispersion (that is, $t_{\text{mod}} = 0$), so in these fibers the rise time is related only to CD and PMD.

16.3.1. Basic rise times

Generally the total transition-time degradation t_{sys} of a digital link should not exceed 70 percent of an NRZ (non-return-to-zero) bit period or 35 percent for RZ (return-to-zero) data, where 1 bit period is defined as the reciprocal of the data rate. Section 16.4 discusses NRZ and RZ data formats in greater detail.

The rise times of the transmitters and receivers generally are known to the link designer. The transmitter rise time is attributable primarily to the speed at which a light source responds to an electric drive current. A *rule-of-thumb estimate* for the *transmitter rise time* is 2 ns for a light-emitting diode (LED) and 0.1 ns for a laser diode source. The receiver rise time results from the photodetector response speed and the 3-dB electrical bandwidth B_{RX} of the receiver front end. The rise time typically is specified as the time it takes the detector output to increase from the 10 percent to the 90 percent point, as shown in Fig. 16.6. If B_{RX} is given in megahertz (MHz), then the *receiver front-end rise time* in nanoseconds is

$$t_{\text{RX}} = \frac{350}{B_{\text{RX}}} \tag{16.7}$$

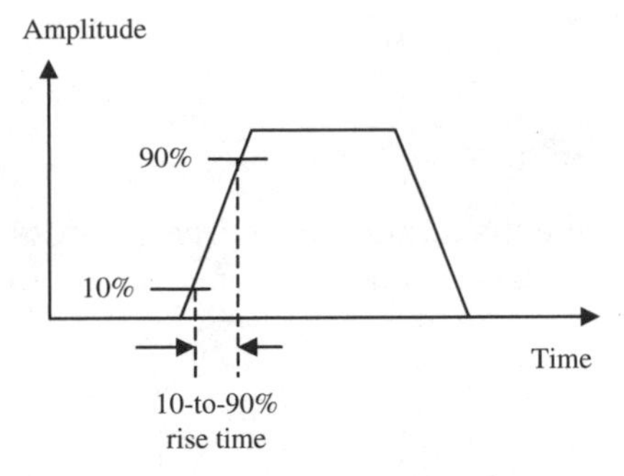

Figure 16.6. Illustration of the 10 to 90 percent rise time of a pulse.

In practice, an optical fiber link seldom consists of a uniform, continuous, jointless fiber. Instead, a transmission link nominally is formed from several concatenated (tandemly joined) fibers which may have different dispersion characteristics. This is especially true for dispersion-compensated links operating at 10 Gbps and higher. In addition, multimode fibers experience modal distributions at fiber-to-fiber joints owing to mechanical misalignments, different core index profiles in each fiber, and/or different degrees of mode mixing in individual fibers. Determining the fiber rise times resulting from chromatic and modal dispersion then becomes more complex than for the case of a single uniform fiber.

The fiber rise time t_{CD} resulting from chromatic dispersion over a length L can be approximated by

$$t_{CD} \approx |D_{CD}| L \Delta\lambda \tag{16.8}$$

where $\Delta\lambda$ is the half-power spectral width of the light source and D_{CD} is the fiber chromatic dispersion. Since the chromatic dispersion value may change from one section of fiber to another in a long link, an average value should be used for D_{CD} in Eq. (16.8).

For a multimode fiber the *bandwidth*, or *information-carrying capacity*, is specified as a bandwidth-distance relationship with units of megahertz times kilometers. Thus the bandwidth needed to support an application depends on the data rate of transmission; that is, as the data rate goes up (MHz), the distance (km) over which signals can be transmitted at that rate goes down. Multimode fibers with a 50-μm core diameter have about 3 times more bandwidth (500 MHz·km) than 62.5-μm fibers (160 MHz·km) at 850 nm. If B_{mod} is the modal dispersion bandwidth (in MHz·km), then the modal rise time t_{mod} (in nanoseconds) over a fiber of length L km is given by

$$t_{mod} = \frac{440L}{B_{mod}} \tag{16.9}$$

As noted in Chap. 15, the pulse spreading t_{PMD} resulting from polarization mode dispersion is given by

$$t_{PMD} = D_{PMD} \sqrt{\text{fiber length}} \tag{16.10}$$

where D_{PMD} is the polarization mode dispersion measured in units of ps/$\sqrt{\text{km}}$.

Let us now calculate the expected rise times for some of the link examples that are given in Sec. 16.2.

16.3.2. Multimode LAN link rise time

First let us consider two examples of LAN links that use an LED and a VCSEL, respectively.

Example of LED-Based LAN Link The following components are used in the 160-m Fast Ethernet LAN link in Sec. 16.2.2:

- An LED with a 2.0-ns rise time and a 75-nm spectral width
- A *pin* photodiode receiver with a front-end bandwidth $B_{RX} = 100\,\text{MHz}$
- A multimode fiber with $D_{CD} = -20\,\text{ps/(nm·km)}$ and $B_{mod} = 400\,\text{MHz·km}$ at 850 nm

Using Eq. (16.7) yields $t_{RX} = 3.5\,\text{ns}$, Eq. (16.8) gives $t_{CD} = 0.24\,\text{ns}$, and from Eq. (16.9) we have $t_{mod} = 0.18\,\text{ns}$. Then, ignoring the negligible PMD effects, the total rise time is

$$\begin{aligned} t_{sys} &= (t_{TX}^2 + t_{mod}^2 + t_{CD}^2 + t_{RX}^2)^{1/2} \\ &= [(2.0)^2 + (0.18)^2 + (0.24)^2 + (3.5)^2]^{1/2} = 4.0\,\text{ns} \end{aligned}$$

Since the Fast Ethernet signal uses an NRZ format, the rise time needs to be less than 0.7/(100 Mbps) = 7.0 ns. Thus the rise-time criterion is well satisfied.

Analogous to power budget calculations, a convenient procedure for keeping track of the various values in the rise time is to use a tabular or spreadsheet form. Table 16.5 shows an example of this for the above calculation.

Example of Maximum Length of Gigabit Ethernet Links Now suppose the LAN consists of Gigabit Ethernet links that use VCSEL sources and 62.5-μm fibers. Also assume that now the link length is 220 m. In this case we have the following conditions:

- A VCSEL with a 0.1-ns rise time and a 1-nm spectral width
- A *pin* photodiode receiver with a front-end bandwidth $B_{RX} = 1000\,\text{MHz}$
- A multimode fiber with $D_{CD} = -20\,\text{ps/(nm·km)}$ and $B_{mod} = 160\,\text{MHz·km}$ at 850 nm

Then $t_{mod} = 0.60\,\text{ns}$, $t_{CD} = 0.01\,\text{ns}$, and $t_{RX} = 0.35\,\text{ns}$, so that (again leaving out the negligible PMD effects) the total rise time is

$$\begin{aligned} t_{sys} &= (t_{TX}^2 + t_{mod}^2 + t_{CD}^2 + t_{RX}^2)^{1/2} \\ &= [(0.10)^2 + (0.60)^2 + (0.01)^2 + (0.35)^2]^{1/2} = 0.70\,\text{ns} \end{aligned}$$

TABLE 16.5. Spreadsheet for Calculating a LAN Link Rise-Time Budget

Component	Rise time	Rise-time budget
Allowed rise-time budget		$t_{sys} < 0.7/B_{NRZ} = 7.0$ ns
LED transmitter rise time	2.0 ns	
Pulse spread from modal dispersion	180 ps	
Chromatic dispersion in fiber	240 ps	
Receiver rise time	3.5 ns	
System rise time [Eq. (16.5)]		4.0 ns

Since the Gigabit Ethernet signal uses an NRZ format, the rise time needs to be less than 0.7/(1000 Mbps) = 0.70 ns. Here the rise-time criterion is just satisfied, so that the maximum link length is 220 m for Gigabit Ethernet operating at 850 nm on 62.5-μm fibers. The maximum length for Gigabit Ethernet running on 50-μm fibers is 550 m due to the higher bandwidth of these fibers.

16.3.3. SONET link rise time

The following components are used in the 2.5-Gbps SONET link in Sec. 16.2.3:

- A laser transmitter with a 0.1-ns rise time and a 1.0-nm spectral width
- An APD receiver with a front-end bandwidth $B_{RX} = 2500$ MHz
- A G.655 single-mode fiber with $D_{CD} = 4$ ps/(nm·km) and $D_{PMD} = 0.1$ ps/$\sqrt{\text{km}}$ at 1310 nm

Then $t_{CD} = 0.12$ ns, $t_{PMD} = 0.001$ ns, and $t_{RX} = 0.14$ ns, so that the total rise time is

$$\begin{aligned} t_{sys} &= (t_{TX}^2 + t_{CD}^2 + t_{PMD}^2 + t_{RX}^2)^{1/2} \\ &= [(0.10)^2 + (0.12)^2 + (0.001)^2 + (0.14)^2]^{1/2} = 0.21\text{ ns} \end{aligned}$$

Since the SONET signal uses an NRZ format, the rise time needs to be less than 0.7/(2500 Mbps) = 0.28 ns, so there is enough rise-time margin in this case.

16.4. Line Coding

In designing a communication link, an important consideration is the format of the transmitted digital signal. This is significant because the receiver must be able to extract precise *timing information* from the incoming signal. The three main purposes of *timing* are

- To allow the signal to be sampled by the receiver at the time the signal-to-noise ratio is a maximum

- To maintain a proper spacing between pulses
- To indicate the start and end of each timing interval

In addition, since errors resulting from channel noise and distortion mechanisms can occur in the signal detection process, it may be desirable for the signal to have an inherent error-detecting capability as well as an error correction mechanism if it is needed or is practical. These features can be incorporated into the data stream by structuring or *encoding* the signal. Generally one does this by introducing extra bits into the raw data stream at the transmitter on a regular and logical basis and extracting them again at the receiver. This process is called *channel coding* or *line coding*. This section presents some examples of generic encoding techniques.

16.4.1. NRZ and RZ signal formats

The simplest method for encoding data is the unipolar *non-return-to-zero* (NRZ) code. *Unipolar* means that a logic 1 is represented by a voltage or light pulse that fills an entire bit period, whereas for a logic 0 no pulse is transmitted, as shown in Fig. 16.7. The coded patterns in this figure are for the data sequence 1010110. If 1 and 0 voltage pulses occur with equal probability and if the amplitude of the voltage pulse is A, then the average transmitted power for this code is $A^2/2$. In optical systems one typically describes a pulse in terms of its optical power level. In this case the average power for an equal number of 1 and 0 pulses is $P/2$, where P is the peak power in a 1 pulse. An NRZ code is simple to generate and decode, but it possesses no inherent error monitoring or correcting capabilities and it contains no timing features.

The lack of timing capabilities in an NRZ code can lead to misinterpretations of the bit stream at the receiver. For example, since there are no level transitions from which to extract timing information in a long sequence of NRZ 1s or 0s, a long string of N identical bits could be interpreted as either $N + 1$ or $N - 1$ bits, unless highly stable (and expensive) clocks are used. This problem can be alleviated with a code that has transitions at the beginning of each bit interval when a binary 1 is transmitted and no transition for a binary 0. This can be

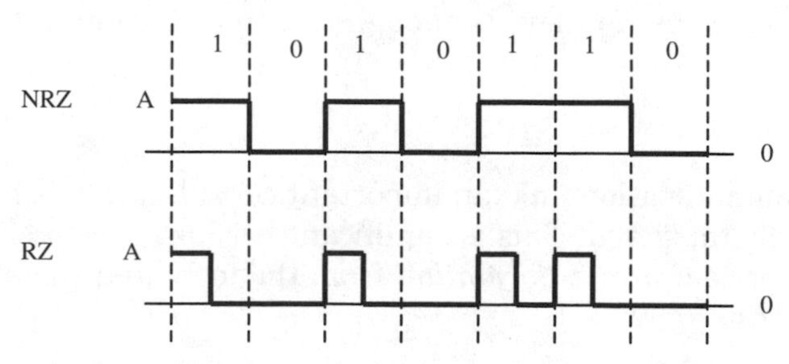

Figure 16.7. NRZ and RZ code patterns for the data sequence 1010110.

achieved with a *return-to-zero* (RZ) code, as shown in Fig. 16.7. Here the pulse for a 1 bit occupies only the first half of the bit interval and returns to zero in the second half of the bit interval. No pulse is used for a 0 bit.

16.4.2. Block codes

Introducing *redundant bits* into a data stream can be used to provide adequate timing and to have error monitoring features. A popular and efficient encoding method for this is the class of mBnB *block codes*. In this class of codes, blocks of m binary bits are converted to longer blocks of $n > m$ binary bits. As a result of the additional redundant bits, the required bandwidth increases by the ratio n/m. For example, in an mBnB code with $m = 1$ and $n = 2$, a binary 1 is mapped into the binary pair 10, and a binary 0 becomes 01. The overhead for such a code is 50 percent.

Suitable mBnB codes for high data rates are the 3B4B, 4B5B, 5B6B, and 8B10B codes. If simplicity of the encoder and decoder circuits is the main criterion, then the 3B4B format is the most convenient code. The 5B6B code is the most advantageous if bandwidth reduction is the major concern. Various versions of Ethernet use either the 3B4B, 4B5B, or 8B10B formats. Fibre Channel employs an 8B10B code.

16.5. Forward Error Correction

For high-speed broadband networks, the data transmission reliability provided by the network may be lower than the reliability requested by an application. In this case, the transport protocol of the network must compensate for the difference in the bit loss rate. Two basic schemes for improving the reliability are *automatic repeat request* (ARQ) and forward error correction (FEC). ARQ schemes have been used for many years and are implemented widely. This technique uses a feedback channel between the receiver and the transmitter to request message retransmission in case errors are detected at the receiver. Since each such retransmission adds at least one round-trip time of latency, ARQ may not be feasible for applications requiring low latency. Among such low-latency applications are voice and video services that involve human interaction, process control, and remote sensing in which data must arrive within a certain time in order to be useful.

Forward error correction avoids the shortcomings of ARQ for high-bandwidth optical networks requiring low delays. FEC is a mathematical signal processing technique that encodes data so that errors can be detected and corrected. In FEC techniques, redundant information is transmitted along with the original information. If some of the original data are lost or received in error, the redundant information is used to reconstruct them. Typically the amount of redundant information is small, so the FEC scheme does not use up much additional bandwidth and thus remains efficient.

The most popular error correction codes are *cyclic codes*. These are designated by the notation (n, m), where n equals the number of original bits m plus the number of redundant bits. Although many types of cyclic codes have been considered over the years in electrical systems, the Reed-Solomon FEC codes are among the best suited for optical signals. They have a low overhead and high coding gain, and they can correct bursts of errors. The RS (255, 239) code is used widely. This code changes 239 data bits into 255 bits, thereby adding about 7 percent overhead. This additional redundancy enables an otherwise unacceptable BER to be sent over a channel by providing a 6- to 10-dB coding gain. A 6-dB coding gain can double the WDM channel count or could increase the transmission distance.

16.6. Modeling and Simulation Tools

With the increased complexity of optical links and networks, computer-based simulation and modeling tools that integrate component, link, and network functions can make the design process more efficient, less expensive, and faster. As described in Chap. 1, such simulation programs are available commercially. The tools typically are based on graphical programming languages that include a library of icons containing the operational characteristics of devices such as optical fibers, couplers, light sources, optical amplifiers, and optical filters, plus the measurement characteristics of instruments such as optical spectrum analyzers, power meters, and bit-error-rate testers. To check the capacity of the network or the behavior of passive and active optical devices, network designers invoke different optical power levels, transmission distances, data rates, and possible performance impairments in the simulation programs.

Associated with this book is an abbreviated version of the software-based tool VPItransmissionMaker from VPIsystems, Inc. The full module is a design and simulation tool for optical devices, components, subsystems, and transmission systems. It enables the user to explore, design, simulate, verify, and evaluate active and passive optical components, fiber amplifiers, dense WDM transmission systems, and broadband access networks. Familiar measurement instruments offer a wide range of settable options when displaying data from multiple runs, optimizations, and multidimensional parameter sweeps.

The abbreviated version of the VPItransmissionMaker simulation tool is called VPIplayer and contains predefined component and link configurations that allow interactive concept demonstrations. Results are shown in a format similar to the displays presented by laboratory instruments. VPIplayer can be downloaded free from the VPIphotonics web site at www.VPIphotonics.com. In addition, at www.PhotonicsComm.com there are numerous interactive examples of optical communication components and links related to topics in this book that the reader can download and simulate.

16.7. Summary

The design of a high-quality transmission link involves a series of tradeoffs among the many interrelated performance variables of each component based on the system operating requirements. Thus, the link analyses may require several iterations before they are completed satisfactorily. Since performance and cost constraints are very important factors in a transmission link, the designer must choose the components carefully to ensure that the link meets the operational specifications over the expected system lifetime without overstating the component requirements.

Key topics related to link design include these:

- A link power budget analysis wherein one first determines the power margin between the optical transmitter output and the minimum receiver sensitivity needed to establish a specified BER and then allocates this margin to fiber, slice, and connector losses, plus any additional margins required for other components, possible device degradations, transmission-line impairments, or temperature effects
- A link margin which is an optical-power safety factor for link design that involves adding extra decibels to the power requirements to compensate for possible unforeseen link degradation factors
- Power penalties which relate to a reduction in the signal-to-noise ratio (SNR) of the system from the ideal case due to signal impairment factors, such as modal noise, chromatic dispersion, polarization mode dispersion, reflection noise in the link, low extinction ratios in the laser, or frequency chirping
- A rise-time budget analysis for determining the dispersion limitation of an optical link
- The format or encoding of the transmitted digital signal so that the receiver can extract precise timing information from the incoming signal
- Forward error correction, which is a mathematical signal processing technique that encodes data so that errors can be detected and corrected. In FEC techniques, redundant information is transmitted along with the original information. If some of the original data are lost or received in error, the redundant information is used to reconstruct them

Associated with this book is an abbreviated version of the VPItransmissionMaker tool from the VPIsystems, Inc. suite of software-based design tools. This simulation engine module is called VPIplayer and has all the tools needed to explore, design, simulate, verify, and evaluate active and passive optical components, fiber amplifiers, dense WDM transmission systems, and broadband access networks. Familiar measurement instruments offer a wide range of settable options when displaying data. The VPIplayer module and a wide variety of interactive demonstration examples are available via Internet access for educational use and concept demonstrations. See Chap. 1 or See. 16.6 for details on how to download the demonstration software.

Further Reading

1. C. D. Poole and J. Nagel, "Polarization effects in lightwave systems," in I. P. Kaminow and T. L. Koch, eds., *Optical Fiber Telecommunications—III*, vol. A, Academic, New York, 1997, Chap. 6.
2. A. B. Carlson, *Communication Systems*, 4th ed., McGraw-Hill, Burr Ridge, Ill., 2002.
3. A. Leon-Garcia and I. Widjaja, *Communication Networks*, McGraw-Hill, Burr Ridge, Ill., 2000.
4. G. Keiser, *Local Area Networks*, 2d ed., McGraw-Hill, Burr Ridge, Ill., 2002.
5. A. Forouzan, *Introduction to Data Communications and Networking*, 2d ed., McGraw-Hill, Burr Ridge, Ill., 2001.
6. S. Haykin, *Communication Systems*, 4th ed., Wiley, New York, 2000.

Chapter

17

Optical Networks

Optical fiber communication technology has been deployed widely all over the world and has become an integral part of telecommunications. This is due to the fact that, compared to copper cables, optical fibers offer a much higher capacity, are smaller, weigh less, and are immune to electromagnetic interference effects. Initially the term *optical networks* referred to collections of optical cable routes that were used for high-capacity point-to-point transmission links. In these networks much of the telecommunication network infrastructure still relied on using electronic signals, particularly in critical functions such as routing and switching of signals. Currently the next generation of optical networks is transitioning some of the routing, switching, and network intelligence into the optical domain.

This chapter first presents some general network concepts in Sec. 17.1, to show what networks consist of and what the related terminology is. Section 17.2 then describes the characteristics and implementations of SONET/SDH networks, which are used for transmission and multiplexing of high-speed signals within the global telecommunications infrastructure. Following this, Sec. 17.3 gives an overview of the meaning and applications of optical Ethernet. A variety of fiber applications to the so-called access networks are the topic of Sec. 17.4. These networks represent the cable segments running from the service provider facility to the business or home user. Finally Sec. 17.5 describes the evolution of optical networks to include concepts such as using all-optical methodologies to reduce the electronic overhead in packet transmissions and the idea of delivering optical services directly to the customer.

17.1. General Network Concepts

To understand the concepts of optical networks better, this section illustrates different types of networks, notes who owns and operates them, defines some network terminology, and describes what the terms *physical layer* and *optical layer* mean.

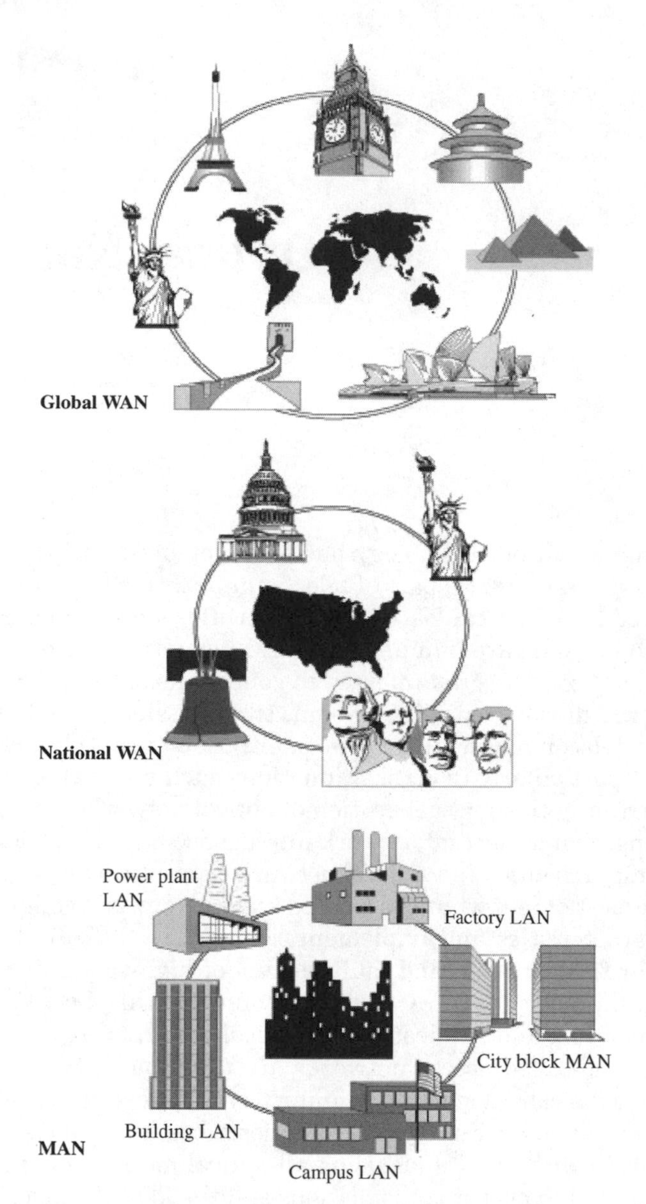

Figure 17.1. Examples of broad categories of networks ranging from LANs to WANs.

17.1.1. Types of networks

Networks can be divided into the following broad categories, as Fig. 17.1 illustrates:

1. *Local-area networks* (LANs) interconnect users in a localized area such as a room, a department, a building, an office or factory complex, or a

campus. Here the word *campus* refers to any group of buildings that are within reasonable walking distance of each other. For example, it could be the collocated buildings of a corporation, a large medical facility, or a university complex. LANs usually are owned, used, and operated privately by a single organization, which is referred to as an *enterprise*.

2. *Metropolitan-area networks* (MANs) span a larger area than a LAN. The size of a MAN could range from interconnections between buildings covering several blocks within a city or it could encompass an entire city and the metropolitan area surrounding it. There is also some means of interconnecting the MAN resources with communication entities located in both LANs and wide-area networks. MANs are owned and operated by many organizations. They commonly are referred to as *metro networks*.
3. *Wide-area networks* (WANs) span a large geographic area. The links can range from connections between switching facilities in neighboring cities to long-haul terrestrial transmission lines running across a country or between countries. WANs are owned and operated by either private enterprises or telecommunication service providers.
4. *Undersea networks* (not shown explicitly in Fig. 17.1) use undersea cables to connect continents. These cables could be several thousand kilometers in length, such as those running across the Atlantic Ocean between North America and Europe or those crossing the Pacific Ocean.

When a network is owned and deployed by a private enterprise, it is referred to as an *enterprise network*. The networks owned by the telecommunication carriers provide services such as leased lines or real-time telephone connections to other users and enterprises. Such networks are referred to as *public networks*.

17.1.2. Network terminology

Before we examine network details, let us define some terms, using Fig. 17.2 for guidance.

- *Stations*. Collections of devices that users employ to communicate are called *stations*. These may be computers, terminals, telephones, or other equipment for communicating. Stations are also referred to as *data terminal equipment* (DTE) in the networking world.
- *Networks*. To establish connections between these stations, one deploys transmission paths running between them to form a collection of interconnected stations called a *network*.
- *Node*. Within this network, a *node* is a point where one or more communication lines terminate and/or where stations are connected. Stations also can connect directly to a transmission line.
- *Trunk*. The term *trunk* normally refers to a transmission line that runs between nodes or networks and that supports large traffic loads.

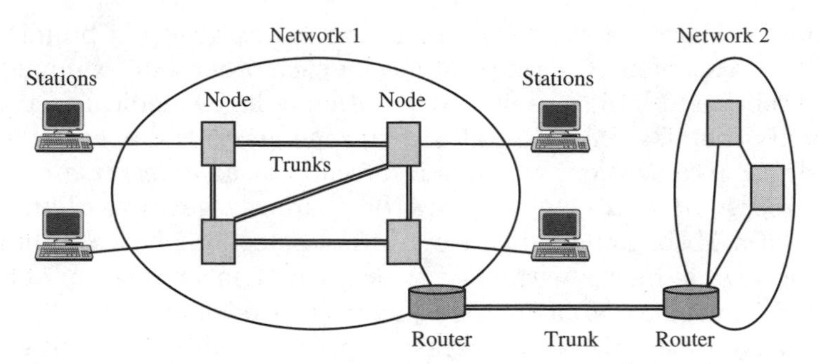

Figure 17.2. Definitions of various elements of a network.

- *Topology*. The *topology* is the logical manner in which nodes are linked together by information transmission channels to form a network.
- *Switching and routing*. The transfer of information from source to destination through a series of intermediate nodes is called *switching*, and the selection of a suitable path through a network is referred to as *routing*. Thus a *switched communication network* consists of an interconnected collection of nodes, in which information streams that enter the network from a station are routed to the destination by being switched from one transmission path to another at a node.
- *Router*. When two networks that use different information-exchange rules (protocols) are interconnected, a device called a *router* is used at the interconnection point to translate the control information from one protocol to another.

To get a better understanding of optical networks, we need to define some terms used in a public network, such as that shown in Fig. 17.3.

- *Central office*. A node in a public network is called a *central office* (CO) or a *point of presence* (POP). The CO houses a series of large switches that establish temporary connections for the duration of a requested connection time between subscriber lines which terminate at the switch.
- *Access network*. The *access network* encompasses connections that extend from the CO to individual businesses, organizations, and homes. Its function is to concentrate the information flows which originate in the access network prior to their entering a long-haul or backbone network.
- *Metro interoffice network*. A metropolitan (typically abbreviated *metro*) network connects groups of central offices within a city or a city-size geographic region. The distances between central offices for this type of network typically range from a few to several tens of kilometers.
- *Backbone network*. The term *backbone* means a network that connects multiple LAN, MAN, or WAN segments. Thus a backbone handles internetwork

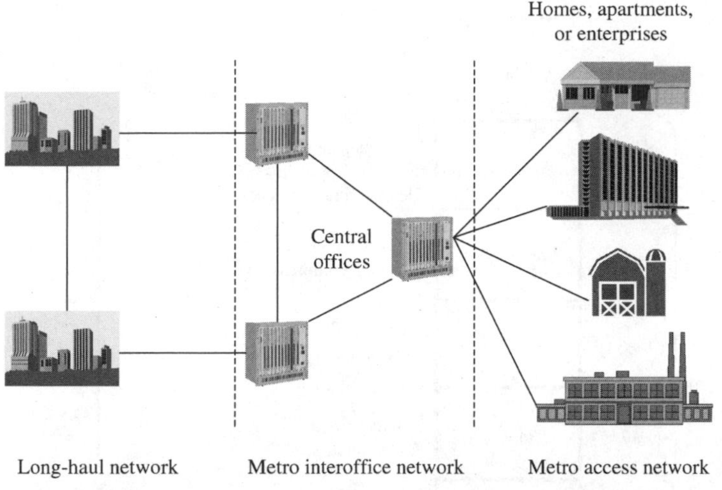

Figure 17.3. Definitions of some terms used in describing a public network.

traffic, that is, traffic that originates in one segment and is transmitted to another segment.

- *Long-haul network.* A *long-haul network* interconnects different cities or widely separated geographic regions and spans hundreds to thousands of kilometers between central offices.

17.1.3. Network layers

To simplify the complexity of modern networks, in the early 1980s the International Organization for Standardization (ISO) developed a model for dividing the functions of a network into seven layers. Each layer performs specific functions using a standard set of protocols, as Fig. 17.4 indicates. A given layer is responsible for providing a service to the layer above it by using the services of the layer below it. In this classical ISO model, the various layers carry out the following functions:

- The *physical layer* refers to a physical transmission medium, such as a wire or an optical fiber, that can handle a certain amount of bandwidth. It provides different types of physical interfaces to equipment, and its functions are responsible for actual transmission of bits across a fiber or wire.
- The purpose of the *data link layer* is to establish, maintain, and release links that directly connect two nodes. Its functions include framing (defining how data are transported), multiplexing, and demultiplexing of data. Examples of data link protocols include the *point-to-point protocol* (PPP) and the *high-level data link control* (HDLC) protocol.
- The function of the *network layer* is to deliver packets from source to destination across multiple network links. Typically, the network layer must find

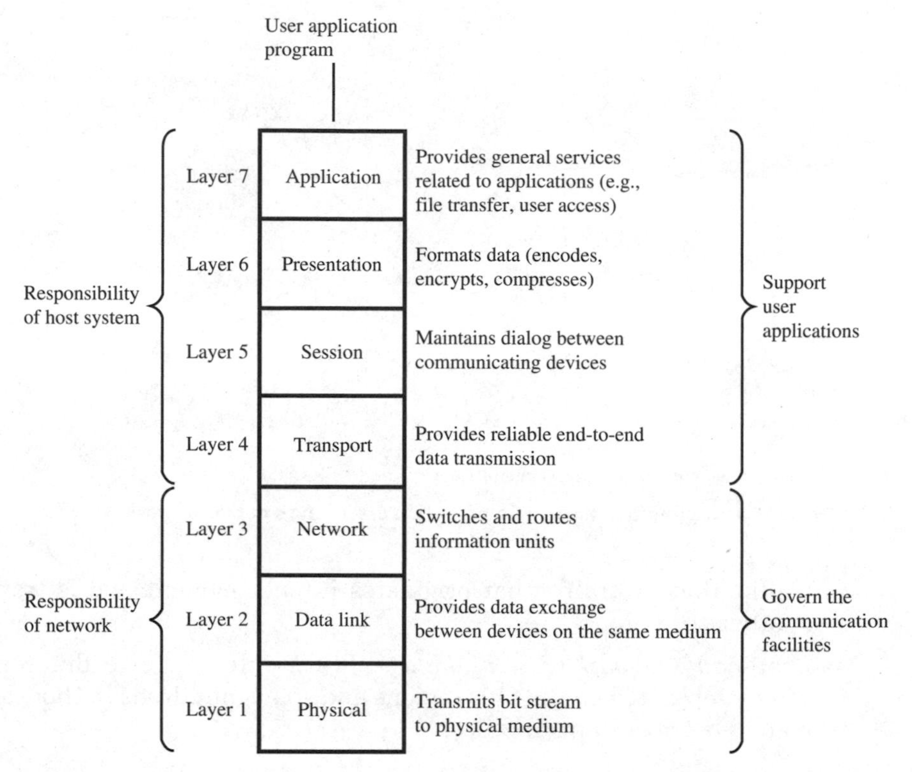

Figure 17.4. General structure and functions of the seven-layer OSI reference model.

a path through a series of connected nodes, and the nodes along this path must forward the packets to the appropriate destination. Currently the dominant network layer protocol is the *Internet Protocol* (IP).

- The *transport layer* is responsible for reliably delivering the complete message from the source to the destination to satisfy a *quality of service* (QoS) requested by the upper layer. The QoS parameters include throughput, transit delay, error rate, delay time to establish a connection, cost, security, and priority. The *transmission control protocol* (TCP) used in the Internet is an example of a transport layer protocol.
- The higher layers (session, presentation, application) support user applications, which are not covered here (see the network books by Keiser[1] or Forouzan[2] for more details).

In actual systems there are many implementation and protocol variations on the classical layering model. A certain layer may work together with lower or higher layers, or a layer may be divided into several sublayers. As an example, consider the widely used IP over SONET architecture. In this case, IP operates at the network and data link layers to format packets in such a way that they

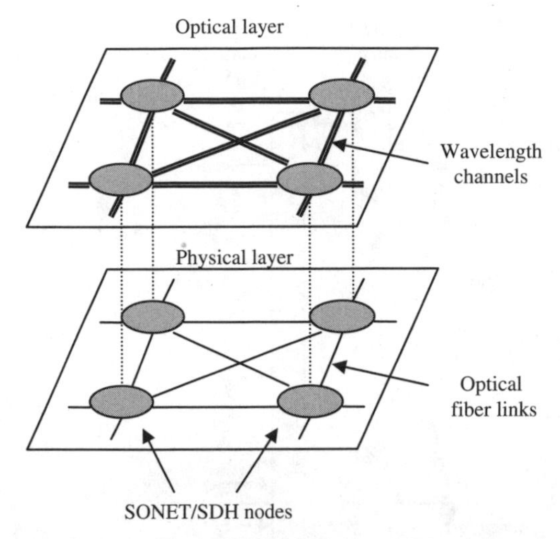

Figure 17.5. The optical layer is a wavelength-based concept and lies just above the physical layer.

can be routed from end to end across a packet-switched network. Here the IP network views the SONET network merely as a set of physical point-to-point links between IP routers. However, as described in Sec. 17.2, the internal switching and routing operations of the SONET itself encompass physical, data link, and network layer functions.

17.1.4. Optical layer

When dealing with optical network concepts, one hears the words *optical layer* used to describe various network functions or services. The *optical layer* is a wavelength-based concept, and it lies just above the physical layer, as shown in Fig. 17.5. This means that whereas the physical layer provides a physical connection between two nodes, the optical layer provides *lightpath services* over that link. A *lightpath* is an end-to-end optical connection that may go through one or more intermediate nodes. For example, in an eight-channel WDM link there are eight lightpaths, which may go over a single physical line. Note that for a specific lightpath the wavelengths between various node pairs may be different.

As shown in Fig. 17.6, the optical layer may carry out processes such as wavelength multiplexing, adding and dropping of wavelengths, and support of optical cross-connects or wavelength switching. Networks which have these optical layer functions are referred to as *wavelength-routed networks*.

17.2. SONET/SDH

With the advent of fiber optic transmission lines, the next step in the evolution of the digital time division multiplexing (TDM) scheme was a standard signal

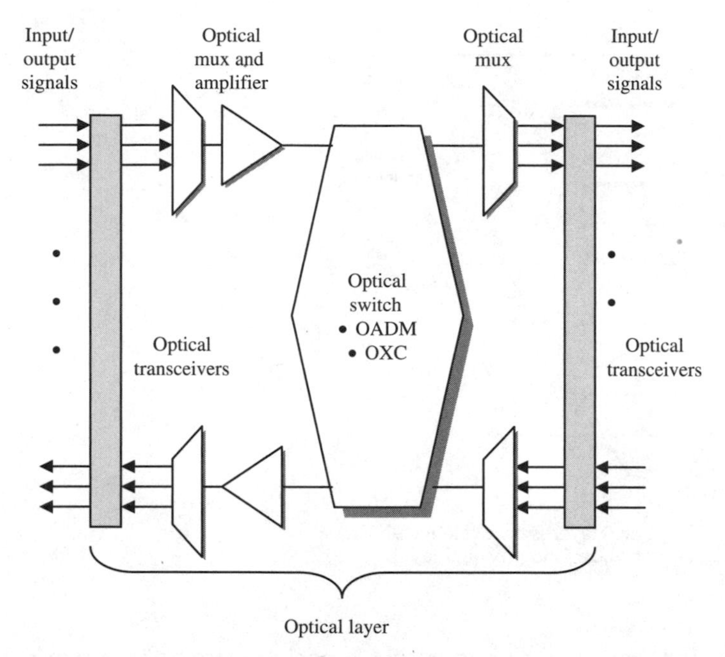

Figure 17.6. The optical layer carries out processes such as multiplexing, adding/dropping of wavelengths, and support of optical cross-connects or wavelength switching.

format called *synchronous optical network* (SONET) in North America and *synchronous digital hierarchy* (SDH) in other parts of the world. This section addresses the basic concepts of SONET/SDH, its optical interfaces, and fundamental network implementations. The aim here is to discuss only the physical layer aspects of SONET/SDH as they relate to optical transmission lines and optical networks. Topics such as the detailed data format structure, SONET/SDH operating specifications, and the relationships of switching methodologies such as asynchronous transfer mode (ATM) with SONET/SDH are beyond the scope of this book.

17.2.1. SONET transmission formats

In the mid-1980s, several service providers in the United States started efforts on developing a standard that would allow network engineers to interconnect fiber optic transmission equipment from various vendors through multiple-owner trunk networks. This soon grew into an international activity, which, after many differences of opinion of implementation philosophy were resolved, resulted in a series of ANSI T1.105 standards for SONET and a series of ITU-T recommendations for SDH. Of particular interest here are the ANSI Standard T1.105.06 and the ITU-T Recommendation G.957. Although there are some implementation differences between SONET and SDH, all SONET specifications conform to the SDH recommendations.

Figure 17.7 shows the basic structure of a SONET frame. This is a two-dimensional structure consisting of 90 columns by 9 rows of bytes, where 1 byte (1 B) is 8 bits (b). This fundamental SONET frame has a 125-μs duration. Thus the transmission bit rate of the basic SONET signal is

$$\text{STS-1} = (90 \text{ bytes/row})(9 \text{ rows/frame})(8 \text{ bits/byte})/(125\ \mu\text{s/frame})$$
$$= 51.84\,\text{Mbps}$$

This is called an STS-1 signal, where STS stands for *synchronous transport signal*.

Higher-rate SONET signals are obtained by byte-interleaving N STS-1 frames, which then are scrambled and converted to an *optical carrier—level N* (OC-N) signal. Thus the OC-N signal will have a line rate exactly N times that of an OC-1 signal; that is, OC-N = $N \times 51.84$ Mbps.

For SDH systems the fundamental building block is the 155.52-Mbps *synchronous transport module—level 1* (STM-1). Again, higher-rate information streams are generated by synchronously multiplexing N different STM-1 signals to form the STM-N signal. Table 17.1 shows commonly used SDH and SONET signal levels and the associated OC rates. For practical purposes, the data rates are abbreviated as shown in the rightmost column.

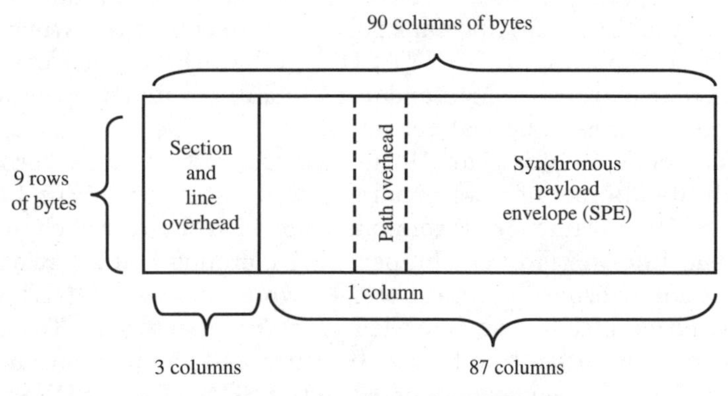

Figure 17.7. Basic structure of an STS-1 SONET frame.

TABLE 17.1. Commonly Used SONET and SDH Transmission Rates

SONET level	Electrical level	SDH level	Line rate, Mbps	Common rate name
OC-1	STS-1	—	51.84	—
OC-3	STS-3	STM-1	155.52	155 Mbps
OC-12	STS-12	STM-4	622.08	622 Mbps
OC-48	STS-48	STM-16	2,488.32	2.5 Gbps
OC-192	STS-192	STM-64	9,953.28	10 Gbps
OC-768	STS-768	STM-256	39,813.12	40 Gbps

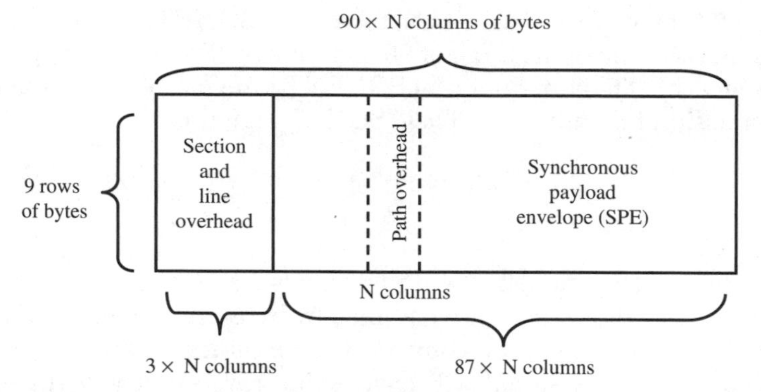

Figure 17.8. Basic format of an STS-N SONET frame.

Referring to Fig. 17.7, the first three columns comprise transport overhead bytes that carry network management information. The remaining field of 87 columns is called the *synchronous payload envelope* (SPE) and carries user data plus 9 bytes of *path overhead* (POH). The POH supports performance monitoring by the end equipment, status, signal labeling, a tracing function, and a user channel. The 9 path-overhead bytes are always in a column and can be located anywhere in the SPE. An important point to note is that the synchronous byte-interleaved multiplexing in SONET/SDH (unlike the asynchronous bit interleaving used in earlier TDM standards) facilitates add/drop multiplexing of information channels in optical networks.

For values of N greater than 1, the columns of the frame become N times wider, with the number of rows remaining at 9, as shown in Fig. 17.8. Thus, an STS-3 (or STM-1) frame is 270 columns wide with the first 9 columns containing overhead information and the next 261 columns being payload data. The line and section overhead bytes differ somewhat between SONET and SDH, so that a translation mechanism is needed to interconnect them. To obtain further details on the contents of the frame structure and the population schemes for the payload field, the reader is referred to the SONET and SDH specifications.

17.2.2. Optical interfaces

To ensure interconnection compatibility between equipment from different manufacturers, the SONET and SDH specifications provide details for the optical source characteristics, the receiver sensitivity, and transmission distances for various types of fibers. Six transmission ranges and the associated fiber types are defined with different terminology for SONET and SDH, as Table 17.2 indicates. The transmission distances are specified for G.652, G.653, and G.655 fibers. The ITU-T Recommendation G.957 also designates the SDH categories by codes such as I-1, S-x.1, L-x.1, and so on, as indicated in the table.

The optical fibers specified in ANSI T1.105.06 and ITU-T G.957 fall into the following three categories and operational windows:

TABLE 17.2. Transmission Distances and Their SONET and SDH Designations, Where x Denotes the STM-x Level

Transmission distance	Fiber type	SONET terminology	SDH terminology
≤ 2 km	G.652	Short-reach (SR)	Intraoffice (I-1)
15 km at 1310 nm	G.653	Intermediate-reach (IR-1)	Short-haul (S-x.1)
15 km at 1550 nm	G.653	Intermediate-reach (IR-2)	Short-haul (S-x.2)
40 km at 1310 nm	G.655	Long-reach (LR-1)	Long-haul (L-x.1)
80 km at 1550 nm	G.655	Long-reach (LR-2)	Long-haul (L-x.3)
120 km at 1550 nm	G.655	Very long-reach (VR-1)	Very long (V-x.3)
160 km at 1550 nm	G.655	Very long-reach (VR-2)	Ultralong (U-x.3)

TABLE 17.3. Wavelength Ranges and Attenuation for Transmission Distances up to 80 km

Distance	Wavelength range at 1310 nm	Wavelength range at 1550 nm	Attenuation at 1310 nm, dB/km	Attenuation at 1550 nm, dB/km
≤15 km	1260–1360 nm	1430–1580 nm	3.5	Not specified
≤40 km	1260–1360 nm	1430–1580 nm	0.8	0.5
≤80 km	1280–1335 nm	1480–1580 nm	0.5	0.3

1. Graded-index multimode in the 1310-nm window
2. Conventional nondispersion-shifted single-mode in the 1310- and 1550-nm windows
3. Dispersion-shifted single-mode in the 1550-nm window

Table 17.3 shows the wavelength and attenuation ranges specified in these fibers for transmission distances up to 80 km.

Depending on the attenuation and dispersion characteristics for each hierarchical level shown in Table 17.2, feasible optical sources include light-emitting diodes (LEDs), multimode lasers, and various single-mode lasers. The system objective in ANSI T1.105.06 and ITU-T G.957 is to achieve a bit error rate (BER) of less than 10^{-10} for rates less than 1 Gbps and 10^{-12} for higher rates and/or higher-performance systems.

The specified receiver sensitivities are the worst-case, end-of-life values. They are defined as the minimum-acceptable, average, received power needed to achieve a 10^{-10} BER. The values take into account the extinction ratio, pulse rise and fall times, optical return loss at the source, receiver connector degradations, and measurement tolerances. The receiver sensitivity does not include power penalties associated with dispersion, jitter, or reflections from the optical path, since these are included in the maximum optical path penalty. Table 17.4 lists the receiver sensitivities for various link configurations up through long-haul distances (80 km). Note that the ANSI and ITU-T recommendations are

TABLE 17.4. Source Output, Attenuation, and Receiver Ranges for Various Rates and Distances up to 80 km (See ITU-T G.957)

Parameter	Intraoffice	Short-haul (1)	Short-haul (2)	Long-haul (1)	Long-haul (3)
Wavelength, nm	1310	1310	1550	1310	1550
Fiber		SM	SM	SM	SM
Distance, km	≤2	15	15	40	80
Designation	I-1	S-1.1	S-1.2	L-1.1	L-1.3
Source range, dBm					
155 Mbps	−15 to −8	−15 to −8	−15 to −8	0 to 5	0 to 5
622 Mbps	−15 to −8	−15 to −8	−15 to −8	−3 to +2	−3 to +2
2.5 Gbps	−10 to −3	−5 to 0	−5 to 0	−2 to +3	−2 to +3
Attenuation range, dB					
155 Mbps	0 to 7	0 to 12	0 to 12	10 to 28	10 to 28
622 Mbps	0 to 7	0 to 12	0 to 12	10 to 24	10 to 24
2.5 Gbps	0 to 7	0 to 12	0 to 12	10 to 24	10 to 24
Receiver sensitivity, dBm					
155 Mbps	−23	−28	−28	−34	−34
622 Mbps	−23	−28	−28	−28	−28
2.5 Gbps	−18	−18	−18	−27	−27

updated periodically, so the reader should refer to the latest version of the documents for specific details.

Longer transmission distances are possible by using higher-power lasers. To comply with eye-safety standards, an upper limit is imposed on fiber-coupled powers. If the maximum total output power (including ASE) is set at the Class-3A laser limit of $P_{3A} = +17$ dBm, then for ITU-T G.655 fiber this allows transmission distances of 160 km for a single-channel link. Using this condition, for M operational WDM channels the maximum nominal channel power P_{chmax} should be limited to

$$P_{chmax} = P_{3A} - 10 \log M \tag{17.1}$$

17.2.3. SONET/SDH rings

A key characteristic of SONET and SDH is that they usually are configured as a ring architecture. This is done to create *loop diversity* for uninterrupted service protection purposes in case of link or equipment failures. The SONET/SDH rings commonly are called *self-healing rings*, since the traffic flowing along a certain path can be switched automatically to an alternate or standby path following failure or degradation of the primary link segment.

Three main features, each with two alternatives, classify all SONET/SDH rings, thus yielding eight possible combinations of ring types. First, there can be either two or four fibers running between the nodes on a ring. Second, the operating signals can travel either clockwise only (which is termed a *unidirectional ring*) or in both directions around the ring (which is called a *bidirectional ring*). Third, protection switching can be performed via either a line-switching or a path-switching scheme. Upon link failure or degradation, *line switching* moves all signal channels of an entire OC-*N* channel to a protection fiber. Conversely, *path switching* can move individual payload channels within an OC-*N* channel (e.g., an STS-1 subchannel in an OC-12 channel) to another path.

Of the eight possible combinations of ring types, the following architectures have become popular for SONET and SDH networks:

- Two-fiber, unidirectional, path-switched ring (called two-fiber UPSR)
- Two-fiber or four-fiber, bidirectional, line-switched ring (called two-fiber or four-fiber BLSR)

The common abbreviations of these configurations are given in parentheses. They also are referred to as a *unidirectional* or a *bidirectional self-healing ring* (USHR or BSHR).

Figure 17.9 shows a two-fiber UPSR network. By convention, in a unidirectional ring the *normal working traffic* travels *clockwise* around the ring, as indicated by the heavy arrows. For example, the connection from node 1 to node 3 uses links 1 and 2, whereas the traffic from node 3 to node 1 traverses links 3 and 4. Thus, two communicating nodes use a specific bandwidth capacity around the entire perimeter of the ring. If nodes 1 and 3 exchange information at an OC-3 rate in an OC-12 ring, then they use one-fourth of the capacity around the ring on all the primary links.

In a unidirectional ring the counterclockwise path is used as an alternate path for protection against link or node failures. To achieve this, the signal from a transmitting node is dual-fed into both the primary and protection fibers. This establishes a designated *protection path* on which traffic flows *counterclockwise*, that is, from node 1 to node 3 via links 4 and 3 (in that order), as shown in Fig. 17.9*a*. A heavy line and a dashed line indicate the primary and protection paths, respectively.

Consequently, two identical signals from a particular node arrive at their destination from opposite directions, usually with different delays, as denoted in Fig. 17.9*b*. The receiver normally selects the signal from the primary path. However, it continuously compares the fidelity of each signal and chooses the alternate signal in case of severe degradation or loss of the primary signal. Thus, each path is switched individually based on the quality of the received signal. For example, if path 2 breaks or equipment in node 2 fails, then node 3 will switch to the protection channel to receive signals from node 1.

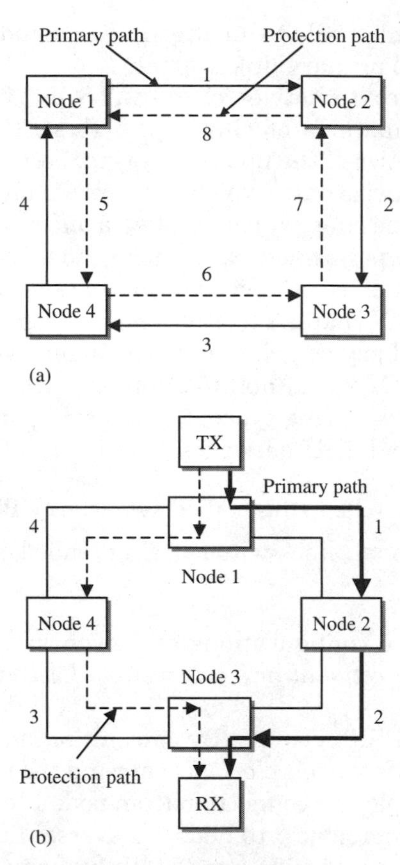

Figure 17.9. (*a*) Generic two-fiber UPSR with a counter-rotating protection path. (*b*) Flow of primary and protection traffic from node 1 to node 3.

Figure 17.10 illustrates the architecture of a four-fiber BLSR. Here two primary fiber loops (with fiber segments labeled 1p through 8p) are used for normal bidirectional communication, and the other two secondary fiber loops are standby links for protection purposes (with fiber segments labeled 1s through 8s). In contrast to the two-fiber UPSR, the four-fiber BLSR has a capacity advantage because it uses twice as much fiber cabling and because traffic between two nodes is sent only partially around the ring. To see this, consider the connection between nodes 1 and 3. The traffic from node 1 to node 3 flows in a clockwise direction along links 1p and 2p. Now, however, in the return path the traffic flows counterclockwise from node 3 to node 1 along links 6p and 5p (in that order). Thus, the information exchange between nodes 1 and 3 does not tie up any of the primary channel bandwidth in the other half of the ring.

To see the function and versatility of the standby links in the four-fiber BLSR, consider first the case where a transmitter or receiver circuit card used

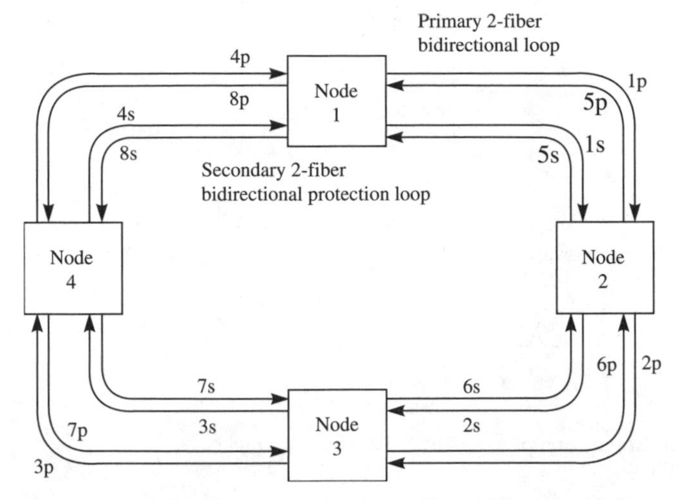

Figure 17.10. Architecture of a four-fiber bidirectional line-switched ring (BLSR).

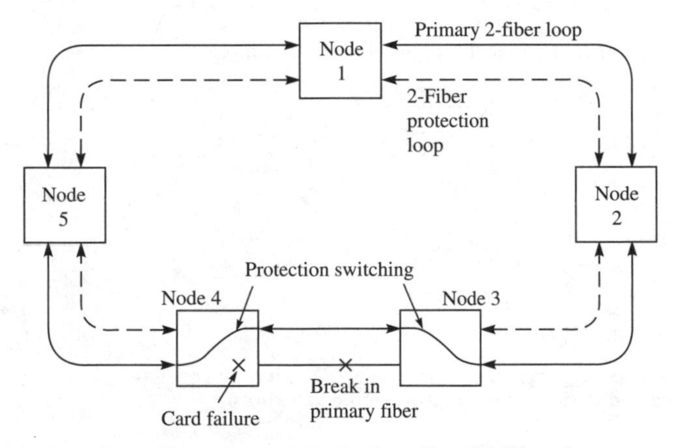

Figure 17.11. Reconfiguration of a four-fiber BLSR under transceiver or line failure.

on the primary ring fails in either node 3 or 4. In this situation the affected nodes detect a loss-of-signal condition and switch both primary fibers connecting them to the secondary protection pair, as shown in Fig. 17.11. The protection segment between nodes 3 and 4 now becomes part of the primary bidirectional loop. The exact same reconfiguration scenario will occur when the primary fiber connecting nodes 3 and 4 breaks. Note that in either case the other links remain unaffected.

Now suppose an entire node fails, or both the primary and the protection fibers in a given span are severed, which could happen if they are in the same cable duct between two nodes. In this case the nodes on either side of the failed

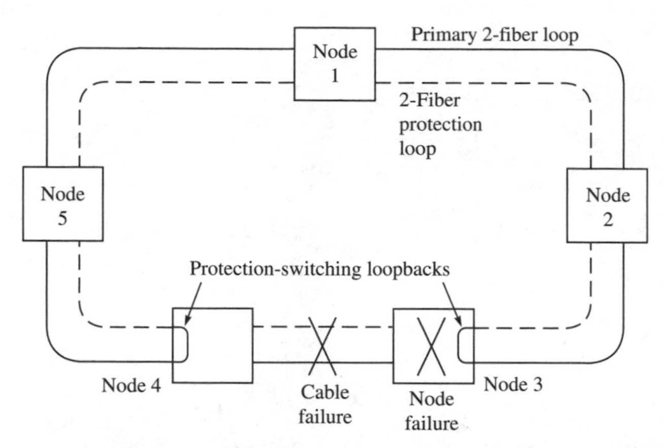

Figure 17.12. Reconfiguration of a four-fiber BLSR under node or fiber cable failure.

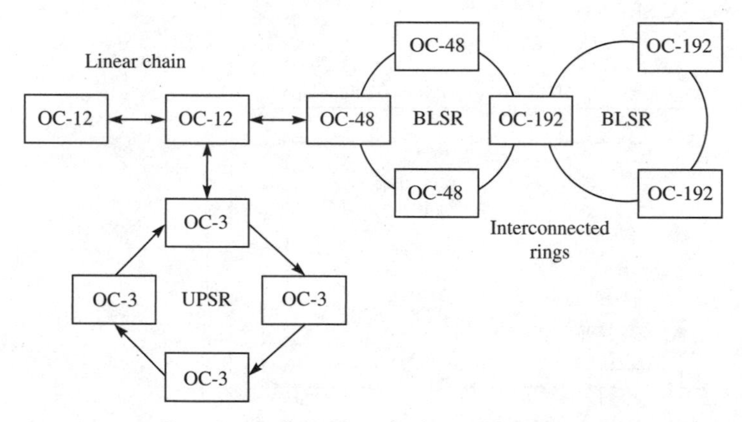

Figure 17.13. Generic configuration of a large SONET network consisting of linear chains and various types of interconnected rings.

interconnecting span internally switch the primary-path connections from their receivers and transmitters to the protection fibers, in order to loop traffic back to the previous node. This process again forms a closed ring, but now with all the primary and protection fibers in use around the entire ring, as shown in Fig. 17.12.

17.2.4. SONET/SDH networks

Commercially available SONET/SDH equipment allows the configuration of a variety of network architectures, as shown in Fig. 17.13. For example, one can build point-to-point links, linear chains, UPSRs, BLSRs, and interconnected rings. The OC-192 four-fiber BLSR could be a large national backbone network with a number of OC-48 rings attached in different cities. The OC-48 rings can have lower-capacity localized OC-12 or OC-3 rings or chains attached to them, thereby providing the possibility of attaching equipment that has an extremely wide range of rates and sizes. Each of the individual rings has its own failure-recovery mechanisms and SONET/SDH network management procedures.

An important SONET/SDH network element is the *add/drop multiplexer* (ADM). This piece of equipment is a fully synchronous, byte-oriented multiplexer that is used to add and drop subchannels within an OC-N signal. Figure 17.14 shows the functional concept of an ADM. Here various OC-12 and OC-3 traffic streams (shown as solid and dashed lines, respectively) are multiplexed into one OC-48 stream. Upon entering an ADM, these subchannels can be dropped individually by the ADM, and others can be added. For example, in Fig. 17.14, one OC-12 and two OC-3 channels enter the leftmost ADM as part of an OC-48 channel. The OC-12 is passed through, and the two OC-3 channels are dropped by the first ADM. Then two more OC-12s and one OC-3 are multiplexed together with the OC-12 channel that is passing through, and the aggregate (partially filled) OC-48 is sent to another ADM node downstream.

The SONET/SDH architectures also can be implemented with multiple wavelengths. For example, Fig. 17.15 shows a dense WDM deployment on an OC-192

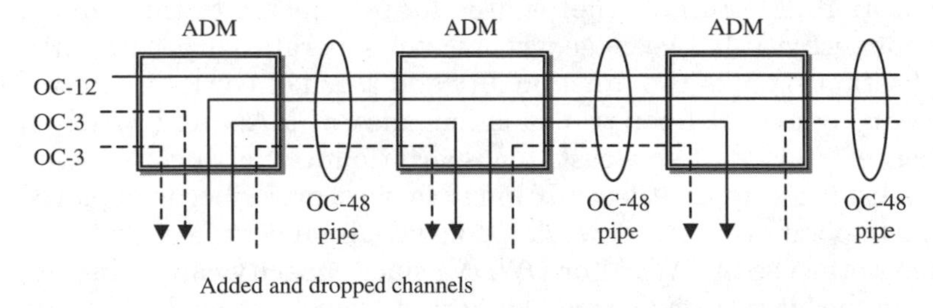

Figure 17.14. Functional concept of an add/drop multiplexer for SONET/SDH applications.

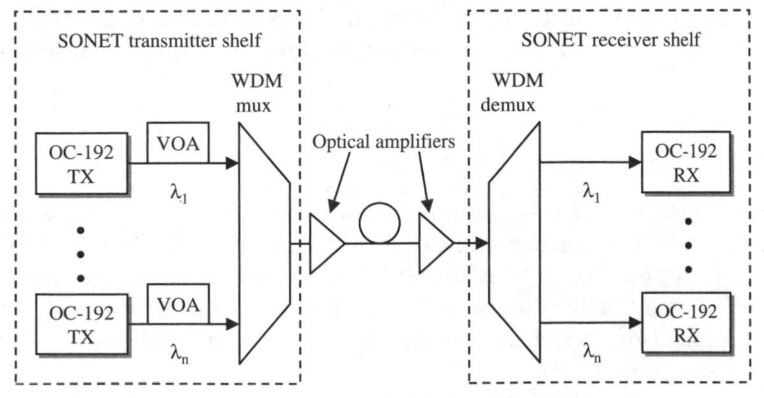

Figure 17.15. Dense WDM deployment of n wavelengths in an OC-192 trunk ring.

trunk ring for n wavelengths (e.g., one could have $n = 16$). The different wavelength outputs from each OC-192 transmitter are passed first through a variable optical attenuator (VOA) to equalize the output powers. These then are fed into a wavelength multiplexer, possibly amplified by a posttransmitter optical amplifier, and sent out over the transmission fiber. Additional optical amplifiers might be located at intermediate points and/or at the receiving end.

17.3. Optical Ethernet

Ethernet is deployed widely in local-area networks, since it is known for its robustness and low cost. Standards-compliant interfaces are available on numerous devices running at line rates ranging from 10 Mbps to 10 Gbps. Therefore Ethernet has matured to become the LAN technology of choice with the best price and performance characteristics.

Ethernet also is being used in metropolitan-area networks and is extending into wide-area networks (see ITU-T Recommendation G.985). In these environments Ethernet can increase network capacity cost-effectively and has the ability to offer a wide range of services in a simple, scalable, and flexible manner. When used in a MAN, Ethernet is referred to as *Metro Ethernet*. In enterprise applications, Metro Ethernet is used for interfacing to the public Internet and for connectivity between geographically separate corporate sites. The latter application extends the functionality and reach of corporate networks.

By using optical fiber transmission lines in MAN and WAN environments, Ethernet provides a low-cost, high-performance networking solution that can span distances up to at least 70 km. Ethernet over fiber is deployed mainly in a point-to-point or mesh network topology. A high degree of scalability is possible through the use of CWDM or DWDM, since capacity can be increased either by raising the bit rate and/or by adding more wavelengths. In addition, with WDM users can lease wavelengths with varying bandwidth and protocol characteristics on a temporary or time-of-day basis. For example, server farms or information storage systems can furnish users with additional bandwidth at specific times of day when standard traffic normally is low. The combined flexibility of Ethernet and WDM on the optical layer allows a rapid activation of such services.

17.4. IP over WDM

The movement in the telecommunications industry toward a greater use of IP is resulting in a dramatic complexity reduction of multiprotocol routing in networks. The popularity of IP is that it has widespread use in enterprise networks and the Internet, it is understood more than any other protocol, gateways for non-IP applications exist, and protocol stacks are available at both the IP and higher levels (e.g., TCP).

As shown in Fig. 17.16, the network layering of a typical wide-area network carries IP on top of ATM, ATM on top of SONET/SDH, and SONET/SDH on top

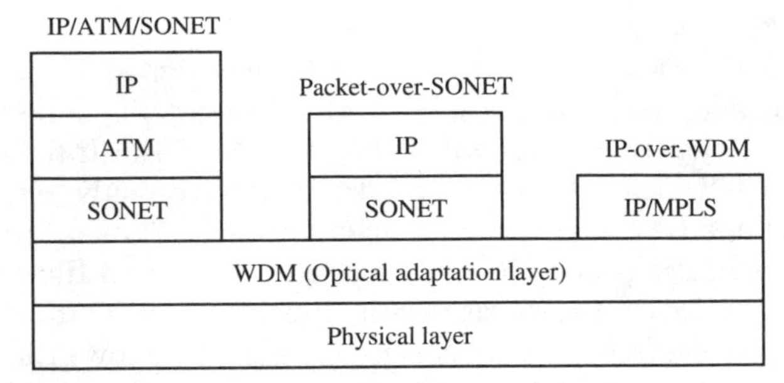

Figure 17.16. Progression of network layering methodologies moving from IP/ATM/SONET to packet-over-SONET to IP-over-MPLS.

of DWDM. However, recently the trend is to format all voice, video, and data as IP packets instead of first encapsulating them in ATM cells. With this "IP-over-SONET" structure, the SONET network protection mechanisms are still in place but the high ATM overhead (at least 5 B in a 53-B cell) is eliminated, thereby reducing the number of management levels from four to three.

A further trend is aimed at bypassing the SONET/SDH layer, thereby combining the IP and SONET layers into one network layer based on *multiprotocol label switching* (MPLS). This "packet-over-WDM" scheme would provide faster provisioning of services and eliminate one electronic bottleneck, so that then there would be only two levels of management. However, there are major framing and fault recovery concerns with this approach. Since an IP packet contains only source and destination IP addresses, to map IP onto a wavelength requires an intermediate step of encapsulating the IP packet into a transport protocol in order to attach a header that contains source and destination physical addresses. This could be a protocol such as Ethernet, ATM, or SONET. After the packet is encapsulated, it is inserted into the modulation format of the wavelength being used. For readers interested in details on this, see the referenced MPLS papers (5 through 7) in *IEEE Communications Magazine* and *IEEE Network*.

17.5. Optical Transport Networks (OTNs)

Emerging next-generation transport networks are referred to as *optical transport networks* (OTNs). In these networks it is envisioned that DWDM-based dynamic optical elements such as optical cross-connect switches and optical add/drop multiplexers (OADMs) will have full control of all wavelengths. In addition they are expected to have full knowledge of the traffic-carrying capacity and the status of each wavelength. With such intelligence these networks are envisioned as being self-connecting and self-regulating. However, there are still many challenges to overcome before such completely intelligent optical networks are feasible.

17.5.1. Optical network services

Traditionally, optical networks such as SONET were designed for backbone networks for enabling a large number of voice and data channels to be multiplexed and transported efficiently and reliably. The growth of the Internet changed that scenario to one in which optical services are delivered directly to the customer. These services include LAN-based packet-oriented services using Ethernet and circuit-oriented applications such as data storage, video, and file transfers. In setting up optical services, a key point to remember is that not all traffic is created equally. Traffic from free Internet services needs only a low grade of service, whereas business traffic such as banking transactions requires a high grade of service. However, no matter who they are, customers demand that their services be provisioned rapidly. Another important factor that needs to be taken into account when establishing the criteria for optical network services is that the dynamic nature of demands from metro area users requires a flexible network that can handle service churns, service mixes, and variable service growths.

17.5.2. OTN standards

Many people are working on OTN concepts, and the ITU-T is establising recommendations. In November 2001 the ITU-T agreed on the following nine new and revised OTN documents (see also G.984.1 and G.984.2).

- G.872, *The Architecture of Optical Transport Networks*
- G.709, *Interface for the OTN*
- G.798, *Characteristics of OTN Hierarchy Equipment*
- G.8251, *The Control of Jitter and Wander within the OTN*
- G.7041, *Generic Framing Procedure (GFP)*
- G.7710, *Common Equipment Management Function Requirements*
- G.874, *Management Aspects of the OTN Element*
- G.874.1, *OTN Protocol-Neutral Management Information Model for the Network Element View*
- G.7712, *Architecture and Specification of Data Communications Network*

17.5.3. Wavelength routing

Recently systems have been devised that use optical wavelength switches or optical cross-connects to enable data to be routed entirely in the optical domain. Wavelength-routed networks need a control mechanism to set up and take down all-optical connections. The functions of the control mechanism are (1) to assign a communications wavelength when a connection request arrives and to configure the appropriate optical switches in the network and (2) to provide information on usage and status of the wavelengths so the nodes can make routing decisions. The wavelength routing can be static or dynamic. In *static routing* a group of lightpaths is set up together and kept in place in the network for long periods of time. In *dynamic routing* a lightpath is set up for each connection request as it arrives and is released after the requested call is over.

17.6. FTTx

As noted in Sec. 17.1.2, an access network refers to the connections that extend from the central office to a neighborhood and further to individual businesses and homes. This network often is called the *last leg* or the *last mile*. Traditionally, copper wires were used as the transmission medium in the access network, since using optical fibers cost-effectively in these transmission spans is a major challenge. However, various means of using fibers in the access network have been explored. These schemes are known by the all-inclusive term *fiber-to-the-x* (FTTx), where x is some letter designating at what point the fiber terminates and copper wires (or wireless links) again take over.

Figure 17.17 shows the reach of various FTTx schemes. These are defined as follows:

- *Fiber to the neighborhood* (FTTN) refers to the optical fiber running from the CO to a main distribution frame (MDF) located in a building or to an outdoor shelter in a neighborhood. FTTN implies that another medium such as coaxial cable will carry the communication signals from the MDF to the users inside buildings within the neighborhood.
- *Fiber to the curb* (FTTC) refers to optical fiber cable running directly from the CO to the outdoor shelters on curbs near homes or any business environment. Another medium (typically copper wires) will carry the signals the very short distance (typically about 100 m) between the curb and the user inside the home or business.
- *Fiber to the building* (FTTB) refers to optical fiber running from the telephone company central office to a specific building such as a business or an apartment house. Inside the building an *optical network unit* (ONU) converts the optical signal to an electrical format for distribution over a wire network to the occupants.

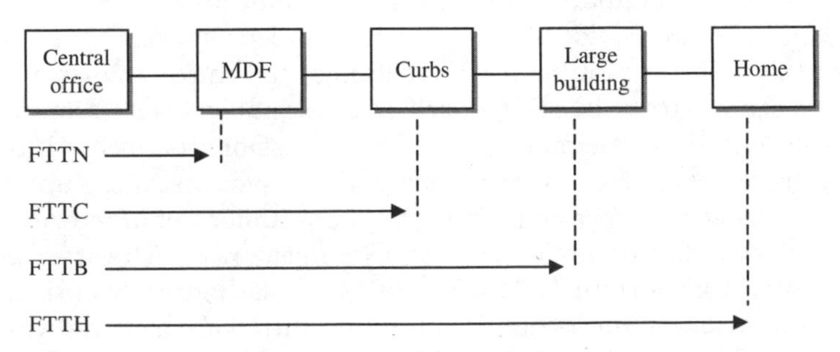

Figure 17.17. The reach of various FTTx schemes.

- *Fiber to the home* (FTTH) is a network technology that deploys fiber optic cable directly to the home or business to deliver voice, video, and data services. Owing to the very high capacity of optical fibers, FTTH can deliver greater capacity than competing copper-based technologies. This is considered the deepest penetration of the fiber optics infrastructure and is the most costly to implement.

17.7. Summary

Networks can be classified broadly as a local-area network (LAN), metropolitan-area network (MAN) or metro network, and wide-area network (WAN). When a network is owned and deployed by a private enterprise, it is referred to as an enterprise network. The networks owned by the telecommunication carriers provide services such as leased lines or real-time telephone connections to other users and enterprises. Such networks are referred to as public networks.

To simplify the complexity of modern networks, the International Organization for Standardization (ISO) developed a model for dividing network functions into seven layers. Each layer performs specific functions using a standard set of protocols. A given layer is responsible for providing a service to the layer above it by using the services of the layer below it. In actual systems there are many implementation and protocol variations on the classical layering model. A certain layer may work together with lower or higher layers, or a layer may be divided into several sublayers.

The words *optical layer* are used to describe various network functions or services. The optical layer is a wavelength-based concept and lies just above the physical layer. Whereas the physical layer provides a physical connection between two nodes, the optical layer provides *lightpath services* over that link. A *lightpath* is an end-to-end optical connection that may go through one or more intermediate nodes. For example, in an eight-channel WDM link there are eight lightpaths, which may go over a single physical line.

The SONET and SDH standards allow network engineers to interconnect fiber optic transmission equipment from various vendors through multiple-owner trunk networks. The ANSI Standard T1.105.06 describes SONET, and ITU-T Recommendation G.957 describes SDH. To ensure interconnection compatibility between equipment from different manufacturers, the SONET and SDH specifications provide details for the optical source characteristics, the receiver sensitivity, and transmission distances for various types of fibers. Table 17.1 shows commonly used SDH and SONET signal levels and the associated OC rates. Table 17.2 lists SDH and SONET transmission distances, Table 17.3 gives wavelength ranges and attenuation for transmission distances up to 80 km, and Table 17.4 lists some representative optical equipment characteristics.

Ethernet is deployed widely in local-area networks with interfaces available at line rates ranging from 10 Mbps to 10 Gbps. Ethernet also is being used in metropolitan-area networks and is extending into wide-area networks. In these environments Ethernet uses optical fiber transmission links to increase network capacity cost-effectively and has the ability to offer a wide range of services in a simple, scalable, and flexible manner. When used in a MAN, Ethernet is referred to as *Metro Ethernet*. In enterprise applications, Metro Ethernet is used for interfacing to the public Internet and for connectivity between geographically separate corporate sites.

Emerging next-generation transport networks are referred to as optical transport networks (OTNs). In these networks it is envisioned that DWDM-based dynamic optical network elements such as optical cross-connect switches and optical add/drop multiplexers (OADMs) will have full control of all wavelengths. In addition, they are expected to have full knowledge of the traffic-carrying capacity and the status of each wavelength. With such intelligence these networks are envisioned as being self-connecting and self-regulating.

Various means of using fibers in the access network have been explored. These schemes are known by the all-inclusive term *fiber-to-the-x* (FTTx), where x is some letter designating at what point the fiber terminates and copper wires (or wireless links) again take over.

Further Reading

1. G. Keiser, *Local Area Networks*, 2d ed., McGraw-Hill, Burr Ridge, Ill., 2002.
2. B. A. Forouzan, *Introduction to Data Communications and Networking*, 2d ed., McGraw-Hill, Burr Ridge, Ill., 2001.
3. American National Standards Institute (ANSI), *ANSI T1.105 - 2001, Telecommunications*, "Synchronous optical network (SONET)—Basic description including multiplex structures, rates, and formats," New York, July 2002.
4. International Telecommunication Union—Telecommunication Standardization Sector, (ITU-T), http://www.itu.int, ITU-T Recommendation G.957, *Optical Interfaces for Equipment and Systems Relating to the Synchronous Digital Hierarchy*, July 1999.
5. P. Bonenfant and A. Rodriguez-Moral, "Framing techniques for IP over fiber," *IEEE Network*, vol. 15, pp. 12–18, July/August 2001.
6. M. Murata and K. I. Kitayama, "A perspective on photonic multiprotocol label switching," *IEEE Network*, vol. 15, pp. 56–63, July/August 2001.
7. D. Benjamin, R. Trudel, S. Shew, and E. Kus, "Optical services over the intelligent optical network," *IEEE Comm Mag*, vol. 39, pp. 73–78, September 2001.
8. R. Ramaswami and K. N. Sivarajan, *Optical Networks*, 2d ed., Morgan Kaufmann, San Francisco, 2002.
9. H. Zang, J. P. Jue, L. Sahasrabuddhe, R. Ramamurthy, and B. Mukherjee, "Dynamic lightpath establishment in wavelength-routed WDM networks," *IEEE Comm Mag*. vol. 39, pp. 100–108, September 2001.
10. ITU-T Recommendation G.985, *100 Mbps Point-to-Point Ethernet Based Optical Access System*, March 2003.
11. ITU-T Recommendation G.984.1, *Gigabit-Capable Passive Optical Network: General Characteristics*, March 2003.
12. ITU-T Recommendation G.984.2, *Gigabit-Capable Passive Optical Network: Physical Layer Specification,* March 2003.

Chapter

18

Network Management

A telecommunications service provider typically will offer a legal contract known as a *service-level agreement* (SLA) to its business customers. The terms of the SLA state that the service provider guarantees a measurable *quality of service* (QoS) to the customer. For example, an SLA may guarantee to a customer that the service will be available 99.999 percent of the time with a designated bit error rate (BER) within a monthly or annual time period. If the SLA guarantees are not met, the customer usually will receive a rebate. Thus there is a financial incentive for the service provider to manage and monitor the key performance parameters of the network very closely. The performance and operations management of a network requires the ability to configure and monitor network devices quickly and easily so that connections and services are always available. Early detection of changes in network status is critical in avoiding potential problems. This requires the use of sophisticated instruments and software-based diagnostic tools.

Thus, once the hardware and software elements of an optical network have been installed properly and integrated successfully, they need to be managed to ensure that the required level of network performance is met. In addition, the network devices must be monitored to verify that they are configured properly and to ensure that corporate policies regarding network use and security procedures are being followed. This is carried out through *network management*, which is a service that uses a variety of hardware and software tools, applications, and devices to assist human network managers in monitoring and maintaining networks. The International Organization for Standardization (ISO) has defined five primary conceptual areas of management for networks. These functional areas are performance, configuration, accounting, fault, and security management.

In an actual system, different groups of network operations personnel normally take separate responsibilities for issues such as administration aspects, performance monitoring, network integrity, access control, and security. There is no special method of allocating the various management functions to particular groups of people, since each organization may take a different approach to fit its own needs.

As the first topic, Sec. 18.1 describes the constituents of a network management architecture and outlines their purposes. Section 18.2 presents the basic concepts of the five generic network management areas outlined by the ISO. The gathering of status information from network devices is done via some type of network management protocol. The *Simple Network Management Protocol* (SNMP) is one example of this, as Sec. 18.3 shows. This section also addresses a set of remote-monitoring standards that extends and improves on the SNMP framework.

To deal with standardized management functions in the optical layer, the ITU-T created document ITU-T Recommendation G.709, *Network Node Interface for the Optical Transport Network (OTN)*, which also is referred to as the *Digital Wrapper* standard. As discussed in Sec. 8.4, whereas the SONET/SDH standard enabled the management of single-wavelength optical networks, the G.709 standard enables the broad adoption of technology for managing multiwavelength optical networks. Section 18.5 gives details on functions that are specific to optical communication elements directly, such as optical line terminals (OLTs), optical amplifiers, optical add/drop multiplexers (OADMs), and optical cross-connects (OXCs). Finally, Sec. 18.6 describes the use of a separate *optical service channel* (OSC) in links that contain optical amplifiers. The OSC operates on a wavelength that is outside of the standard WDM transmission grid being used. This allows the OSC to control and manage traffic without deploying a separate Ethernet control connection to each active device in the network.

18.1. Management Architecture

Figure 18.1 shows the components of a typical network management system and their relationships. The *network management console* is a specialized workstation that serves as the interface for the human network manager. There can be several of these workstations that perform different functions in a network. From such a console a network manager can view the health and status of the network to verify that all devices are functioning properly, that they are configured correctly, and that their application software is up to date. A network manager also can see how the network is performing, for example, in terms of traffic loads and fault conditions. In addition, the console allows control of the network resources.

The *managed devices* are network components, such as optical line terminals, optical amplifiers, optical add/drop multiplexers, and optical cross-connects. Each such device is monitored and controlled by its *element management system* (EMS). Management software modules, called *agents*, residing in a microprocessor within the elements continuously gather and compile information on the status and performance of the managed devices. The agents store this information in a *management information base* (MIB) and then provide this information to *management entities* within a *network management system* (NMS) that resides in the management workstation. A MIB (typically pronounced *mib*) is a logical base of information that defines data elements and their appropriate syntax and

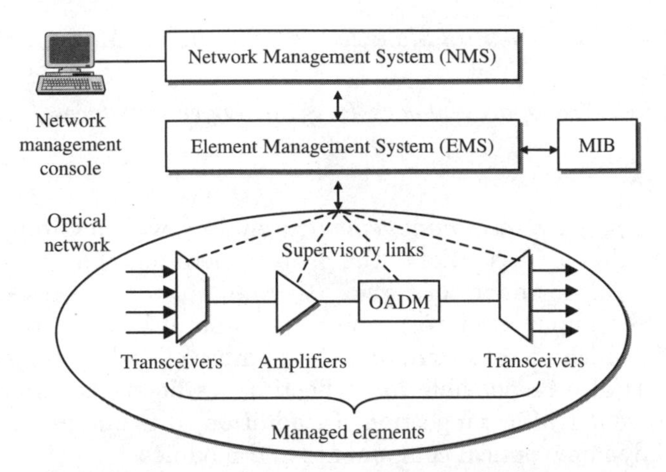

Figure 18.1. The components of a typical network management system and their relationships.

identifier, such as the fields in a database. This information may be stored in tables, counters, or switch settings. The MIB does not define how to collect or use data elements; it only specifies what the agent should collect and how to organize these data elements so that other systems can use them. The information transfer from the MIB to the NMS is done via a *network management protocol* such as SNMP (see Sec. 18.3).

When agents notice problems in the element they are monitoring (e.g., link or component faults, wavelength drifts, reduction in optical power levels, or excessive bit error rates), they send alerts to the management entities. Upon receiving an alert, the management entities can initiate one or more actions such as operator notification, event logging, system shutdown, or automatic attempts at fault isolation or repair. The EMS also can query or poll the agents in the elements to check the status of certain conditions or variables. This polling can be automatic or operator-initiated. In addition, there are *management proxies* that provide management information on behalf of devices that are not able to host an agent.

18.2. Basic Management Functions

The ISO has grouped network management functions into five generic categories: performance, configuration, accounting, fault, and security management. The principles for applying these functions to managing networks in general are described in ITU-T Recommendation X.701, *System Management Overview*. This section defines each of these categories and shows how they relate to managing optical networks. Other related ITU Recommendations for optical systems include

- G.7710, *Common Equipment Management Function Requirements*
- G.874, *Management Aspects of the OTN Element*

- G.874.1, *OTN Protocol-Neutral Management Information Model for the Network Element View*
- G.959.1, *Optical Transport Network Physical Layer Interfaces*

18.2.1. Performance management

In carrying out *performance management*, a system will monitor and control key parameters that are essential to the proper operation of a network in order to guarantee a specific quality of service (QoS) to network users. Among these are network throughput, user response times, line utilization, the number of seconds during which errors occur, and the number of bad messages delivered. This function is also responsible for collecting traffic statistics and applying controls to prevent traffic congestion. In addition, it examines the operating parameters of dynamic optical components and modules.

Examples of parameters that may be monitored at the physical level in an optical network are wavelength stability, bit error rate, and optical power levels. The performance management includes assigning threshold values to such parameters and informing the management system or generating alarms when these thresholds are exceeded.

18.2.2. Configuration management

The goal of *configuration management* is to monitor both network setup information and network device configurations, in order to track and manage the effects on network operation of the various constituent hardware and software elements. Configuration management allows a system to provision network resources and services, monitor and control their state, and collect status information. This provisioning may include remote provisioning of specific wavelengths to a user, automatically maintaining optical power-level settings in remote equipment as wavelengths are added or removed from the network, assigning special features requested by a user, distributing software upgrades to agents, and reconfiguring equipment to isolate faults.

Configuration management stores all this information in a readily accessible database, so that when a problem occurs, the database can be searched for assistance in solving the problem.

18.2.3. Accounting management

The purpose of *accounting management* is to measure network utilization parameters so that individuals or groups of users on the network can be regulated and billed for services appropriately. This regulation maximizes the fairness of network access across all users, since network resources can be allocated based on their capacities. Thus, accounting management is responsible for measuring, collecting, and recording statistics on resource and network usage. In addition, accounting management may examine current usage patterns in order to allocate network usage quotas. From the gathered statistics, the service provider then can generate a bill or a tariff for the usage of the service as well as ensure continued fair and optimal resource utilization.

18.2.4. Fault management

Faults in a network, such as physical cuts in a fiber transmission line or failure of a circuit card or optical amplifier, can cause portions of a network to be inoperable. Since network faults can result in system downtime or unacceptable network degradation, *fault management* is one of the most widely implemented and important network management functions. With the growing dependence of people on network resources for carrying out their work and communications, users expect rapid and reliable resolutions of network fault conditions. As Fig. 18.2 illustrates, fault management involves the following processes:

- Fault or degradation symptoms are detected, usually through alarm surveillance. *Alarm surveillance* involves reporting alarms that may have different levels of severity and indicating possible causes of these alarms. Fault management also provides a summary of unresolved alarms and allows the network manager to retrieve and view the alarm information from an alarm log.
- The origin and possible cause of faults are determined either automatically or through the intervention of a network manager. To determine the location or

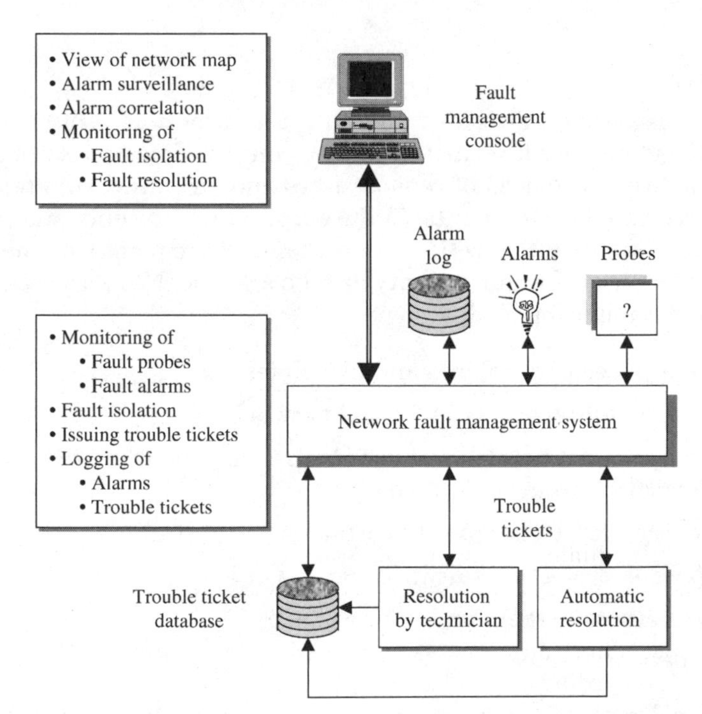

Figure 18.2. Functions and interactions of a network fault management system.

origin of faults, the management system might use *fault isolation* techniques such as alarm correlation from different parts of the network and diagnostic testing.

- Once the faults are isolated, the system issues *trouble tickets* that indicate what the problem is and possible means of how to resolve the fault. These tickets go to either a technician for manual intervention or an automatic fault correction mechanism. When the fault or degradation is corrected, this fact and the resolution method are indicated on the trouble ticket, which then is stored in a database.
- Once the problem has been fixed, the repair is operationally tested on all major subsystems of the network. *Operational testing* involves requesting performance tests, tracking the progress of these tests, and recording the results. The classes of tests that might be performed include echo tests and connectivity examinations.

An important factor in troubleshooting faults is to have a comprehensive physical and logical map of the network. Ideally this map should be part of a software-based management system that can show the network connectivity and the operational status of the constituent elements of the network on a display screen. With such a map, failed or degraded devices can be viewed easily, and corrective action can be taken immediately.

18.2.5. Security management

The ability of users to gain worldwide access to information resources easily and rapidly has made network security a major concern among network administrators. In addition, the need of remote users and personnel who telecommute to access corporate data from outside the corporation presents another dimension to network security. Figure 18.3 shows some of the points in a network and at its external interfaces where security may be an issue. Network security covers a number of disciplines, including

- Development of security policies and principles
- Creating a security architecture for the network
- Implementing special firewall software to prevent unauthorized access of corporate information from the Internet
- Applying encryption techniques to certain types of traffic
- Setting up virus protection software
- Establishing access authentication procedures
- Enforcing network security

The principal goal of network *security management* is to establish and enforce guidelines to control access to network resources. This control is needed to

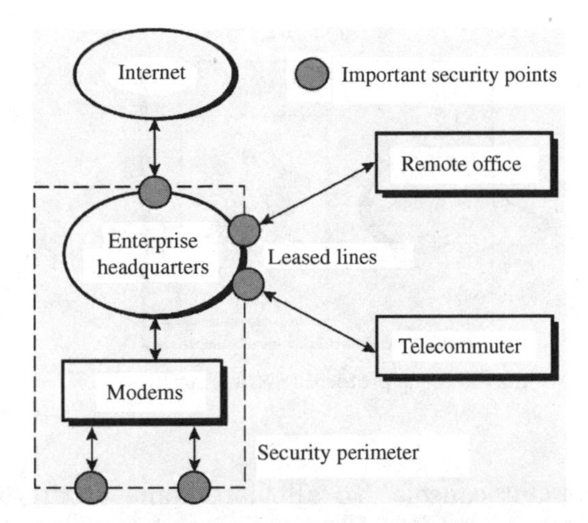

Figure 18.3. Points in a network and at its external interfaces where security may be an issue.

prevent intentional or unintentional sabotage of network capabilities and to prevent viewing or modification of sensitive information by people who do not have appropriate access authorization. For example, a security management system can monitor users attempting to log on to a particular network resource and can prevent access to those who do not have an authorized password.

18.3. Management Protocols

A number of communication protocols exist for gathering information from network devices. This section describes the widely used *simple network management protocol* (SNMP) and some of the enhancements and extensions that have been added to increase its scope and flexibility.

18.3.1. SNMP

SNMP is applicable in all types of networking environments. As shown in Fig. 18.4, each network device hosts an agent that gathers information about the status of that device and sends it to the management console. SNMP is the protocol that provides the query language for gathering the information and for sending it to the console. In general, the SNMP management system will discover the topology of the network automatically and will display it on the management console in the form of a graph. From this display the human network manager can select a particular segment of the network to view its status in greater detail.

18.3.2. RMON

Although SNMP is a simple and robust protocol, the information-gathering procedures increase network traffic and put a large management burden on the

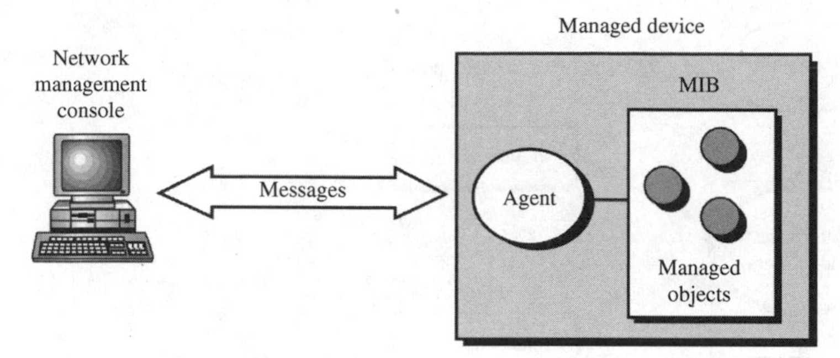

Figure 18.4. Basic simple network management protocol (SNMP) architecture.

central network management console. To alleviate some of this stress, the Internet Engineering Task Force (IETF) developed the *remote monitoring* (RMON) scheme.

The RMON set of standards extends and improves the SNMP framework. The main document is the RFC 1757 Ethernet RMON MIB, which describes the framework for remote monitoring of a MIB. The basic purpose of the RMON MIB specification is to readily implement a distributed management system consisting of an *RMON manager* and an *RMON probe*. The probe is responsible for collection of management data. It may be a stand-alone piece of equipment, or it could be a software application that is embedded within the managed device. The function of the manager is to retrieve the RMON information, process it, and then present it to the system administrator.

The RMON specifications give network administrators greater freedom to select network monitoring probes and consoles that meet the needs of a particular network. By defining a set of statistics and functions that can be exchanged between RMON-compliant consoles and network probes, the specifications provide the administrators with comprehensive fault diagnosis, planning, and performance-tuning information.

Management applications do not communicate directly with the managed device itself, but instead go through an RMON agent in the probe by using SNMP. This setup makes it easier to share information among multiple management stations. The RMON MIB specifications define how the information should be categorized in a common format so that the manager and the probe can exchange data. They do not define explicitly how the probe should collect the data or how the manager should format these data for presentation.

18.4. Optical Layer Management

To deal with standardized management functions in the optical layer, in February 2001 the ITU-T defined a three-layer model. The document is ITU-T Recommendation G.709, *Network Node Interface for the Optical Transport Network* (OTN), which also is referred to as the *Digital Wrapper standard*. Just as the SONET/SDH standard enabled the management of single-wavelength optical networks using equipment from many different vendors, the G.709 standard enables the broad adoption of technology for managing multiwavelength optical networks. The structure and layers of the OTN closely parallel the path, line, and section sublayers of SONET.

The model is based on a client/server concept. The exchange of information between processes running in two different devices connected through a network may be characterized by a *client/server interaction*. The terms *client* and *server* describe the functional roles of the elements in the network, as Fig. 18.5 illustrates. The process or element that requests or receives information is called the *client* (here the browser), and the process or element that supplies the information is called the *server*.

Figure 18.6 illustrates the three-layer model for a simple link. Client signals such as IP, Ethernet, or OC-*N*/STM-*M* are mapped from an electrical digital format into an optical format in an optical channel (OCh) layer. The OCh deals with single wavelength channels as end-to-end paths or as subnetwork connections between routing nodes. As shown in Fig. 18.7, the *optical multiplex section* (OMS) layer represents a link carrying groups of wavelengths between

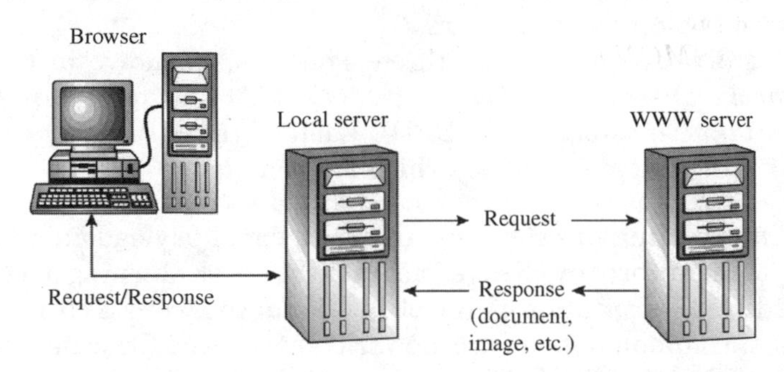

Figure 18.5. The terms *client* and *server* describe the functional roles of communicating elements in the network. Here the browser is the client.

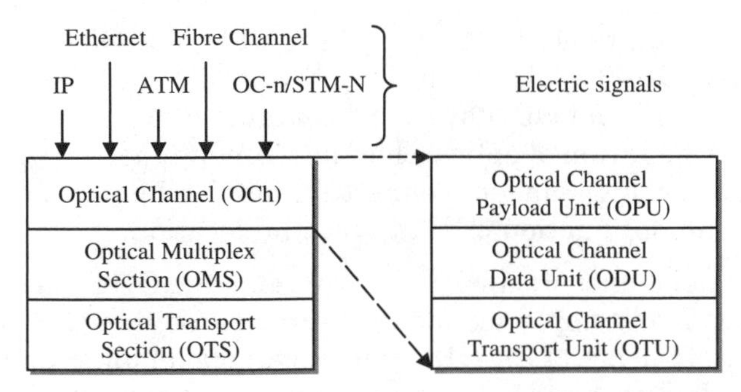

Figure 18.6. Three-layer model for a simple link in an OTN. The OCh is divided further into three sublayers.

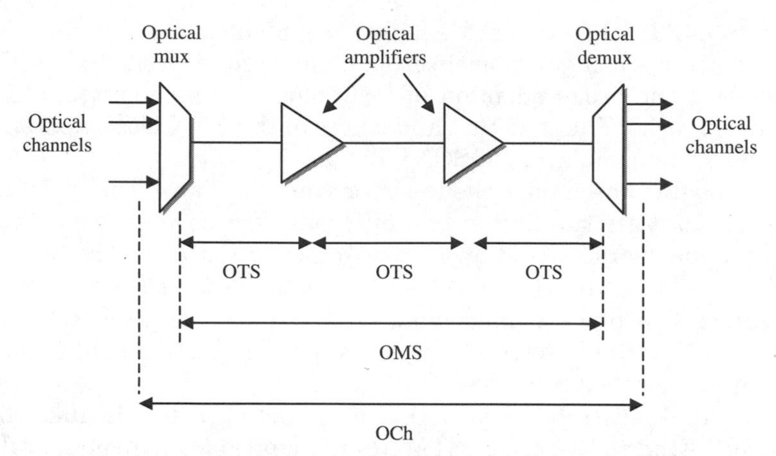

Figure 18.7. The OMS layer represents a link carrying wavelengths between multiplexers or OADMs. The OTS layer relates to a link between two optical amplifiers.

multiplexers or OADMs. The *optical transmission section* (OTS) layer relates to a link between two optical amplifiers.

The OCh is divided further into three sublayers, as shown in Fig. 18.6: the *optical channel transport unit* (OTU), the *optical channel data unit* (ODU), and the *optical channel payload unit* (OPU). Each of these sublayers has its own functions and associated overhead, which are as follows:

- The OPU frame structure contains the client signal payload and the overhead necessary for mapping any client signal into the OPU. Mapping of client signals may include rate adaptation of the client signal to a constant-bit-rate signal. Examples of common signals are IP, various forms of Ethernet, ATM, Fibre Channel, and SONET/SDH. The three payload rates associated with the OPU sublayer are 2.5, 10, and 40 Gbps. These correspond to standard SONET/SDH data rates (OC-48/STM-16, OC-192/STM-64, and OC-768/STM-256, respectively), but may be used for any client signal.
- The ODU is the structure used to transport the OPU. The ODU consists of the OPU and the associated ODU overhead and provides path-layer-connection monitoring functions. The ODU overhead contains information that enables maintenance and operation of optical channels. Among these are maintenance signals, path monitoring, tandem connection monitoring, automatic protection switching, and designation of fault type and location.
- The optical channel transport unit (OTU) contains the ODU frame structure, the OTU overhead, and appended forward error correction (FEC). The OTU changes the digital format of the ODU into a light signal for transport over an optical channel. It also provides error detection and correction and section layer connection monitoring functions.

18.5. Element Management

This section addresses some of the concepts and functions related to monitoring the overall network performance together with managing the various active elements in an optical network and checking on their health and status. The topics covered include BER measurements, wavelength assignment management, monitoring the performance of various network elements, fault detection and recovery techniques, and the implementation of separate wavelengths for monitoring active devices.

18.5.1. Error monitoring

From the overall system point of view, the bit error rate (BER) is the main performance parameter for any specific lightpath. The BER is calculated by the receiving equipment after the optical signal has been converted back to an electric signal.

This BER calculation process is well established in networking equipment that uses SONET or SDH as the underlying transport protocol. This protocol contains an inherent set of parity-check bytes that are used for continuously monitoring the information stream traveling from one SONET/SDH terminal to another for errors. The ITU-T Recommendation G.709, or Digital Wrapper, uses the same error-monitoring technique as is employed in SONET/SDH. The performance metrics that are calculated in these protocols include coding violations in the incoming bit stream, the number of seconds in which at least one error occurs, the number of seconds in which multiple errors occur (called *severally errored seconds*), and the total number of seconds in which service is not available.

In local-area networks and other communication environments that use Gigabit Ethernet (GigE) or Fibre Channel (FC) in place of SONET/SDH as the underlying transport protocol, the SLA metrics need to be slightly different. As described in Sec. 16.4, GigE and FC utilize 8B10B encoding in which 8 bits are encoded into 10 bits. In this scheme only certain groups of bit patterns (called *code groups*) are allowed. Thus, the 8B10B decoding process can be used for error monitoring by detecting invalid code groups. In addition a process called *cyclic redundancy check* (CRC) that makes use of a standard preset polynomial normally is implemented in GigE and FC to check for errors. In this case the CRC calculation is based on a binary division method involving 32 bits that consist of the data portion of a packet plus a sequence of redundant bits.

Since BER monitoring has been implemented worldwide for many years, numerous software packages of varying degrees of complexity are available commercially. Often they are part of a larger network management software system running on a management terminal. By means of a Java-based *graphical user interface* (GUI), a network manager using such programs has the ability to carry out full system monitoring and control via the Internet. The performance data gathered and calculated by these management packages can provide an overall graphical view of the network, which typically consists of a variety of hardware technologies from different vendors. In addition, the BER information can be integrated with other *operations support system* (OSS) functions such as order processing, billing, and service provisioning.

Note that since the BER measurement is done after the received signal is transformed back to the electrical domain, this metric only gives an indication of the overall performance of a link. It does not tell whether a change in link quality was caused by optical power reduction, degradation in the optical signal-to-noise ratio, or component aging or failure. To measure these effects, one needs to invoke the element management procedures described in Sec. 18.5.3.

18.5.2. Wavelength management

To deploy new services and applications rapidly to their customers, service providers must possess a complete end-to-end management capability of the optical access link. For example, the SLA can specify that certain wavelengths will be available during specific time periods or only on weekends when normal traffic flow is low. Thus, network managers need the ability to reassign wavelengths or portions of the capacity of a wavelength quickly, thereby temporarily or permanently increasing bandwidth allocation according to customer requirements.

Wavelength management is especially important in metro networks where there typically is a high degree of dynamic change in customer bandwidth requirements. By using dynamic lightpath reconfiguration rather than depending on a fixed physical infrastructure, network management personnel can provision for these needs by means of a point-and-click feature on their management console. Such capabilities are part of large network management programs that are available commercially. Among the features in these programs are capabilities such as path tracing for individual wavelengths, power tracing and remote power adjustment for individual wavelengths, and optical-layer topology discovery that permits fault isolation to specific network sections.

18.5.3. Element monitoring

Since the signal quality of an optical network depends critically on the proper operation of all its constituent elements, monitoring techniques that can be performed directly in the optical domain are a key requirement. The three main parameters for any element are wavelength, optical power, and OSNR (optical signal-to-noise ratio). The measurement instruments are based on spectrum analysis techniques (see Chap. 19) and are known by a variety of names. For example, one may see the names *optical channel monitor* (OCM), *optical performance monitor* (OPM), or *optical channel analyzer* (OCA). The original intent was to designate slightly different monitoring functions, but the differences among these functions are becoming blurred. Therefore for simplicity we will refer to them as *optical performance monitors*.

An OPM taps off a small portion of the light signals in a fiber and separates the wavelengths or scans them onto a detector or detector array. This enables the measurement of individual channel powers, wavelength, and OSNR. These

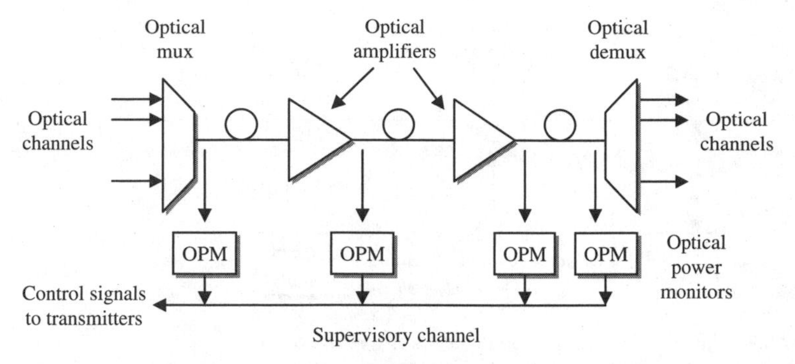

Figure 18.8. DWDM networks use an automated OPM to measure the light level of each wavelength at various network points and to adjust the individual laser outputs at the transmitter.

devices have an important role in controlling DWDM networks. For example, as shown in Fig. 18.8, most long-haul DWDM networks incorporate automated end-to-end power-balancing algorithms that use a high-performance OPM to measure the optical power level of each wavelength at optical amplifiers and at the receiver and to adjust the individual laser outputs at the transmitter. This information is exchanged by means of a separate supervisory channel, which is described in Sec. 18.6. In addition, manufacturers may embed an OPM function into dynamic elements such as an EDFA, an OADM, or an OXC to provide feedback for active control of total output power and to balance the power levels between channels. Other functions of an OPM include determining whether a particular channel is active, verifying whether wavelengths match the specified channel plan, and checking whether optical power and OSNR levels are sufficient to meet the QoS requirements.

An OPM may have the following operational characteristics:

- Measures absolute channel power to within ±0.5 dBm
- Identifies channels without prior knowledge of the wavelength plan
- Makes full S-, C-, or L-band measurements in less than 0.5 s
- Measures center wavelength accuracy to better than ±50 pm
- Determines OSNR with a 35-dB dynamic range to a ±0.1-dB accuracy

18.6. Optical Service Channel

ITU-T Recommendation G.692 describes the use of a separate *optical service channel* (OSC) in links that contain optical amplifiers. The OSC operates on a wavelength that is outside of the standard WDM transmission grid being used. For example, in a C-band DWDM link (1530 to 1565 nm) the OSC might operate at 1310, 1480, 1510, or 1620 nm. Of these the ITU has adopted 1510 nm as the preferred wavelength. In a 32-channel system this would be referred to as

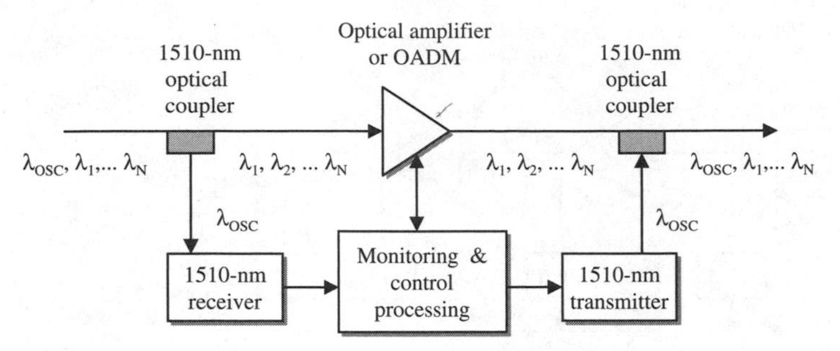

Figure 18.9. The OSC bypasses the device being monitored.

using a *33rd wavelength* (or channel 0), which allows the OSC to control and manage traffic without deploying a separate Ethernet control connection to each active device in the network.

As shown in Fig. 18.9, the OSC bypasses the device being monitored and always terminates on a neighboring node. This is in contrast to data channels which do not necessarily terminate on a given node (such as an optical amplifier or an OADM).

An OSC carries out the following types of functions:

- *Discovery*. This function sends packets over the OSC to discover the logical topology of the network
- *Monitoring*. With this keep-alive function, nodes exchange packets that allow them to determine the operational status of their neighbors.
- *Management*. IP packets are carried over the OSC to support SNMP and Telnet sessions

A variety of vendors offer 1510-nm channel couplers and lasers that operate from the 2- to 155-Mbps data rates used on the OSC.

18.7. Summary

System maintenance and system control are necessary and important functions in any network. To meet the requirements of the service-level agreements with their customers, service providers must manage and monitor the key performance parameters of the network very closely. The performance and operations management of a network requires the ability to configure and monitor network devices quickly and easily so that connections and services are always available. Early detection of changes in network status is critical in avoiding potential problems. This requires the use of sophisticated instruments and software-based diagnostic tools.

Once the hardware and software elements of an optical network have been installed properly and integrated successfully, they need to be managed to ensure that they are configured correctly and operating properly, and that corporate policies regarding network use and security procedures are being followed. This is carried out through *network management*, which is a service that uses a variety of hardware and software tools, applications, and devices to assist human network managers in monitoring and maintaining networks. The ISO has defined five primary conceptual areas of management for networks. These functional areas are performance, configuration, accounting, fault, and security management.

The gathering of status information from network devices is done via some type of network management protocol, such as the *Simple Network Management Protocol* (SNMP). This protocol provides the query language for gathering performance information and sending it to a management console. In general, the SNMP system will discover the topology of the network automatically and will display it on the management console in the form of a graph. From this display the human network manager can view the status of a particular network segment in greater detail.

To deal with standardized management functions in the optical layer, the ITU-T issued the document G.709, *Network Node Interface for the Optical Transport Network* (*OTN*), which describes a three-layer model. This also is referred to as the *Digital Wrapper standard*. This standard enables the broad adoption of technology for managing multiwavelength optical networks. The model described in G.709 is based on a client/server concept. Client signals such as IP, Ethernet, or OC-*N*/STM-*M* are mapped from an electrical digital format into an optical format in an optical channel (OCh) layer. The OCh deals with single-wavelength channels as end-to-end paths or as subnetwork connections between routing nodes.

ITU-T Recommendation G.692 describes the use of a separate *optical service channel* (OSC) in links that contain optical amplifiers. The OSC operates on a wavelength that is outside of the standard WDM transmission grid being used. This allows the OSC to control and manage traffic without deploying a separate Ethernet control connection to each active device in the network.

Further Reading

1. L. Raman, "OSI systems and network management," *IEEE Commun. Mag.*, vol. 36, pp. 46–53, March 1998.
2. G. Keiser, *Local Area Networks*, 2d ed., McGraw-Hill, Burr Ridge, Ill., 2002, Chap. 11.
3. ITU-T Recommendation X.701, *Information Technology—Open Systems Interconnection—Systems Management Overview*, August 1997.
4. E. Park, "Error monitoring for optical metropolitan network services," *IEEE Commun. Mag.*, vol. 40, pp. 104–109, February 2002.
5. C. A. Armiento and Y. A. Yudin, "Wavelength managers monitor reconfigurable DWDM networks," *WDM Solutions*, vol. 4, pp. 45–49, February 2002.
6. R. Ramaswami and K. N. Sivarajan, *Optical Networks*, 2d ed., Morgan Kaufmann, San Francisco, 2002, Chap. 9.
7. L. Berthelon, "Management of WDM networks," *Proc. 25th European Conf. on Optical Communications (ECOC)*, pp. II.94–II.97, September 1999.

Chapter

19

Test and Measurement

The installation and powering up of an optical fiber communication system requires measurement techniques for verifying the link has been configured properly and that its constituent components are functioning correctly. Of particular importance are accurate and precise measurements of the optical fiber, since this component cannot be replaced readily once it has been installed. In addition, various test methods are needed for continually monitoring the link condition to verify that the performance requirements are being met during operation.

During the link design phase an engineer can find the operational parameters of many components on vendor data sheets. These include fixed parameters for fibers (e.g., core and cladding diameters, refractive index profile, mode-field diameter, cutoff wavelength); passive splitters, connectors, and couplers; and electrooptic components such as sources, photodetectors, and optical amplifiers. Once these parameters are known, there is no need to measure them again.

However, the attenuation and dispersion of a fiber can change during fiber cabling and cable installation. In single-mode fibers, chromatic and polarization mode dispersions are important factors that can limit the transmission distance or data rate. Chromatic dispersion effects are of particular importance in high-speed WDM links, and polarization mode dispersion ultimately can limit the highest achievable data rate in single-mode links. Measurement procedures for these parameters thus are of interest to the user, as are methods for locating breaks and faults in optical fiber cables.

When a link is being installed and tested, the operational parameters of interest include the bit error rate, timing jitter, and signal-to-noise ratio as indicated by the eye pattern. During actual operation, measurements are needed for maintenance and monitoring functions to determine factors such as fault locations in fibers and the status of remotely located optical amplifiers and other active devices.

This chapter discusses measurements and performance tests of interest to installers and operators of fiber optic links and networks. Of particular interest

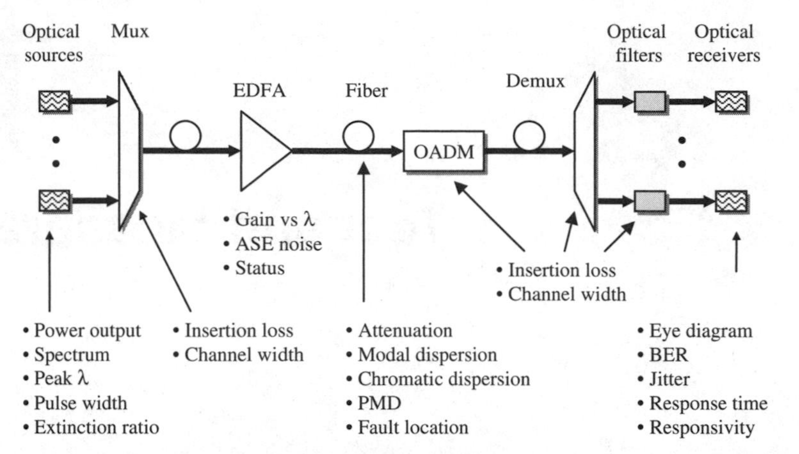

Figure 19.1. Some of the relevant test parameters and at what points in a WDM link they are of importance.

here are measurements for WDM links. Figure 19.1 shows some of the relevant test parameters and at what points in a WDM link they are of importance. The chapter first addresses measurement standards in Sec. 19.1 and basic test equipment for optical fiber communication links in Sec. 19.2. Next, Sec. 19.3 discusses optical power and its measurement with optical power meters. Section 19.4 describes tunable laser sources, which are important support instruments for wavelength-dependent tests. The optical spectrum analyzer (OSA) is a key instrument since it measures optical power as a function of wavelength. This instrument and some its applications are described in Sec. 19.5. The long-term workhorse measurement system in fiber optic systems is the optical time-domain reflectometer (OTDR). This instrument is mainly for field evaluations of fibers, as Sec. 19.6 explains.

19.1. Measurement Standards

Before we examine measurement techniques, let us look at what standards exist for fiber optics. As summarized in Table 19.1, there are three basic classes: primary standards, component testing standards, and system standards.

Primary standards refer to measuring and characterizing fundamental physical parameters such as attenuation, bandwidth, mode-field diameter for single-mode fibers, and optical power. In the United States the main organization involved in primary standards is the National Institute of Standards and Technology (NIST). This organization carries out fiber optic and laser standardization work, and it sponsors an annual conference on optical fiber measurements. Another goal is to support and accelerate the development of emerging technologies. Other national organizations include the National Physical Laboratory (NPL) in the United Kingdom and the Physikalisch-Technische Bundesanstalt (PTB) in Germany.

TABLE 19.1. Summary of Key Standards Organizations and Their Functions

Standards class	Key organizations	Functions
Primary	• NIST (U.S.) • NPL(UK) • PTB(Germany)	• Characterize physical parameters • Support and accelerate development of emerging technologies (NIST)
Component testing	• TIA/EIA • ITU-T • IEC	• Define component evaluation tests • Establish equipment calibration procedures
System testing	• ANSI • IEEE • ITU-T	• Define physical-layer test methods • Establish measurement procedures for links and networks

Component testing standards define relevant tests for fiber optic component performance, and they establish equipment calibration procedures. Several different organizations are involved in formulating testing standards, some very active ones being the Telecommunication Industries Association (TIA) in association with the Electronic Industries Association (EIA), the Telecommunication Standardization Sector of the International Telecommunication Union (ITU-T), and the International Electrotechnical Commission (IEC). The TIA has a list of over 120 fiber optic test standards and specifications under the general designation TIA/EIA-455-XX-YY, where XX refers to a specific measurement technique and YY refers to the publication year. These standards are also called Fiber Optic Test Procedures (FOTPs), so that TIA/EIA-455-XX becomes FOTP-XX. These include a wide variety of recommended methods for testing the response of fibers, cables, passive devices, and electrooptic components to environmental factors and operational conditions. For example, TIA/EIA-455-60A-2000, or FOTP-60, is a method revised in 2000 for measuring fiber or cable length using an OTDR.

System standards refer to measurement methods for links and networks. The major organizations involved here are the American National Standards Institute (ANSI), the Institute for Electrical and Electronic Engineers (IEEE), and the ITU-T. Of particular interest for fiber optics systems are test standards and recommendations from the ITU-T that are aimed at all aspects of optical networking. Within the TIA, the FO-2 Committee develops physical-layer test procedures, system design guides, and system specifications to assist both suppliers and users of fiber optic communications technology.

Interoperability and compatibility between different vendor equipment are important concerns. The committee addresses the performance and reliability of active components and systems, such as transmitters, receivers, amplifiers, and modulators. Systems include single-mode digital and analog systems, optically amplified systems with dense wavelength division multiplexing (DWDM), point-to-point multimode systems, and local-area network (LAN) applications.

19.2. Basic Test Equipment

As optical signals pass through the various parts of an optical link, they need to be measured and characterized in terms of the three fundamental areas of optical power, polarization, and spectral content. The basic pieces of test equipment for carrying out such measurements on optical fiber components and systems include optical power meters, attenuators, tunable laser sources, spectrum analyzers, and time-domain reflectometers. These come in a variety of capabilities, with sizes ranging from portable, handheld units for field use to sophisticated briefcase-size bench-top or rack-mountable instruments for laboratory and manufacturing applications. In general, the field units do not need to have the extremely high precision of laboratory instruments, but they need to be more rugged to maintain reliable and accurate measurements under extreme environmental conditions of temperature, humidity, dust, and mechanical stress. However, even the handheld equipment for field use has reached a high degree of sophistication with automated microprocessor-controlled test features and computer interface capabilities.

More sophisticated instruments, such as polarization analyzers and optical communication analyzers, are available for measuring and analyzing polarization mode dispersion (PMD), eye diagrams, and pulse waveforms. These instruments enable a variety of statistical measurements to be made at the push of a button, after the user has keyed in the parameters to be tested and the desired measurement range.

Table 19.2 lists some widely used optical system test equipment and their functions. This chapter covers a selection of test equipment used for manufacturing, installation, and operation. Test equipment and measurement

TABLE 19.2. Some Widely Used Optical System Test Equipment and Its Functions

Test equipment	Function
Optical power meter	Measures total power over a selected wavelength band
Optical spectrum analyzer (OSA)	Measures optical power as a function of wavelength
Test-support laser (multiple-wavelength or broadband)	Assists in tests that measure the wavelength-dependent response of an optical component or link
Optical power attenuator	Reduces power level to prevent instrument damage or to avoid overload distortion in the measurements
OTDR (field instrument)	Measures attenuation, length, connector/splice losses, and reflectance levels; helps locate fiber breaks
Multifunction optical test system	Factory or field instruments with exchangeable modules for performing a variety of measurements
Polarization analyzer	Measures polarization-dependent loss (PDL) and polarization mode dispersion (PMD); see Chap. 15
BER test equipment	Uses standard eye pattern masks to evaluate the data handling ability of an optical link; see Chap. 14

methodologies for analyzing optical fibers and various passive and active components are not covered here but can be found in the book by Derickson.

19.3. Optical Power Measurements

Optical power measurement is the most basic function in fiber optic metrology. However, this parameter is not a fixed quantity and can vary as a function of other parameters such as time, distance along a link, wavelength, phase, and polarization.

19.3.1. Definition of optical power

To get an understanding of optical power, let us look at its physical basis and how it relates to other optical quantities such as energy, intensity, and radiance.

- As described in Chap. 3, light particles are known as *photons*, which have a certain energy associated with them. The relationship between the energy E of a photon and its wavelength λ is given by the equation $E = hc/\lambda$, which is known as *Planck's law*. In terms of wavelength (measured in units of micrometers), the energy in electron volts (eV) is given by the expression E (eV) = $1.2406/\lambda$ (µm). Note that 1 eV = 1.60218×10^{-19} J (joules).
- *Optical power P* measures the rate at which photons arrive at a detector; that is, it is a measure of energy transfer per time. Since the rate of energy transfer varies with time, the optical power is a function of time. It is measured in *watts* or joules per second (J/s).
- As noted in Chap. 6, *radiance* (or *brightness*) is a measure, in watts, of how much optical power radiates into a unit solid angle per unit of emitting surface.

Since optical power varies with time, its measurement also changes with time. As shown in Fig. 19.2, which plots the power level in a signal pulse as a function of time, different instantaneous power-level readings are obtained depending on the precise time when the measurement is made. Therefore, two standard classes of power measurements can be specified in an optical system. These are the peak power and the average power. The *peak power* is the maximum power level in a pulse, which might be sustained for only a very short time.

The *average power* is a measure of the power level averaged over a relatively long time period compared to the duration of an individual pulse. For example, the measurement time period could be 1s, which contains many signal pulses. As a simple example, in a non-return-to-zero (NRZ) data stream (see Chap. 16) there will be an equal probability of 1 and 0 pulses over a long time period. In this case, as shown in Fig. 19.2, the average power is one-half of the peak power. If a return-to-zero (RZ) modulation format is used, the average power over a long sequence of pulses will be one-fourth of the peak power since there is no pulse in a 0 time slot and a 1 time slot is only half filled.

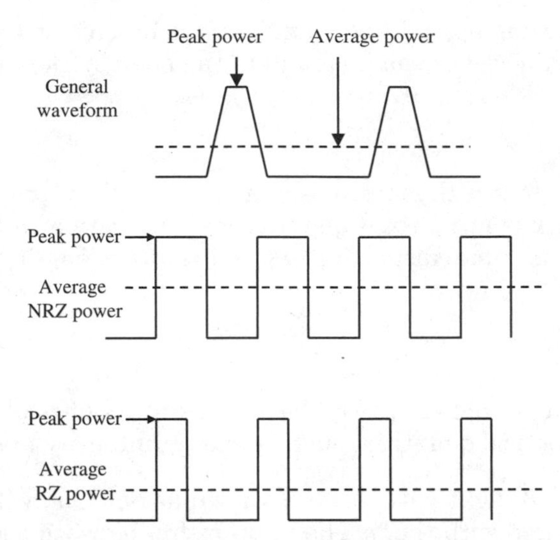

Figure 19.2. Peak and average powers in a series of general, NRZ, and RZ optical pulses.

The sensitivity of a photodetector normally is expressed in terms of the average power level impinging on it, since the measurements in an actual fiber optic system are done over many pulses. However, the output level for an optical transmitter normally is specified as the peak power. This means that the average power coupled into a fiber, and the power level which a photodetector measures, is at least 3 dB lower than if the link designer incorrectly used the peak source output in power budget calculations as the light level entering the fiber.

19.3.2. Optical power meters

The function of an *optical power meter* is to measure total power over a selected wavelength band. Some form of optical power detection is in almost every piece of lightwave test equipment. Handheld instruments come in a wide variety of types, with different levels of capabilities. Multiwavelength optical power meters using photodetectors are the most common instrument for measuring optical signal power levels. Usually the meter outputs are given in dBm (where 0 dBm = 1 mW) or dBμ (where 0 dBμ = 1 μW).

As an illustration, Fig. 19.3 shows a handheld model FOT-90A fiber optic power meter from EXFO. In this versatile instrument, various photodetector heads having different performance characteristics are available. For example, using a Ge photodetector allows a measuring range of +18 to −60 dBm in the 780- to 1600-nm wavelength band, whereas an InGaAs photodetector allows a measuring range of +3 to −73 dBm in the 840- to 1650-nm wavelength band. In each case, the power measurements can be made at 20 calibrated wavelengths with a ±20-dB accuracy. An RS-232 interface together with application software allows a user to download measurements and view, export, or print them

Figure 19.3. Example of a versatile handheld optical power meter. (*Model FOT-90A, provided courtesy of EXFO; www.exfo.com.*)

in either tabular or graphic form. The permanent memory registers can store 512 readings manually or 400 readings automatically at a programmable time interval.

Figure 19.4 shows another handheld tester that also contains optical sources to carry out more sophisticated optical power measuring. For example, this instrument can function as a power meter, an optical-loss tester for automatically measuring loss in a fiber in two directions at two wavelengths, an optical return-loss tester for measuring the quality of optical patch cords, a visual fault indicator for locating breaks and failures in a fiber cable, and a talk set for full-duplex communications between field personnel.

19.3.3. Optical power attenuators

In many laboratory or production tests, the characteristics of a high optical signal level may need to be measured. If the level is very high, such as a strong output from an optical amplifier, the signal may need to be attenuated precisely before being measured. This is done to prevent instrument damage or to avoid overload distortion in the measurements. An *optical attenuator* allows a user to reduce an optical signal level up to, for example, 60 dB (a factor of 10^6) in precise steps at a specified wavelength, which is usually 1310 or 1550 nm. The capabilities of attenuators range from simple tape-cassette-size devices for quick field measurements that may only need to be accurate to 0.5 dB to laboratory instruments that have an attenuation precision of 0.001 dB.

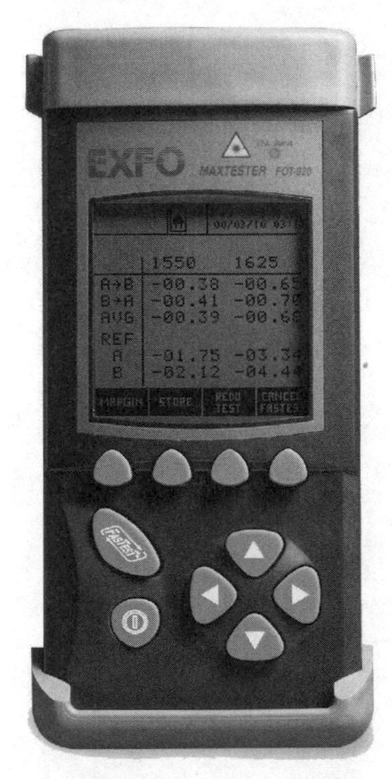

Figure 19.4. Compact, portable multipurpose test instrument for use in field environments. (*Model FOT-920, provided courtesy of EXFO; www.exfo.com.*)

TABLE 19.3. Characteristics of Laser Source Instruments Used for Test Support

Parameter	Tunable source	Broadband source
Spectral output range	Selectable: for example, 1370–1495 or 1460–1640 nm	Peak wavelength ± 25 nm
Total optical output power	Up to 8 dBm	>3.5 mW (5.5 dBm) over a 50-nm range
Power stability	$<\pm 0.02$ dB	$<\pm 0.05$ dB
Wavelength accuracy	$<\pm 10$ pm	Not applicable

19.4. Test-Support Lasers

Specialized light sources are desirable for testing optical components. Table 19.3 lists the characteristics of two such laser source instruments used for test support.

Tunable laser sources are important instruments for measurements of the wavelength-dependent response of an optical component or link. A number of vendors offer such light sources that generate a true single-mode laser line for every selected wavelength point. Typically the source is an external-cavity semiconductor laser. A movable diffraction grating may be used as a tunable filter for wavelength selection. Depending on the source and grating combination, an instrument may be tunable over (for example) the 1280- to 1330-nm, the 1370- to 1495-nm, or the 1460- to 1640-nm band. Wavelength scans, with an output power that is flat across the scanned spectral band, can be done automatically. The minimum output power of such an instrument usually is -10 dBm, and the absolute wavelength accuracy is typically ± 0.01 nm (± 10 pm).

A *broadband incoherent light source* with a high output power coupled into a single-mode fiber is desirable to evaluate passive DWDM components. Such an instrument can be realized by using the *amplified spontaneous emission* (ASE) of an erbium-doped fiber amplifier (EDFA). The power spectral density of the output is up to 100 times (20 dB) greater than that of edge emitting LEDs and up to 100,000 times (50 dB) greater than white-light tungsten lamp sources. The instrument can be specified to have a total output power of greater than 3.5 mW (5.5 dBm) over a 50-nm range with a spectral density of 13 dBm/nm (-50 μW/nm). The relatively high-power spectral density allows test personnel to characterize devices with medium or high insertion loss. Peak wavelengths might be 1200, 1310, 1430, 1550, or 1650 nm.

19.5. Optical Spectrum Analyzer

The widespread implementation of WDM systems calls for making optical spectrum analyses to characterize the spectral behavior of various telecommunication network elements. One widely used instrument for doing this is an *optical spectrum analyzer* (OSA), which measures optical power as a function of wavelength. The most common implementation uses a diffraction-grating-based optical filter, which yields wavelength resolutions to less than 0.1 nm. Higher wavelength accuracy (± 0.001 nm) is achieved with wavelength meters based on Michelson interferometry.

Figure 19.5 illustrates the operation of a grating-based optical spectrum analyzer. Light emerging from a fiber is collimated by a lens and is directed onto a diffraction grating that can be rotated. The exit slit selects or filters the spectrum of the light from the grating. Thus, it determines the *spectral resolution* of the OSA. The term *resolution bandwidth* describes the width of this optical filter. Typical OSAs have selectable filters ranging from 10 nm down to 0.1 nm. The optical filter characteristics determine the *dynamic range*, which is the ability of the OSA to simultaneously view large and small signals in the same sweep. The bandwidth of the amplifier is a major factor affecting the sensitivity and sweep time of the OSA. The photodiode is usually an InGaAs device.

The OSA normally sweeps across a spectral band, making measurements at discretely spaced wavelength points. This spacing depends on the bandwidth resolution capability of the instrument and is known as the *trace-point spacing*.

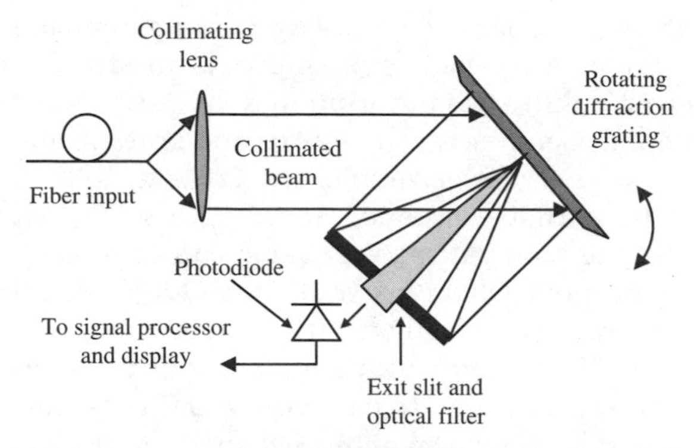

Figure 19.5. Operation of a grating-based optical spectrum analyzer.

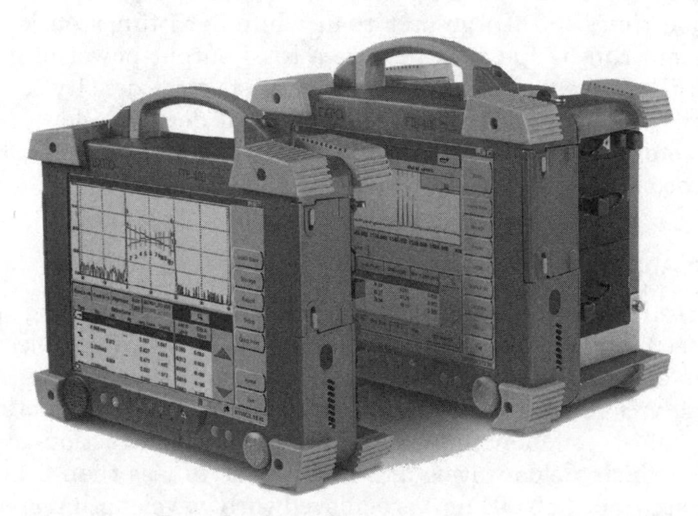

Figure 19.6. Example of portable universal test sets that can be used for OTDR analysis, optical spectrum analysis, dispersion analysis, and other test functions. (*Model FTB-400, provided courtesy of EXFO; www.exfo.com.*)

19.6. Optical Time-Domain Reflectometer

The long-term workhorse instrument in fiber optic systems is the *optical time-domain reflectometer* (OTDR). In addition to locating faults within an optical link, this instrument measures parameters such as attenuation, length, connector and splice losses, and reflectance levels. A typical OTDR consists of an optical source and receiver, a data acquisition module, a central processing unit (CPU), an information storage unit for retaining data either in the internal memory or on an external disk, and a display. As an example of an OTDR instrument, the portable unit shown in Fig. 19.6 can perform tests for outside plant installation, maintenance, and troubleshooting.

An OTDR is fundamentally an optical radar. It operates by periodically launching narrow laser pulses into one end of a fiber under test by using either a directional coupler or a beam splitter. The properties of the optical fiber link then are determined by analyzing the amplitude and temporal characteristics of the waveform of the backscattered light.

19.6.1. OTDR trace

Figure 19.7 shows a typical trace as would be seen on the display screen of an OTDR. The scale of the vertical axis is logarithmic and measures the returning (back-reflected) signal in decibels. The horizontal axis denotes the distance between the instrument and the measurement point in the fiber. The backscattered waveform has four distinct features:

- A large initial pulse resulting from Fresnel reflection at the input end of the fiber.
- A long decaying tail resulting from Rayleigh scattering in the reverse direction as the input pulse travels along the fiber. In Fig. 19.7 the different slopes of the three curves mean that the three fibers have different attenuations.
- Abrupt shifts in the curve caused by optical loss at joints, at connectors, or because of sharp bends in the fiber line.
- Positive spikes arising from Fresnel reflection at the far end of the fiber, at fiber joints, and at fiber imperfections.

Fresnel reflection and Rayleigh scattering principally produce the backscattered light. *Fresnel reflection* occurs when light enters a medium having a different index of refraction. For a glass-air interface, when light of power P_0 is incident perpendicular to the interface, the reflected power P_{ref} is

$$P_{\mathrm{ref}} = P_0 \left(\frac{n_{\mathrm{fiber}} - n_{\mathrm{air}}}{n_{\mathrm{fiber}} + n_{\mathrm{air}}} \right)^2 \tag{19.1}$$

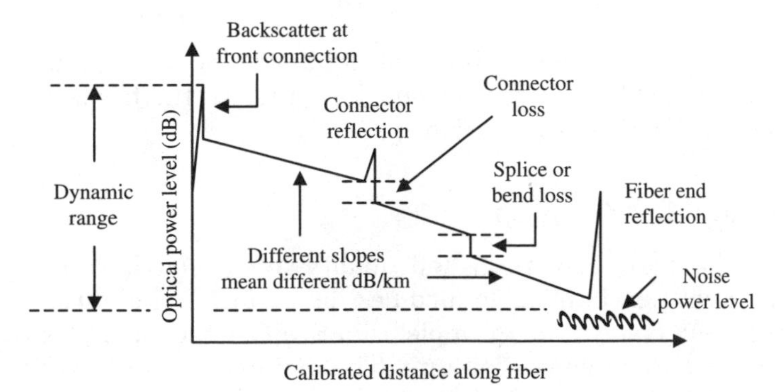

Figure 19.7. Representative trace of backscattered optical power as displayed on an OTDR screen and the meanings of various trace features.

where n_{fiber} and n_{air} are the refractive indices of the fiber core and air, respectively. A perfect fiber end reflects about 4 percent of the power incident on it. However, since fiber ends generally are not polished perfectly and perpendicular to the fiber axis, the reflected power tends to be much lower than the maximum possible value.

Two important performance parameters of an OTDR are dynamic range and measurement range. *Dynamic range* is defined as the difference between the initial backscattered power level at the front connector and the noise level after 3 min of measurement time. It is expressed in decibels of one-way fiber loss. Dynamic range provides information on the maximum fiber loss that can be measured and denotes the time required to measure a given fiber loss. Thus it often is used to rank the capabilities of an OTDR. A basic limitation of an OTDR is the tradeoff between dynamic range and resolution. For high spatial resolution, the pulse width has to be as small as possible. However, this reduces the signal-to-noise ratio and thus lowers the dynamic range. For example, a 100-ns pulse width allows a 24-dB dynamic range, whereas a 20-μs pulse width increases the dynamic range to 40 dB.

Measurement range deals with the capability of identifying events in the link, such as splice points, connection points, or fiber breaks. It is defined as the maximum allowable attenuation between an OTDR and an event that still enables the OTDR to accurately measure the event. Normally, for definition purposes, a 0.5-dB splice is selected as the event to be measured.

19.6.2. Fiber fault location

In addition to measuring attenuation and component losses, an OTDR can be used to locate breaks and imperfections in an optical fiber. The fiber length L (and hence the position of the break or fault) can be calculated from the time difference between the pulses reflected from the front and far ends of the fiber. If this time difference is t, then the length L is given by

$$L = \frac{ct}{2n_1} \tag{19.2}$$

where n_1 is the core refractive index of the fiber. The factor 2 accounts for the fact that light travels a length L from the source to the break point and then another length L on the return trip.

19.7. Multifunction Optical Test Systems

For laboratory, manufacturing, and quality-control environments, there are instruments with exchangeable modules for performing a variety of measurements. Figure 19.8 shows an example from EXFO, which includes a basic modular mainframe and an expansion unit. The mainframe is a Pentium-based unit that coordinates data compilation and analyses from a variety of test instruments. This test system can control external instruments having RS-232

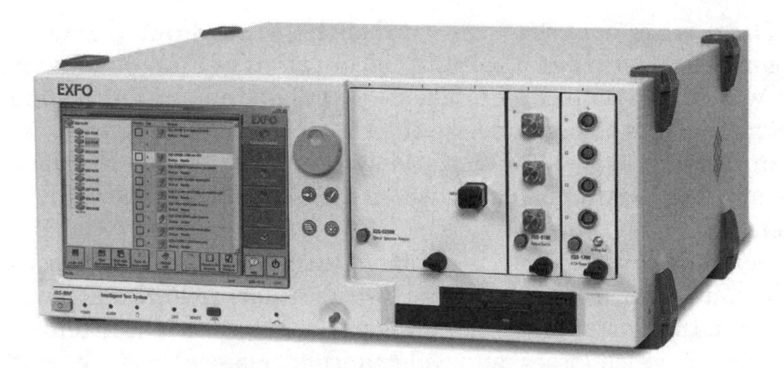

Figure 19.8. Example of a modular instrument for laboratory, manufacturing, and quality-control environments. (*Model IQS-500, provided courtesy of EXFO; www.exfo.com.*)

communication capability, and it has a networking capability for remote access from a computer. The plug-in modules cover a wide range of test capabilities. Example functions include single-channel or multichannel power meter, ASE broadband source, C-band or C+L-band tunable laser sources, variable attenuator, optical spectrum analyzer, return-loss meter, and PMD analyzer.

19.8. Summary

The installation and powering up of an optical fiber communication system requires measurement techniques for verifying that the link has been configured properly and that its constituent components are functioning correctly. The basic test equipment for measurements on optical fiber components and systems includes optical power meters, attenuators, tunable laser sources, spectrum analyzers, and time-domain reflectometers. These come in a variety of capabilities, with sizes ranging from portable, handheld units for field use to sophisticated briefcase-size bench-top or rack-mountable instruments for laboratory and manufacturing applications. More sophisticated instruments, such as polarization analyzers and optical communication analyzers, are available for measuring and analyzing polarization mode dispersion, eye pattern diagrams, and pulse waveforms.

Most test and measurement instruments enable a variety of statistical performance readings to be made at the push of a button, after the user has keyed in the parameters to be tested and the desired measurement range. Table 19.2 lists some widely used optical system test equipment and its functions.

Optical power measurement is the most basic function in fiber optic metrology. However, this parameter is not a fixed quantity and can vary as a function of other parameters such as time, distance along a link, wavelength, phase, and polarization. Therefore, two standard classes of power measurements in an optical system are the peak power and the average power. The *peak power* is the maximum power level in a pulse, which might be sustained for only a very short time. The *average power* is a measure of the power level averaged over a relatively long time period compared to the duration of an individual pulse.

The widespread implementation of WDM systems calls for making optical spectrum analyses to characterize the spectral behavior of various telecommunication network elements. One widely used instrument for doing this is an optical spectrum analyzer (OSA), which measures optical power as a function of wavelength.

The long-term workhorse instrument in fiber optic systems is the optical time-domain reflectometer (OTDR). In addition to locating faults within an optical link, this instrument measures parameters such as attenuation, length, connector and splice losses, and reflectance levels.

Further Reading

1. D. Derickson, ed., contributors from Hewlett-Packard Co., *Fiber Optic Test and Measurement*, Prentice Hall, Upper Saddle River, N.J., 1998.
2. National Institute of Standards and Technology (NIST), http://www.nist.gov.
3. Telecommunication Industries Association (TIA), http://www.tiaonline.org.
4. Electronic Industries Association (EIA), Washington.
5. Telecommunication Standardization Sector of the International Telecommunication Union (ITU-T), Geneva, Switzerland, http://www.itu.int.
6. American National Standards Institute (ANSI), New York, http://www.ansi.org.
7. F. Caviglia, V. C. Di Biase, and A. Gnazzo, “Optical maintenance in PONs,” *Optical Fiber Technology*, vol. 5, pp. 349–362, October 1999.
8. C. Mas and P. Thiran, “A review on fault location methods and their application to optical networks,” *Optical Networks Mag.*, vol. 2, pp. 73–87, July/August 2001.
9. Circadiant Systems, white papers on measurement (http://www.circadiant.com). See measurement white papers on this website or on websites of other test equipment manufacturers.

Chapter

20

Manufacturing Issues

A key issue in the successful widespread application of any emerging technology is the implementation of cost-efficient component production. Whereas it is acceptable to use highly expensive custom-made devices for proving that the technology works, the cost of components must be reduced significantly when bringing systems into the field. This is especially challenging in optical communications where new concepts have been appearing rapidly and standards are not fully mature in many areas. Ways of achieving this include adapting semiconductor manufacturing techniques, setting up automation processes, and devising reliable optoelectronic packaging methods.

This chapter looks at a few of the many issues associated with optical fiber and component production. First, Sec. 20.1 describes methods for producing optical fibers. Next, Sec. 20.2 discusses some component design issues, such as integration of various functions on planar lightwave circuits and athermal package designs. An important step in manufacturing is the connector and component polishing process, which is critical for high light coupling efficiency and low optical return loss. Section 20.3 describes some factory automation equipment to achieve this. Finally, Sec. 20.4 addresses various component packaging issues such as package designs, bonding of metallized fibers, and hermetic sealing.

20.1. Fiber Fabrication

Modern optical fibers are made of highly pure *silica* (SiO_2) to which certain impurities, such as germanium or boron, have been added to induce slight changes in the value of the refractive index. These fibers are produced principally by what is called a *vapor-phase oxidation* process. In this method, highly pure vapors of metal halides (e.g., $SiCl_4$ and $GeCl_4$) react with oxygen in a high-temperature environment to form a white powder of SiO_2 particles called *soot*. These particles then are collected on the surface of a bulk glass by one of three different commonly used processes. During or immediately after the collection

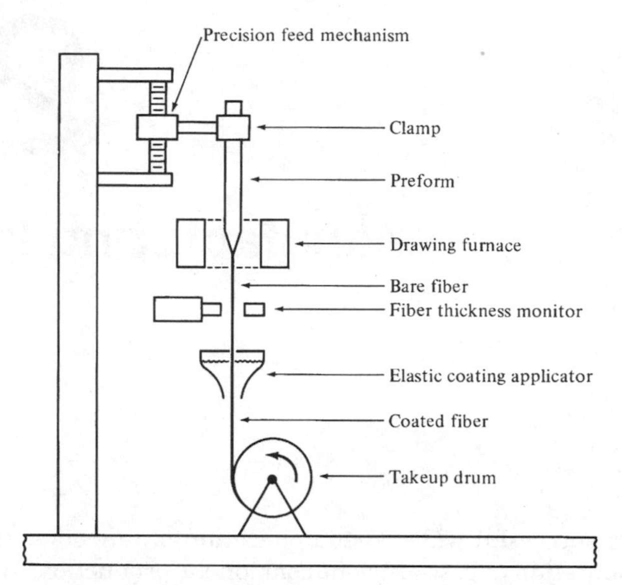

Figure 20.1. Illustration of an optical fiber-drawing process.

process, the particles are *sintered* (transformed to a homogeneous glass mass by heating without melting) by one of a variety of techniques to form a clear glass rod or tube (depending on the fabrication process). The chlorine gas produced in this process is evacuated from the system.

The glass rod or tube is called a *preform*. It is typically around 10 to 25 mm in diameter and 60 to 120 cm long. Fibers are made from the preform by using the *draw tower* equipment shown in Fig. 20.1. The preform is precision-fed into a circular heater called the *drawing furnace*. Here the preform end is softened to the point where it can be drawn into a very thin filament, which becomes the optical fiber. The turning speed of the takeup drum at the bottom of the draw tower determines how fast the fiber is drawn. This in turn will determine the thickness of the fiber, so that a precise rotation rate must be maintained. An optical fiber thickness monitor is used in a feedback loop for this speed regulation. To protect the bare glass fiber from external contaminants, such as dust and water vapor, an elastic coating is applied immediately after the fiber is drawn.

20.1.1. Outside vapor-phase oxidation

Figure 20.2 shows the *outside vapor-phase oxidation* (OVPO) process. Here a layer of SiO_2 particles is deposited from a burner onto a rotating graphite or ceramic rod called a *mandrel*. The glass soot adheres to this bait rod, and layer by layer, a porous cylindrical glass preform is built up. By properly controlling the constituents of the metal halide vapor stream during the deposition process, the glass composition and dimensions desired for the core and cladding can be incorporated into the preform.

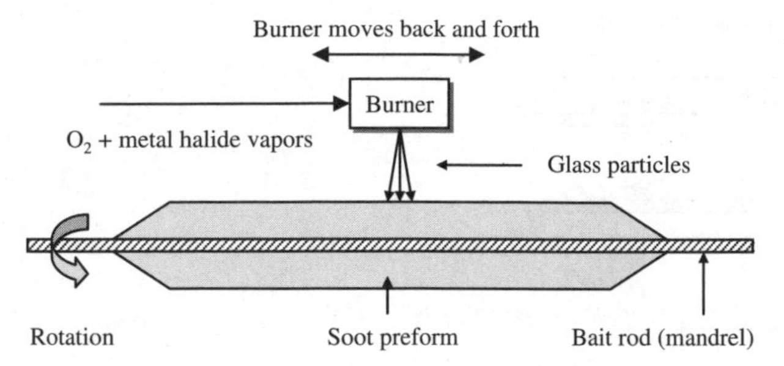

Figure 20.2. Illustration of the outside vapor-phase oxidation (OVPO) process.

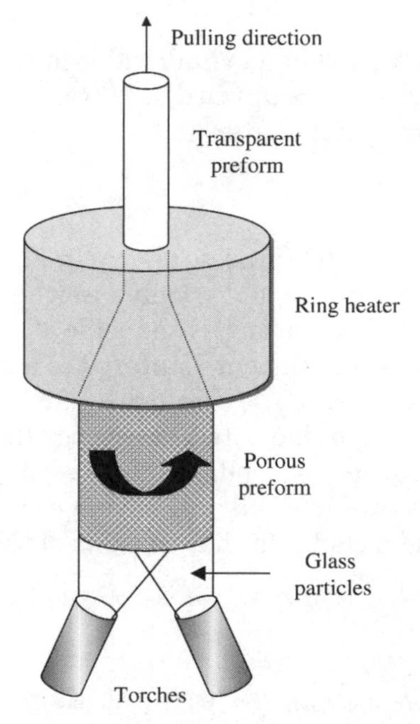

Figure 20.3. Illustration of the vapor-phase axial deposition (VAD) method.

When the deposition process is complete, the mandrel is removed and the porous tube is vitrified to a clear glass preform. The central hole in the tube preform collapses during the fiber drawing process.

20.1.2. Vapor-phase axial deposition

In the *vapor-phase axial deposition* (VAD) method, shown in Fig. 20.3, the soot particles are formed in the same way as in the OVPO process. However, in this case the particles are deposited on the end of a silica glass rod, which acts as a seed. A porous glass preform is grown in the axial direction by moving the rod

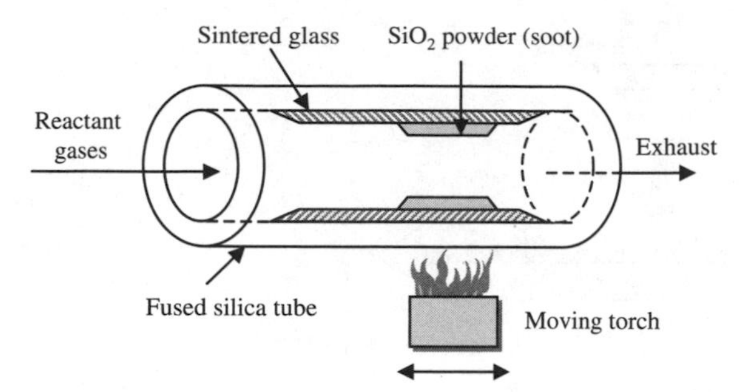

Figure 20.4. Illustration of the modified chemical vapor deposition (MCVD) process.

upward and rotating it continuously to maintain a cylindrical symmetry of the particle deposition. As the porous preform moves upward, it is transformed into a solid transparent glass rod preform by a ring heater.

20.1.3. Modified chemical vapor deposition

The *modified chemical vapor deposition* (MCVD) process shown in Fig. 20.4 was pioneered at Bell Laboratories and has been adopted widely elsewhere. In this technique, metal halide gases and oxygen flow through the inside of a revolving silica tube. As the SiO_2 particles are deposited, they are sintered to a clear glass layer by an oxyhydrogen torch which travels back and forth along the tube. When the desired thickness of glass has been deposited, the vapor flow is shut off and the tube is strongly heated to cause it to collapse into a solid rod preform. The fiber that is drawn from this preform rod will have a core that consists of the vapor-deposited material and a cladding that consists of the original silica tube.

20.1.4. Plasma-activated chemical vapor deposition

The *plasma-activated chemical vapor deposition* (PCVD) process is similar to the MCVD method in that deposition occurs within a silica tube. However, a moving microwave plasma operating at low pressure initiates the chemical reaction within the tube. This process deposits clear glass material directly on the tube wall without going through a soot deposition step. Thus, no extra sintering step is required. Just as in the MCVD case, when the desired thickness of glass has been deposited, the vapor flow is shut off and the tube is strongly heated to cause it to collapse into a solid rod preform.

20.2. Component Designs

Significant cost reductions and enhanced performance are two key factors for component design related to the growing use of dense WDM systems. These are great challenges, since performance improvement often leads to more expensive devices. However, this is being addressed through developments in areas such as planar lightwave circuits and inexpensive passive athermal designs for thermal wavelength stability. This section addresses those two issues. Other cost reduction techniques are the use of manufacturing automation, as described in Sec. 20.3, and improved packaging methods, which is the topic of Sec. 20.4.

20.2.1. Planar lightwave circuits

A significant saving in manufacturing cost may be realized by integrating the functions of several optical functions (such as filtering, amplification, coupling, and attenuation) on a single substrate. This substrate then can be placed in a single package, thereby reducing the size required on a circuit card. This is the aim of *planar lightwave circuit* (PLC) technology, which also is known as *integrated optics* technology. Since PLC technology increases the density and functionality of photonic devices per square centimeter, it leads to optical transmission equipment designs that are simpler, cost less, and are not as expensive to operate since they consume less electric power.

The manufacture of PLC devices is achieved by using tools and techniques from the semiconductor industry, which allow high-performance components to be made in high volume at low cost. At the core of this technology are traditional silicon-wafer processes and silica-on-silicon photonic circuit designs. These processes are extremely flexible and allow a modular design approach. Circuit configuration modifications within established design rules may be implemented rapidly through mask changes. In addition to using waveguides for passive optical functions, active control of light signals can be achieved through the incorporation of thermooptic elements. Naturally, similar to the fabrication of electronic components, extremely precise control must be exercised in every manufacturing step.

Components that may be integrated include variable optical attenuators (VOAs), optical switches, optical power splitters, optical couplers, arrayed waveguide gratings (AWGs), and rare earth–doped waveguides. As one example, suppose an engineer wants to construct an eight-channel wavelength multiplexer that has a variable output control on each channel. By using separately packaged components, this can be done with an AWG and eight VOAs plus some control circuitry, as shown in Fig. 20.5. Here the discrete packages need to be interconnected by short fiber jumpers. Integrating the AWG and VOA functions into a single PLC device greatly reduces the size and cost of the final module. In addition, the PLC design eliminates the external optical connections that normally are needed to interconnect the AWG and the VOAs, which results in a much lower optical power insertion loss.

As another example, consider the replacement of the erbium-doped fiber in a conventional EDFA with an integrated erbium-doped waveguide. The resultant *erbium-doped waveguide amplifier* (EDWA) is significantly smaller than its EDFA counterpart. In the example from NKT Integration shown in Fig. 20.6, eight EDWA devices are integrated on a single substrate. Each amplifier has

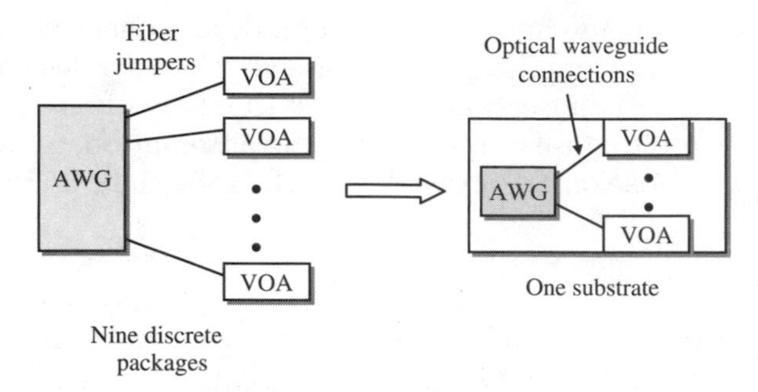

Figure 20.5. Example of planar lightwave circuit (PLC) technology. Here an AWG and eight VOAs are combined into a single PLC device.

Figure 20.6. Example of PLC technology showing the integration of eight EDWA devices plus associated signal and pump multiplexers on a single substrate. (*Photo courtesy of NKT Integration; www.nktintegration.com.*)

TABLE 20.1. Representative Specifications of an EDWA (Specification from NKT Integration; www.nktintegration.com)

Parameter	Specification
Wavelength range	1528–1562 nm
Pump wavelength	980 nm
Small-signal gain	>20 dB @ 100-mW pump power
Output power	>10 dBm @ 100-mW pump power and 0-dBm input signal
Noise figure	<5.0 dB @ −20-dBm input power
PDL (PDG)	<0.3 dB
PMD	<0.3 ps
Size of bare chip	25 × 55 × 1 mm

individual on-chip multiplexers for C-band and 980-nm pump wavelengths on both the input and the output for either codirectional or counterdirectional pumping. Table 20.1 gives some specifications of such an amplifier. Here the acronyms *PDL* and *PDG* refer to polarization-dependent loss and gain, respectively. This module may be used in metro applications where a few channels are added or dropped from a high-capacity DWDM trunk line.

20.2.2. Athermal designs

The performance of a passive optical component may change significantly with temperature. Of particular concern is wavelength drift in a DWDM application. Among thermally sensitive components are standard arrayed waveguide gratings, fiber Bragg gratings, and bulk-grating-based optical products. To maintain reliable performance of a DWDM communication link, it is essential that the wavelength characteristics of such system components be as invariant as possible.

Conventional AWGs consist of lightwave circuits made of quartz glass waveguides. Since the index of refraction of quartz glass changes with temperature, the wavelengths of light transmitted through such an AWG also change. Thus a conventional AWG typically will have a thermoelectric cooler-based temperature controller built into the package to maintain wavelength stability. However, by using different materials for the waveguides, an AWG can be fabricated without the need for a temperature control device and, consequently, without the need for a source of electric power. For example, an athermal AWG made by NTT Electronics uses a special silicon resin in part of the lightwave circuit that has a different temperature coefficient from that of quartz glass. This design cuts the temperature dependence of the wavelengths of transmitted light to less than one-tenth of its original value, which makes using a temperature control device unnecessary.

A different approach based on a passive temperature compensation method can be taken for a fiber Bragg grating (FBG). In this case the FBG is attached to a platform consisting of a material which has the opposite thermal expansion property from that of the FBG. With this method a wavelength stability of better than 0.5 pm/°C can be achieved together with a temperature-dependent loss below 0.01 dB/°C over an operating range of 0 to 70°C.

Likewise, bulk-grating-based optical products may be designed to be passively athermal, so that no active temperature control is required. In contrast to components that require active temperature control, for devices with passive athermal compensation there is no need for electric power, alarms, and monitoring circuitry.

20.3. Integrated Automation

Sometimes what appears to be the simplest step in a fabrication process can take a long time if individual pieces are processed manually. Among these are connector attachment, preparation of optical fiber pigtails for use in devices, and polishing of optical components. This section examines integrated-automation machines from two different vendors for these functions.

First consider the connector preparation system from Sagitta, which is shown in Fig. 20.7. This system has the integrated capabilities of batch polishing,

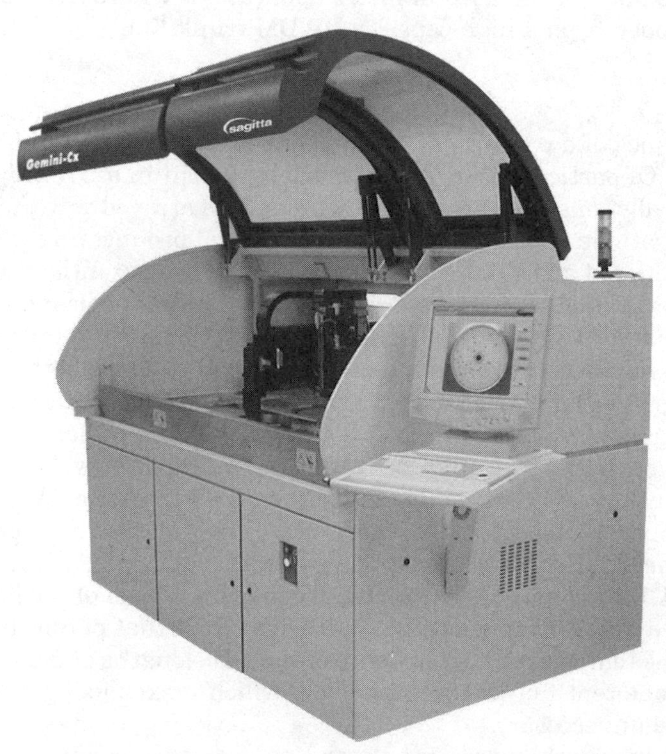

Figure 20.7. Example of an automated connector preparation system. (*Photo courtesy of Sagitta; www.sagitta.com.*)

cleaning, and optical-surface inspection of various types of fiber optic connectors and cable assemblies. These include the SC, FC, ST, LC, MU, MT-RJ, and MPO connectors described in Chap. 8. Connector properties, such as surface quality, end-face curvature, fiber height, and apex offset, can be controlled through a servo-control polishing head. The system includes an integrated microscope and CCD camera for image capture and the manual or automatic inspection of optical surfaces for scratches, pits, and material defects. The areas to be examined may be defined by regions in the core, cladding, or ferrule. Imaging software enables automatic surface-defect recognition, and a pass/fail sorting criterion can be set. The images can be seen on a display screen that is connected to the equipment. The machine allows a throughput of 100 to 300 connectors per hour, depending on the connector style.

As another example, consider the system shown in Fig. 20.8 from kSaria which fully automates the processes of fiber pigtail fabrication. When done manually, the fiber pigtail preparation process normally requires five processes and takes up to 10 min to complete. With this system a fiber preparation module automatically pays out a predetermined length of fiber (ranging from 0.75 to 3 m) and prepares the end. First, the spooling/pay-out tool dispenses a preprogrammed length of fiber, coils it, and inserts it into a holding tray with the two ends secured for further processing. The spooled trays then are transported to the strip, clean, and cleave tools. When the fiber preparation process is

Figure 20.8. Example of a system that fully automates the processes of fiber pigtail fabrication. (*Photo courtesy of kSaria; www.ksaria.com.*)

complete, a ferrule is loaded automatically and attached to the prepared fiber. Before attachment, a metered amount of thermally curable epoxy is dispensed into the ferrule. The fiber is inserted into the ferrule, and the assembly is heated to cure the epoxy. After the ferrule attachment process is complete, pigtails are unloaded into an empty cassette as a fiber pigtail. The entire process is controlled by a personal computer. The system software provides all operator controls, including multilevel user access, error logging, process statistics, and diagnostics. The throughput capability is 120 fibers per hour.

20.4. Packaging

The term *packaging* refers to the encapsulation of a device in a form that takes into account optical power coupling, thermal management, mechanical support, and hermetic sealing for environmental protection. In addition there are the standard concerns related to electrical connections. These concerns are the same as for ordinary microelectronics packages and are relevant only for active devices, since passive components do not have electrical connections.

20.4.1. Optical connection

The optical connection is the most critical aspect of a photonics package. Since light must be coupled efficiently into very small fiber cores, the optical connections require extremely precise alignment. They are especially sensitive to axial misalignments, as noted in Sec. 8.3. In some cases alignments need to be held to a fraction of a micrometer in order to have an acceptable component. To make matters more complicated, passive devices often have multiple couplings, all of which need to be precision-aligned simultaneously.

Micropositioning machines are available commercially that allow six-axis alignments to a resolution precision of 0.1 nm. Such systems usually can be software-controlled to allow their use for both development techniques and manufacturing procedures. Typically included with such machines is a video magnification system with illumination.

20.4.2. Thermal management

Management of *thermal effects* is an important factor for both active and passive devices. Thermal gradients within or across a device can lead to misalignments of components within a device package, which may cause a reduction in optical power coupling efficiency. In addition, heat generated within active components needs to be dissipated in order to avoid degradation of the device. In most cases this is done by means of a thermoelectric (TE) cooler. Typically the TE cooler is soldered to the package base, which later will be mounted on an external heat sink.

A TE cooler (which also is known as a *Peltier cooler*, since its function is based on the Peltier effect) is a semiconductor-based electronic component that functions as a small heat pump. When a low voltage is applied to a TE module, heat moves from one side of the module to the other in proportion to the applied voltage. Thus, one module face will be cooled while the other is heated simultaneously. As noted in Chap. 6, with appropriate control circuitry TE coolers can stabilize temperatures to ±0.02°C.

In packaging devices such as laser diodes, careful design procedures must be followed to ensure that the heat generated by the laser chip during operation will flow away from the laser out of the package. Creating the interface between a laser chip and the carrier it is mounted on is a standard procedure known as *eutectic die attachment*. This bonding traditionally is done by means of a gold/tin solder (80% Au/ 20% Sn), which can be done in a fairly straightforward manner in an automated assembly environment.

20.4.3. Mechanical concerns

The mechanical structure that houses active components usually is the industry-standard hermetically sealed 14-pin *butterfly package* or some variation thereof, as illustrated in Fig. 20.9. The housing is made of either Kovar (a low-expansion iron-nickel-cobalt alloy) or Kovar and a copper-tungsten (Cu-W) alloy. Once the components are placed inside and are bonded and interconnected properly, the package may be heated to reduce or even eliminate any moisture and other contaminants (see Sec. 20.4.4). Next the package is sealed, evacuated, and then backfilled with helium to a very low pressure.

To mount the optoelectronic components within the package, manufacturers use a metallic subassembly or platform that fits inside the butterfly package, as illustrated in Fig. 20.10. Included on the subassembly may be bonding pads for attaching wires, an alignment block for precise component placement, miniature circuit boards, and special inserts for mounting items such as fiber alignment

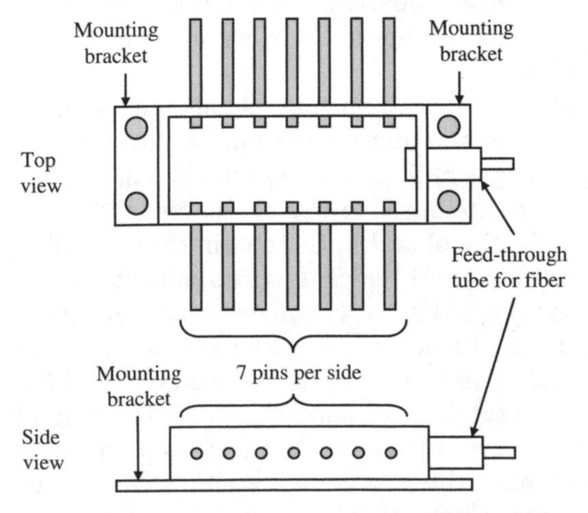

Figure 20.9. Schematic of an industry-standard hermetically sealed 14-pin butterfly package.

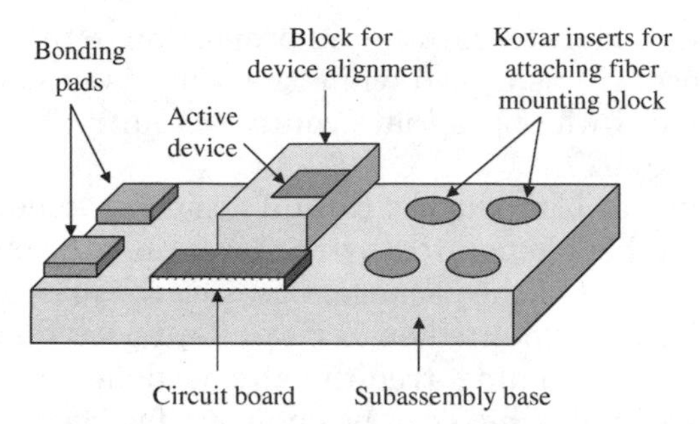

Figure 20.10. Subassembly or platform and some of its associated parts for mounting optoelectronic components within a butterfly package.

blocks. The basic properties of these complex metallic subassemblies include a heat dissipation ability, a thermal expansion match with the internal devices, and an ability to weld or bond items such as metallized fibers, semiconductor die carriers, and high-speed electronic circuitry. The basic platform structure nominally consists of a Cu-W alloy and also may include other metals or metallic alloys such as copper, nickel, a copper-molybdenum alloy, and Kovar.

Manufacturers of these butterfly packages often supply the housings with a TE cooler already attached inside.

20.4.4. Hermetic sealing

The purpose of *hermetic sealing* is to prevent contaminants from entering a device package. The items inside a package that require such protection include electronics, electrodes, microelectromechanical system (MEMS) devices, epoxy and die bonds, laser chips, and component surfaces. A key culprit is moisture, which over time can corrode electrodes, cause mechanical shifts in items bonded with water-sensitive epoxy, and damage the surfaces of fiber ends, mirrors, optical filters, or MEMS devices.

Hermetic packaging designs basically use the same packaging principles that have proved very successful in the microelectronics and microwave industries. The additional challenge in optical fiber components lies in how to create a hermetic seal at the point where the optical fiber passes through the package wall. When a fiber is inside a cable, the plastic buffer coating surrounding the cladding is sufficient to protect the fiber from typical environmental effects and mechanical disturbances that it may encounter. However, this coating is not adequate to provide a tight moisture seal when the fiber traverses a package wall.

To provide such a seal, the elastic coating is stripped off the fiber down to the cladding, and then a thin metal film is bonded to the bare fiber. The metal film nominally consists of two layers. First the fiber is coated with a layer of nickel to provide a strong adhesion to the glass and a stable soldering base. Then the fiber is plated with an overcoat layer of gold to provide resistance to oxidization. This creates a strong metal coating that is capable of withstanding the rigors of

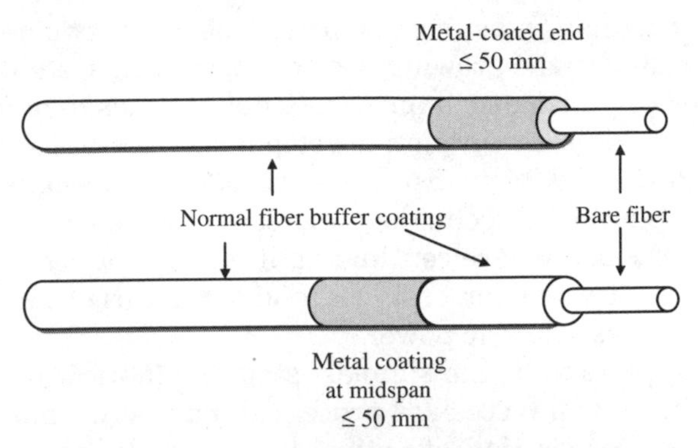

Figure 20.11. Two possible locations for creating a metal coating on a fiber to allow hermetic sealing to device packages.

soldering and hermetic sealing. A typical coating length ranges up to 50 mm, and the thickness is less than 2 μm. Metallic coatings can be put on either single fibers or ribbon fibers.

As shown in Fig. 20.11, the metal coating can be at any location on the fiber. For metallic coatings on the end, the metal can go completely to the fiber end, or a short final length of fiber can remain bare so that it may be terminated with ceramic ferrules, fusion-spliced to another fiber, or connected directly to a device inside a package. In the *midspan metallization* only a short section near the end of the fiber is metallized. When such a fiber passes through a package, it can be soldered to the package at the midspan point, thereby leaving a short pigtail within the package for connecting to a device inside.

20.5. Emerging Package Technology

Although most devices for optical fiber communications are enclosed in hermetically sealed metallic packages, there is a trend to produce both hermetic and non-hermetic ceramic packages. This development is being driven by the need to reduce manufacturing costs and to produce packaged devices in very high volumes. The mechanical structure that houses active components usually is the industry-standard hermetically sealed 14-pin butterfly package or some variation thereof.

20.6. Summary

The successful application of any emerging technology lies in the implementation of cost-efficient component production. This is especially challenging in optical communications where new concepts have been appearing rapidly and standards are not fully mature in many areas. Ways of achieving this include adapting semiconductor manufacturing techniques, setting up automation processes, and devising reliable optoelectronic packaging methods.

Significant cost reductions and enhanced performance are two key factors for component design related to the growing use of dense WDM systems. These are great challenges, since performance improvement often leads to more expensive devices. This is being addressed through developments in areas such as planar lightwave circuits (PLCs) and inexpensive passive athermal designs for thermal wavelength stability. Since PLC technology increases the density and functionality of photonic devices per square centimeter, it leads to optical transmission equipment designs that are simpler, cost less, and are not as expensive to operate since they consume less electric power.

Sometimes what appears to be the simplest step in a fabrication process can take a long time if individual pieces are processed manually. Among these are connector attachment, preparation of optical fiber pigtails for use in devices, and polishing of optical components. Several companies are offering integrated-automation machines for such functions.

Packaging of devices for optical fiber communications must take into account issues such as optical power coupling, thermal management, mechanical support, and hermetic sealing for environmental protection. Hermetic packaging designs have the challenge of creating a hermetic seal at the point where the optical fiber passes through the package wall. This can be achieved by stripping the fiber down to the cladding and then bonding a thin metal layer to the glass. This creates a strong metal coating that allows the fiber to be soldered to the package at the point where it passes through the enclosure wall, thereby providing a hermetic sealing.

Further Reading

1. J. G. Bornstein, "Semiconductor techniques enhance production of planar lightwave circuits," *Optical Manufacturing*, vol. 1, no. 1, pp. 25–29, March 2002.
2. R. Cisneros and G. Shechter, "Integrated automation promises improved planar lightwave circuits," *Optical Manufacturing*, vol. 1, no. 1, pp. 31–34, March 2002.
3. J. Sepulveda and L. Valenzuela, "Integrated subassemblies improve optoelectronic package performance," *Optical Manufacturing*, vol. 1, no. 2, pp. 27–29, May 2002.
4. B. Benton, "Automating optical and IC module assembly and thermal management," *Optical Manufacturing*, vol. 1, no. 2, pp. 23–25, May 2002.
5. G. Ogura, "Hermetic packaging," *Lightwave*, vol. 18, pp. 28–30, October 15, 2001 (www.lightwave.com).
6. S. J. Horowitz and D. I. Amey, "Ceramics meet next-generation fiber optic packaging requirements," *Photonics Spectra*, pp. 123–126, November 2001 (www.photonics.com).

Appendix

A

Units, Physical Constants, and Conversion Factors

A.1. International System of Units

Parameter	Unit	Symbol	Dimension
Length	meter	m	
Mass	kilogram	kg	
Time	second	s	
Temperature	kelvin	K	
Electric current	ampere	A	
Frequency	hertz	Hz	1/s
Force	newton	N	$kg \cdot m/s^2$
Pressure	pascal	Pa	N/m^2
Energy	joule	J	$N \cdot m$
Power	watt	W	J/s
Electric charge	coulomb	C	$A \cdot s$
Potential	volt	V	J/C
Conductance	siemens	S	A/V
Resistance	ohm	Ω	V/A

A.2. Physical Constants

Constant	Symbol	Value, mks units
Speed of light in vacuum	c	2.99793×10^{8} m/s
Electron charge	q	1.60218×10^{-19} C
Planck's constant	h	6.6256×10^{-34} J · s
Boltzmann's constant	k_B	1.38054×10^{-23} J/K
k_B/q at $T = 300\,\text{K}$	—	0.02586 eV
Electron volt	eV	1.60218×10^{-19} J
Base of natural logarithms	e	2.71828
Pi	π	3.14159
2π radians	—	360°

A.3. Conversion Factors

From	To	Conversion
cm	inch	cm × 0.3937
km	mile	km × 0.6214
inch	cm	inch × 2.5400
mile	km	mile × 1.6093
eV	J	eV × 1.60218×10^{-19}
radian	degree	radian × 57.296
°C	°F	°C × 1.8 + 32
K	°C	K + 273

Appendix

B

ITU-T Frequency and Wavelength Grid

The table below gives the ITU-T frequency grid and the corresponding wavelengths for 200 channels in the L- and C-bands for 50-GHz and 100-GHz channel spacing. The column labeled "50-GHz offset" means that for the 50-GHz grid one uses the 100-GHz spacings with these 50-GHz values interleaved. For example, the 50-GHz channels in the L-band would be at 186.00 THz, 186.05 THz, 186.10 THz, and so on.

	L-band				C-band			
	100-GHz		50-GHz offset		100-GHz		50-GHz offset	
Unit	THz	nm	THz	nm	THz	nm	THz	nm
1	186.00	1611.79	186.05	1611.35	191.00	1569.59	191.05	1569.18
2	186.10	1610.92	186.15	1610.49	191.10	1568.77	191.15	1568.36
3	186.20	1610.06	186.25	1609.62	191.20	1576.95	191.25	1567.54
4	186.30	1609.19	186.35	1608.76	191.30	1567.13	191.35	1566.72
5	186.40	1608.33	186.45	1607.90	191.40	1566.31	191.45	1565.90
6	186.50	1607.47	186.55	1607.04	191.50	1565.50	191.55	1565.09
7	186.60	1606.60	186.65	1606.17	191.60	1564.68	191.65	1564.27
8	186.70	1605.74	186.75	1605.31	191.70	1563.86	191.75	1563.45
9	186.80	1604.88	186.85	1604.46	191.80	1563.05	191.85	1562.64
10	186.90	1604.03	186.95	1603.60	191.90	1562.23	191.95	1561.83
11	187.00	1603.17	187.05	1602.74	192.00	1561.42	192.05	1561.01
12	187.10	1602.31	187.15	1601.88	192.10	1560.61	192.15	1560.20

	L-band				C-band			
	100-GHz		50-GHz offset		100-GHz		50-GHz offset	
Unit	THz	nm	THz	nm	THz	nm	THz	nm
13	187.20	1601.46	187.25	1601.03	192.20	1559.79	192.25	1559.39
14	187.30	1600.60	187.35	1600.17	192.30	1558.98	192.35	1558.58
15	187.40	1599.75	187.45	1599.32	192.40	1558.17	192.45	1557.77
16	187.50	1598.89	187.55	1598.47	192.50	1557.36	192.55	1556.96
17	187.60	1598.04	187.65	1597.62	192.60	1556.55	192.65	1556.15
18	187.70	1597.19	187.75	1596.76	192.70	1555.75	192.75	1555.34
19	187.80	1596.34	187.85	1595.91	192.80	1554.94	192.85	1554.54
20	187.90	1595.49	187.95	1595.06	192.90	1554.13	192.95	1553.73
21	188.00	1594.64	188.05	1594.22	193.00	1553.33	193.05	1552.93
22	188.10	1593.79	188.15	1593.37	193.10	1552.52	193.15	1552.12
23	188.20	1592.95	188.25	1592.52	193.20	1551.72	193.25	1551.32
24	188.30	1592.10	188.35	1591.68	193.30	1550.92	193.35	1550.52
25	188.40	1591.26	188.45	1590.83	193.40	1550.12	193.45	1549.72
26	188.50	1590.41	188.55	1589.99	193.50	1549.32	193.55	1548.91
27	188.60	1589.57	188.65	1589.15	193.60	1548.51	193.65	1548.11
28	188.70	1588.73	188.75	1588.30	193.70	1547.72	193.75	1547.32
29	188.80	1587.88	188.85	1587.46	193.80	1546.92	193.85	1546.52
30	188.90	1587.04	188.95	1586.62	193.90	1546.12	193.95	1545.72
31	189.00	1586.20	189.05	1585.78	194.00	1545.32	194.05	1544.92
32	189.10	1585.36	189.15	1584.95	194.10	1544.53	194.15	1544.13
33	189.20	1584.53	189.25	1584.11	194.20	1543.73	194.25	1543.33
34	189.30	1583.69	189.35	1583.27	194.30	1542.94	194.35	1542.54
35	189.40	1582.85	189.45	1582.44	194.40	1542.14	194.45	1541.75
36	189.50	1582.02	189.55	1581.60	194.50	1541.35	194.55	1540.95
37	189.60	1581.18	189.65	1580.77	194.60	1540.56	194.65	1540.16
38	189.70	1580.35	189.75	1579.93	194.70	1539.77	194.75	1539.37
39	189.80	1579.52	189.85	1579.10	194.80	1538.98	194.85	1538.58
40	189.90	1578.69	189.95	1578.27	194.90	1538.19	194.95	1537.79
41	190.00	1577.86	190.05	1577.44	195.00	1537.40	195.05	1537.00
42	190.10	1577.03	190.15	1576.61	195.10	1536.61	195.15	1536.22
43	190.20	1576.20	190.25	1575.78	195.20	1535.82	195.25	1535.43
44	190.30	1575.37	190.35	1574.95	195.30	1535.04	195.35	1534.64
45	190.40	1574.54	190.45	1574.13	195.40	1534.25	195.45	1533.86

	L-band				C-band			
	100-GHz		50-GHz offset		100-GHz		50-GHz offset	
Unit	THz	nm	THz	nm	THz	nm	THz	nm
46	190.50	1573.71	190.55	1573.30	195.50	1533.47	195.55	1533.07
47	190.60	1572.89	190.65	1572.48	195.60	1532.68	195.65	1532.29
48	190.70	1572.06	190.75	1571.65	195.70	1531.90	195.75	1531.51
49	190.80	1571.24	190.85	1570.83	195.80	1531.12	195.85	1530.72
50	190.90	1570.42	190.95	1570.01	195.90	1530.33	195.95	1529.94

Appendix C

Acronyms

ADSS	All-dielectric self-supporting
ANSI	American National Standards Institute
A-NZDSF	Advanced nonzero dispersion-shifted fiber
APD	Avalanche photodiode
ARQ	Automatic repeat request
ASE	Amplified spontaneous emission
ASK	Amplitude shift keying
ATM	Asynchronous transfer mode
AWG	Arrayed waveguide grating
BER	Bit error rate
BLSR	Bidirectional line-switched ring
CATV	Cable television
CD	Chromatic dispersion
CGM	Cross-gain modulation
CMEMS	Compliant MEMS
CNR	Carrier-to-noise ratio
CO	Central office
CPM	Cross-phase modulation
CPU	Central processing unit
CRC	Cyclic redundancy check
CWDM	Coarse WDM
DBR	Distributed Bragg reflector
DCE	Dynamic channel equalizer
DCF	Dispersion-compensating fiber
DCM	Dispersion-compensating module

DFA	Doped-fiber amplifier
DFB	Distributed-feedback (laser)
DGE	Dynamic gain equalizer
DSF	Dispersion-shifted fiber
DTE	Data terminal equipment
DWDM	Dense WDM
EAM	Electroabsorption modulator
EDFA	Erbium-doped fiber amplifier
EDWA	Erbium-doped waveguide amplifier
EIA	Electronic Industries Association
EM	Electromagnetic
EMS	Element management system
EO	Electrooptical
FBG	Fiber Bragg grating
FC	Fibre Channel
FDM	Frequency division multiplexing
FEC	Forward error correction
FOTP	Fiber Optic Test Procedure
FP	Fabry-Perot
FPM	Four-photon mixing
FRPE	Flame-retardant polyethylene
FSR	Free spectral range
FTTB	Fiber to the building
FTTC	Fiber to the curb
FTTH	Fiber to the home
FTTN	Fiber to the neighborhood
FTTx	Fiber-to-the-x
FWHM	Full-width half-maximum
FWM	Four-wave mixing
GFF	Gain-flattening filter
GigE	Gigabit Ethernet
GUI	Graphical-user interface
HASB	High-airspeed-blown
HDLC	High-level data link control (protocol)
IEC	International Electrotechnical Commission
IEEE	Institute for Electrical and Electronic Engineers
IETF	Internet Engineering Task Force
IP	Internet Protocol

ISO	International Organization for Standardization
ITU	International Telecommunication Union
LAN	Local-area network
LC	Lucent (connector)
LED	Light-emitting diode
MAN	Metropolitan-area network
MCVD	Modified chemical vapor deposition
MDF	Main distribution frame
MEMS	Microelectromechanical system
MIB	Management information base
MPLS	Multiprotocol label switching
MPO	Multiple-fiber push-on/pull-off (connector)
MSA	Multisource agreement
MT-RJ	Media termination—recommended jack (connector)
MU	Miniature unit (connector)
MZI	Mach-Zehnder interferometer
NA	Numerical aperture
NEC	National Electrical Code
NIC	Network interface card
NIST	National Institute of Standards and Technology
NMS	Network management system
NPL	National Physical Laboratory
NRZ	Non-return-to-zero
NZDSF	Nonzero dispersion-shifted fiber
OADM	Optical add/drop multiplexer
OCA	Optical channel analyzer
OCh	Optical channel (layer)
OCM	Optical channel monitor
OC-*N*	Optical carrier—level *N*
OCPM	Optical channel performance monitoring
ODU	Optical channel data unit
OFC	Optical fiber conductive
OFCP	Optical fiber conductive plenum
OFCR	Optical fiber conductive riser
OFN	Optical fiber nonconductive
OFNP	Optical fiber nonconductive plenum
OFNR	Optical fiber nonconductive riser
OLT	Optical line terminal

OMS	Optical multiplex section
ONU	Optical network unit
OOK	On/off keying
OPGW	Optical ground wire
OPM	Optical performance monitor
OPU	Optical channel payload unit
OSA	Optical spectrum analyzer
OSC	Optical service channel
OSNR	Optical signal-to-noise ratio
OSS	Operations support system
OTDR	Optical time-domain reflectometer
OTN	Optical transport network
OTS	Optical transmission section
OTU	Optical channel transport unit
OVPO	Outside vapor-phase oxidation
OXC	Optical cross-connect
PDG	Polarization-dependent gain
PDH	Plesiochronous digital hierarchy
PDL	Polarization-dependent loss
PE	Polyethylene
PLC	Planar lightwave circuit
PMD	Polarization mode dispersion
POH	Path overhead
PON	Passive optical network
POP	Point of presence
PPP	Point-to-point protocol
PTB	Physikalisch-Technische Bundesanstalt
PU	Polyurethane
PVC	Polyvinyl chloride
QoS	Quality of service
RF	Radio-frequency
RIN	Relative intensity noise
RMON	Remote monitoring
RS	Reed-Solomon (code)
RZ	Return-to-zero
SBS	Stimulated Brillouin scattering
SC	Subscriber connector or square connector
SDH	Synchronous digital hierarchy

SFF	Small form factor
SLA	Service-level agreement
SMA	SubMiniature version A (connector)
SNMP	Simple Network Management Protocol
SNR	Signal-to-noise ratio
SOA	Semiconductor optical amplifier
SONET	Synchronous optical network
SOP	State of polarization
SPE	Synchronous payload envelope
SPM	Self-phase modulation
SRS	Stimulated Raman scattering
ST	Straight-tip (connector)
STM-1	Synchronous transport module—level 1
STS-1	Synchronous transport signal—level 1
SWP	Spatial walk-off polarizer
TCP	Transmission control protocol
TDM	Time-division multiplexing
TE	Transverse electric (modes)
TE	Thermoelectric
TEC	Thermoelectric cooler
TFF	Thin-film filter
TIA	Telecommunication Industries Association
TW	Traveling-wave (amplifier)
UL	Underwriters Laboratories
UPSR	Unidirectional, path-switched ring
VAD	Vapor-phase axial deposition
VCSEL	Vertical cavity surface emitting laser
VOA	Variable optical attenuator
WAN	Wide-area network
WDM	Wavelength division multiplexing
WSC	Wavelength-selective coupler
XPM	Cross-phase modulation
YIG	Yttrium iron garnet

Index

ABOUT THE AUTHOR

Gerd Keiser is founder and president of PhotonicsComm Solutions, Inc., a firm specializing in consulting and education for the optical communications industry. (Visit www.PhotonicsComm.com.) He had extensive experience at Honeywell, GTE, and General Dynamics in optical networking technology for telecommunications applications, has served as an Adjunct Professor of Electrical Engineering at Northeastern University and Tufts University, and is a Fellow of the IEEE. The world-known author of McGraw-Hill's *Optical Fiber Communications* and *Local Area Networks*, he is also an Associate Editor of the technical journal *Optical Fiber Technology*. His popular books have been translated into Chinese, Japanese, and Italian. Gerd can be contacted at gerd.keiser@ieee.org.